Concepts of
Probability

Concepts of Probability

E. R. Mullins, Jr., and David Rosen

Swarthmore College

Bogden & Quigley, Inc., Publishers
Tarrytown-on-Hudson, New York / Belmont, California
1972

Cover design by Winston G. Potter
Text design by Craven, Evans and Stone, NYC

Library of Congress Card No.: 76-164947
Standard Book No.: 0-8005-0015-6

to Julie

Preface

The concepts of probability are important to the understanding of many current issues and problems in the behavioral and biological sciences. This is so because probability is the underlying justification for statistical applications. Because probability theory sheds light on how and why statistical techniques work, this book attempts to present a reasonable and understandable theory of probability which will enable the reader to see more easily what is going on in a statistical analysis.

The book is divided into forty-two sections, not necessarily of equal length or intensity, but certainly sufficient for a one-semester course meeting three times a week. The prerequisite for the first seven chapters is standard high-school mathematics through two years of algebra. Chapter 8 requires some knowledge of sequences and Chapters 9 and 10 use some basic calculus concepts. The book is primarily an axiomatic treatment of finite probability (the first seven chapters). In Chapter 8 the theory is extended to the countably infinite case and to the continuous model in Chapters 9 and 10. Although more demanding, this material is included for completeness and gives students some idea of the mathematical background needed to pursue the subject further. At the same time, these chapters make available useful facts about the Poisson and normal probability functions.

Our goal has been to organize the material and to present it so that students and instructors may work through the book selectively. Indeed, an instructor could not possibly discuss every exercise or proof in class. The presentation, therefore, is adaptable to self-study, encouraging students to read some mathematics by themselves. It is also designed to give instructors flexibility in the classroom, freeing them to pursue the more difficult concepts in detail.

Care has been taken to introduce those mathematical topics which are needed in probability theory so that the reader will also develop an appreciation of the power and utility of mathematics itself. When one applies mathematics to non-mathematical situations, one makes assumptions and constructs hypotheses about the phenomena under consideration. In short, one constructs a model. We have, therefore, interwoven theory and applications to provide insight into this model-building process.

Proofs and examples designated by a solid square (■) may by omitted without loss of continuity. A solid bar (▌) appears at the end of each proof and each example. The appendixes on the real number system, mathematical induction, and analytic geometry can be incorporated into the course or simply used as a reference or review.

There are numerous problems at the end of each section, many similar to the worked-out examples in the section. The more theoretical problems appear near the end of each set and on occasion are extensions of the theory itself.

Statistical applications are found in many of the worked-out examples and many of the exercises, which come from such diverse fields as sports, politics, physics, biology, economics, business, and sociology.

This book could not have been developed without the advice and influence of many, particularly those authors whose texts we have used in the classroom. Hopefully we have acknowledged this assistance everywhere. There have also been many students who have freely offered helpful and useful suggestions which we acknowledge happily.

Special thanks must go to Beth Christiansen, Charles Jepsen, and J. Edward Skeath, all of whom cheerfully used multilithed versions in their classes and then offered sensible, constructive criticism. We appreciate the suggestions made by Professor Robert Kozelka of Williams College and Professor S. K. Katti of the University of Missouri, both of whom read the manuscript. Our deep thanks go to Dorothy Blythe, our principal typist and to Joann Lauritzen and Jane Mullins, who have typed various versions or parts of the text, sometimes under trying circumstances. Robert Katz worked all problems for the first three chapters and the appendixes. Our publisher, Bogden & Quigley, Inc., deserves our special praise for steadfast faith in our project and for the kind of editorial assistance that makes writing pleasant.

E. R. Mullins, Jr.
Swarthmore, Pennsylvania

David Rosen
Swarthmore, Pennsylvania

Contents

Contents

chapter one

Introduction to Probability

1.1   Basic Concepts

A primary aim of our book is to involve the reader in mathematics so that he can understand and use the mathematician's perspective in handling problems that arise in a wide variety of fields.   We begin this by studying the theory of probability, for it underlies the technical justification that is needed to create appropriate mathematical models.

Although the theory of probability owes its origins to the analysis of classical games of chance, its applications now range from analyzing games of chance to determining the likelihood that Madison was an author of the *Federalist* papers or to calculating the likelihood that two molecules will collide.   Such diverse problems as predicting the outcome of an election and predicting the efficiency of a new vaccine lend themselves to similar mathematical and probabilistic analysis.

Technically, the mathematical theory of probability is a body of axioms, definitions, and theorems that form a mathematical system just as (ordinary) Euclidean geometry is a mathematical system.   As in every mathematical system, the axioms and theorems are statements about elements or objects that are undefined. The undefined elements in geometry are point and line, whereas in probability theory the undefined objects are the occurrences of *random phenomena*.   To describe this term and to prepare the way for the formal presentation, we present some examples.

Physical occurrences that are commonly considered to be random phenomena are the toss of a coin, the number of cars that will pass a given corner in 1 hour, the color of a ball drawn from an urn that contains balls of different colors, and whether a given human egg will be fertilized (also whether by a male-determining or a female-determining sperm). These phenomena are related in the sense that, intuitively at least, the results are not predictable.

Examples of physical phenomena that are *not* considered to be random are the temperature of boiling water in Chicago, the time of sunrise in Philadelphia on the day this book is published, or the date the first man will land on Mars. These outcomes can be determined. Consider some other examples.

Suppose a stone is dropped from a bridge at a point directly over a stream. Will the stone hit the water? Yes or no? Suppose it is dropped from the bridge while a rowboat is moving gently at anchor under the bridge. Will the stone land in the rowboat? Yes or no? These two questions emphasize the difference between nonrandom and random phenomena. In the first the law of gravity assures us that the stone will hit the water. In the second it assures us only that the stone will fall—but whether it will hit the boat or the water we do not know. The phenomenon of the stone hitting the boat is considered a random phenomenon. If the bridge is close to the water and the rowboat comparatively large, quite possibly we can predict where the stone will land; then the experiment no longer would be considered a random experiment. Incidentally, for our purposes, random phenomenon and random experiment are synonymous.

If in an automated manufacturing process 20 items are picked out of the hopper, these constitute a sample of the manufactured items, and the percentage of faulty products in the sample is an estimation of the level of faulty production in the entire output. This process of examining a sample and making an assertion about the original collection is referred to as *statistical inference*. Since the assertion is a nondeterministic inference (that is, if we had drawn a different sample of 20 we might have a different assertion), there is a degree of uncertainty about the accuracy of the assertion. It is just this uncertainty that links statistical inference to mathematics.

Observations such as preferences in a voters' poll are characterized by the fact that although the behavior of each individual is unpredictable, it is possible to make a predictive statement about the whole voting population by interviewing only a relatively few voters. The basic assumption is that there will be some regularity and consistency in the way the population behaves, and the behavior of a small sample of the population mirrors the whole population. Any phenomenon or experiment, wherever it occurs, that has the feature that the outcomes are unpredictable will be called a *random phenomenon* or a *random experiment*.

Conceivably we may find it difficult to decide whether a given physical occurrence is a random phenomenon. A guiding rule is that if on different trials or occasions a physical occurrence yields different results with no apparent change in the conditions of the experiment, it is safe to assume that it is a random physical phenomenon. Our primary concern in studying probability will be to formulate abstract models of physical phenomena that can be assumed to be random.

Problems

1. Give an example of a random phenomenon that would be studied by a
   - (a) postmaster
   - (b) traffic engineer
   - (c) quality-control engineer
   - (d) economist
   - (e) psychologist
   - (f) retail merchant
   - (g) farmer

2. Discuss whether or not each of the following occurrences would be classified as a random phenomenon:
   - (a) a hole-in-one at the golf course
   - (b) a strike thrown by a particular pitcher in a baseball game
   - (c) the average daily temperature in New York City
   - (d) the total gallons of gasoline purchased in 1 day at a particular gas station
   - (e) the total of shares traded in 1 day on the New York Stock Exchange

3. At a track meet, usually three to five people are assigned to measure the elapsed time in any running event. Seldom will all the timers record the same time. Discuss the problem of measuring the elapsed time of a race as a random phenomenon.

4. Let each student in a chemistry-laboratory class of 20 perform the same experiment to determine a physical constant, such as the boiling point of water. Discuss this situation as a possible random phenomenon.

5. Discuss whether or not each of the following occurrences would be classified as random phenomena:
   - (a) A light bulb manufactured by a machine is tested.
   - (b) A man selects a girl and asks her for a date.
   - (c) A traveling salesman selects an itinerary of 10 cities.
   - (d) A person is questioned by a national public opinion poll.
   - (e) A winner and a runner-up for a beauty contest are selected from among 10 girls.

6. Discuss whether or not each of the following experiments would be classified as a random experiment:
   - (a) Stop a car on the highway and count the number of persons riding in the car.
   - (b) Measure the difference in the number of man-hours required to complete a job and the number of man-hours estimated for the job.

(c) Select a family of size 4 and record the total income.

(d) Record the grades received by the same person in three different tests.

(e) Ask a person a question and measure the time lapse between the end of the question and the beginning of the response.

1.2 The Sample Space

Consider the following three random experiments:

1. Tossing a coin.

2. Rolling a die with three faces painted yellow and three faces painted green.

3. Drawing a chip from an urn with 14 red chips and 14 blue chips, with the chips indistinguishable except for color.

In each experiment we can determine two possible, observable results: head or tail, yellow or green face, red or blue chip. In each experiment, intuitively at least, either outcome is equally likely. These three physical phenomena have a common characteristic which we want to identify precisely and formally as we create a mathematical model for the experiment.

Usually many models will describe an experiment; choosing one is part of the skill that is needed to apply probability theory to real problems. At the outset there must be agreement as to possible outcomes. For example, in the toss of the coin it is quite possible that the result "coin on edge" can occur, and if this is important for the experiment it can be included with "head" and "tail." Then there would be three thinkable outcomes. Our own idealization of the coin toss will be to regard "head" and "tail" as the only possible outcomes of the experiment.

By collecting all outcomes in one set, we obtain a *sample space*.

Definition 1: Sample space

The collection of all possible outcomes of a random phenomenon is called the sample space.

A synonym for sample space is the word *set*, which is a primitive mathematical notion, usually undefined, but generally understood to mean a collection of objects or items. Space is used here in the sense of a set or collection. In no way does its use imply a physical expanse.

Since we are concerned with the whole collection of all possible outcomes of an experiment, our set is in a sense a full set or a universal set of outcomes. Everything that can occur will be represented in the sample space. Thus the sample space provides a model of an ideal experiment, as in the following example:

Example 1

(a) A random experiment consists of tossing a coin twice. If not only the results are important, but also which came first and which last, then the four pairs HH, HT, TH, and TT can be used to represent the thinkable outcomes for a sample space S. The first letter in each pair represents the result of the first toss; the second letter, the second toss:

$$S = \{HH, HT, TH, TT\}.$$

(b) Suppose that the only thing of interest is the number of times a head (alternatively, the number of times a tail) appears. Now, the thinkable outcomes are 0, 1, or 2 heads (or tails), and a sample space can be represented as

$$S = \{0, 1, 2\}.$$

(c) Still another model can be described if the important concern is simply whether or not the two tosses produce like or different results. By letting A denote "like" and D "different" results, a sample space can be represented as

$$S = \{A, D\}.$$

No other sample space or set of outcomes is possible, since we have ruled out "coin on edge." ∎

The symbol ∎ is used to denote the end of an example; later it will also be used to indicate the end of a proof.

Example 2

An ordinary die when rolled will come to rest with one of six possible faces uppermost. If the faces are numbered 1, 2, 3, 4, 5, and 6, then six possible outcomes make up the sample space. If two of the faces are numbered 1 and the remaining four are numbered 3, then the sample space has only two outcomes $\{1, 3\}$. Other sample spaces obtain if the faces are specified differently. ∎

Example 3

For any specific telephone, the number of calls that come through in 1 hour is a random phenomenon with no upper limit (theoretically) on the number of calls, so a sample space will consist of the nonnegative integers. ∎

Example 4

The time of arrival of a parcel to be delivered by the department store where it was purchased will be a random phenomenon with the times 9 A.M.–6 P.M. of the promised day (and perhaps the next!) as a sample space. ∎

Example 5

If a pair of ordinary dice is rolled, the sum of the number of spots on the uppermost face will be one of the numbers 2, 3, 4, 5, 6, 7, 8, 9, 10, 11, or 12. The sample space is this set of 11 integers. If the dice are distinguishable, for instance, if one is red and the other white, then there are 6 different possibilities for each die and a total of 36 different outcomes for the pair. A red 4 and a white 2 is not the same as a white 4 and a red 2. The difference can be represented by pairs of numbers written (4, 2) and (2, 4), where the first number of each pair represents the number of spots on the red die, and the second number, the spots on the white die. The order of the numbers is important, and the symbol (x, y) is called an *ordered pair*. The 36 possible outcomes collected in a square array in Table 1 comprise the sample space.

Table 1

(1, 1)	(2, 1)	(3, 1)	(4, 1)	(5, 1)	(6, 1)
(1, 2)	(2, 2)	(3, 2)	(4, 2)	(5, 2)	(6, 2)
(1, 3)	(2, 3)	(3, 3)	(4, 3)	(5, 3)	(6, 3)
(1, 4)	(2, 4)	(3, 4)	(4, 4)	(5, 4)	(6, 4)
(1, 5)	(2, 5)	(3, 5)	(4, 5)	(5, 5)	(6, 5)
(1, 6)	(2, 6)	(3, 6)	(4, 6)	(5, 6)	(6, 6) ∎

Example 6

A family of four is selected in some arbitrary way and the total income is calculated to the nearest dollar. A suitable sample space is all the nonnegative integers. ∎

Example 7

The time lapse between the end of a question and the beginning of an answer is called the response time. A sample space of all nonnegative numbers is appropriate. ∎

Example 8

Students are asked to rate statements such as "I have spent more time in the dormitory lounge this year" on a scale from -5 (strongly disagree) to $+5$ (strongly agree), letting 0 signify "I can't decide." Thus there are 11 different ratings for each question. A sample space for two such questions would be the 121 pairs of numbers with the first number the rating for the first question and the second number the rating for the second question. ∎

Problems

In the problems calling for the description of a sample space, use appropriate symbols and explicitly list the set of elements.

1. Three coins are tossed. Give two sample spaces for this experiment.

2. A coin is tossed and then a die is rolled. Give a sample space for this experiment.

3. A survey of families with three children is made. Describe the sample space for the experiment of selecting one family and recording the number of boys and girls.

4. A political candidate makes a speech on some or all of the major issues of his platform. If there are four such issues, construct a sample space for the different speeches.

5. Two large tribes and three smaller tribes live in the same geographical region. An anthropologist has noted that the smaller tribes ally themselves with one of the larger tribes. Devise a sample space for the possible alignments of tribes.

6. Two balls are drawn from an urn containing six balls, of which four are white, one is red, and one is black.
   (a) If the experiment is conducted by drawing one ball, recording the color, and replacing the ball in the urn before the second draw, list a sample space.
   (b) If the experiment is conducted by drawing one ball, recording the color, and then drawing the second without replacing the first ball, list a sample space.

7. Two people are shown an abstract painting and asked to rate the painting unpleasing, no preference, or pleasing. If the ratings are coded 0, 1, and 2, describe a sample space for the experiment.

8. The direction of change in the Dow–Jones averages of the New York Stock Exchange for three consecutive days is observed, where it is assumed that the average always changes either up or down. Devise a sample space for the experiment.

9. An urn contains six balls identical except for the numbers 1, 2, 3, 4, 5, and 6. A ball is drawn, its number noted, and the ball is replaced before another ball is drawn and its number noted. Devise a sample space for this experiment.

10. Refer to Problem 9. This time perform the experiment without returning the first ball to the urn before the second ball is drawn.

11. An ordinary deck of 52 playing cards is shuffled thoroughly. The number, but not the value, of cards down to and including the first ace is recorded. Give a sample space for this random experiment.

12. Consider the following pairs of numerical quantities for the students in your class and list sample spaces that would represent these quantities.
    (a) Height in inches, weight in pounds.
    (b) Age in months, grade on last hour exam.
    (c) Number of movies seen in the last month, number of hours spent watching television in the last week.
    (d) Shoe size, number of siblings.

13. Two separate production lines read the same product into the same stockpile. The productive capacity of the first line is five units per day and that of the other is three units per day. If it is assumed that the output of each production line is a random experiment with values from zero to capacity each day, list
    (a) sample spaces that describe the output of each of the lines
    (b) a sample space that describes the stockpile condition.

14. A certain variety of plant can be one of three distinct genetic types. Ten plants are observed, and the genetic type of each is determined and recorded. Give two sample spaces for this experiment.

15. A jar contains a large, but unknown, number of balls, identical except for the colors red, white, and yellow. Three balls are drawn simultaneously and the colors are recorded. Describe a sample space for this experiment.

16. A coin is tossed. If the coin falls heads, a die is thrown, but if the coin falls tails, the coin is tossed twice more. Describe a sample space for this experiment.

17. A penny is tossed twice and a nickel is tossed three times. Set up an appropriate sample space for this experiment.

18. Consider four objects, *a*, *b*, *c*, and *d*. Suppose that the order in which these objects are listed represents the outcome of an experiment. Describe a sample space for this experiment.

19. Each of three distinguishable balls can be placed in any of three cells. List a sample space for this experiment. (You may have more than one ball in each cell.)

20. Each of three indistinguishable balls can be placed in any of four cells. List a sample space for this experiment.

21. Three individuals are randomly selected from a group of 500 people. Describe a sample space for this experiment.

22. A coin is tossed n times, where n is a given positive integer. Describe a sample space for this experiment.

23. Toss a coin until it falls heads the first time. Give a sample space for this experiment.

24. Toss a coin until two successive tosses are identical. Give a sample space for this experiment.

1.3 Events

The definition of sample space as the collection of possible outcomes of a random experiment is the first step in developing a mathematical model for the experiment. The second step is to define the terms *event* and *occurrence of an event*.

Suppose we put twenty identical chips numbered from 1 to 20 in a jar and draw one chip. A sample space for this experiment would be the integers from 1 to 20 inclusive. Having drawn a chip we can ask questions such as

1. Is the number on the chip a multiple of 3?
2. Is the number on the chip greater than 10?
3. Is the number on the chip 10?

If the number on the chip is any of the six outcomes in the set

$$\{3, 6, 9, 12, 15, 18\},$$

the answer to question 1 is affirmative. For question 2 the appropriate set of outcomes is $\{11, 12, 13, 14, 15, 16, 17, 18, 19, 20\}$. For question 3 the set is the single outcome $\{10\}$. The following definitions take these ideas into account.

Definition 1: Event

An event is a set of outcomes of the sample space.

Definition 2: Elementary event

An elementary event is a set consisting of a single outcome of the sample space.

From Definition 1 it is apparent that an event is described by a set of outcomes from the sample space. But event also means happening and in this context we use the phrase "occurrence of an event." To say that an event has occurred is to say that the actual outcome of the random phenomenon is an outcome in the set of outcomes associated with that event. Thus, if on the draw of a chip one of the six numbers in the set {3, 6, 9, 12, 15, 18} occurs, then we say that the event "the number is a multiple of 3" has occurred; otherwise the event does not occur.

To summarize (and emphasize) this second phase in the development of a mathematical model, the intuitive notion of an event as a part of a random phenomenon has been precisely and unambiguously defined as a set of outcomes among all the thinkable outcomes in the sample space. Moreover, every set of outcomes is an event, and the event is said to have occurred only if the particular observed outcome is in the set of outcomes for that event.

Example 1

An animal is being conditioned to respond in a certain way to a given stimulus. For example, the response might be to push (or not to push) a lever, and at each stage of the conditioning, the same stimulus is presented to the animal. A reward is given to the animal based on its response. If the response of the animal is observed for three consecutive trials, a sample space for the random experiment would have eight outcomes, as shown in Table 1. The event "more pushes than not

Table 1

Outcome	Trial 1	Trial 2	Trial 3
1	push	push	push
2	push	push	not push
3	push	not push	push
4	not push	push	push
5	push	not push	not push
6	not push	push	not push
7	not push	not push	push
8	not push	not push	not push

pushes" is represented by the set {1, 2, 3, 4}. The event "alternating push and not push" is described by the set {3, 6}. The event "at least two consecutive pushes" is the set of outcomes {1, 2, 4}. ▌

In the example of twenty chips in a jar, the event "number is a multiple of 3" was described by one set of outcomes and the event "number greater than 10" by a second set. The statement "number a multiple of 3 and greater than 10" defines a new event {12, 15, 18}, and consists of outcomes that are common to the two original events. The implied relationship between the sets of outcomes involves the language and notation of set theory. In Section 1.4 we review briefly the pertinent concepts that will be needed in our discussion of probability theory and calculus.

Problems

1. For the experiment of tossing a die twice, with the sample space described in Example 5 of Section 1.2, list each of the following events.
   (a) The total score is even.
   (b) The total score is 4.
   (c) Each throw results in an even score.
   (d) Each throw is greater than 4.
   (e) The first throw is odd.

2. A marksman fires at 25 targets and records a hit or a miss each time. Set up an appropriate sample space for this experiment and list each of the following events.
   (a) He hits an even number of targets.
   (b) He hits exactly 10 targets.
   (c) He hits more targets than he misses.
   (d) He hits no targets.
   (e) He hits more than 10 and less than 15 targets.

3. An urn contains a large, but unknown, number of colored balls, some of which are red, others blue, and the rest yellow. Three balls are drawn one after another and the color of each is recorded. Describe a suitable sample space and list each of the following events.
   (a) The colors are all different.
   (b) Exactly one color is yellow.
   (c) The first and third colors are the same.
   (d) Exactly two of the colors are the same.
   (e) Only two different colors appear.

4. A coin is tossed four times in succession with the result recorded each time. Set up an appropriate sample space for this experiment and list each of the events.
   (a) Heads and tails alternate.
   (b) The first toss is a head.
   (c) The first and last toss are the same.
   (d) The number of heads is 2.
   (e) There are more heads than tails.

5. A lot contains items weighing 5, 10, 15, 20, and 25 pounds, with at least two items of each weight in the lot. Two items are chosen from the lot one after another. Describe a suitable sample space and list each of the events.
   (a) The second item is twice as heavy as the first.
   (b) The first item weighs 10 pounds less than the second.
   (c) The total weight is less than 35 pounds.
   (d) The average weight is 15 pounds.
   (e) The first item weighs more than the second.

1.4 Definitions and Operations for Sets

The mathematical concept of a set, which is extremely fundamental to much of mathematics, is characterized by its simplicity and its primitive nature. The theory originated with the German mathematician, George Cantor (1845–1918), and the results and applications to mathematics have been profound.

To avoid logical pitfalls we shall accept the notion of set (Section 1.2) as a primitive concept, just as point or line is a primitive or undefined concept in geometry. But just as a line is known by its points, *a set is known by its members*. A set is known when it can be decided whether or not an object belongs to it. The usual notation for showing that an object is in a set is defined next.

Definition 1: Set membership

The individual objects of a set are called elements of the set and are said to belong to the set, or to be members of the set. Set membership is denoted by the symbol $\in$. If S represents the set and x represents an element of the set, then $x \in S$ is read "the element x belongs to the set S," or "x is an element of S."

The simple act of making a list of the members of a set (if it is possible to list them all) involves the idea of counting. To count a set means to associate the set of objects with the consecutive integers 1, 2, 3, and so on, in order. (Three dots will denote "and so on"; thus 1, 2, 3, ... means "the consecutive integers 1, 2, 3, and so on.") The set has a "last" member if we run out of elements to associate with integers, and the last integer matched up tells us how many elements are in the set. Such a set is said to be *finite*.

A set is *infinite* if it is not finite which means that in the matching process there is no last integer used. However, if for each member of the set we can associate a positive integer, and conversely, if for each positive integer we can assign one member of the set, the set is infinite and *countable*. All other infinite sets are said to be *uncountable*. We shall frequently use the four terms *finite, infinite, countable*, and *uncountable* in describing sets.

Although uncountable sets are not easy to demonstrate, it can be shown [31] that the set of rational numbers and irrational numbers is uncountable. The set of points on a line segment of any length is also an uncountable set.

One way of representing a set is simply to list the elements, if this is possible. Another is to give a rule for membership in the set. Some sets can be conveniently identified both ways. For example, the set {4, 8, 12, 16, 20, 24} can also be represented by the statement "the set of all positive integers between 1 and 25 that are multiples of 4." More often the set membership method is preferred, and we shall use the notation

$$S = \{x \mid x \text{ satisfies the membership condition}\}.$$

The curly brackets indicate a set; the vertical bar stands for "such that." The statement is read "S is the set of x's such that x satisfies the membership condition." The set defined by "the set of all positive integers between 1 and 25 that are multiples of 4" can be written $S = \{x \mid x$ a positive integer, $1 < x < 25$, x is divisible by 4}.

Much has been said about objects that satisfy some sort of membership condition. In all the examples so far, every set had at least one member, but there is no guarantee that every set will have members. For example, the set of American mathematicians now on Mars is a case that to our knowledge has no members.

Definition 2: Null set

A null set or empty set is a set with no elements.

If one were to describe a null set by listing the elements, one could use the set symbol "{ }." By convention, however, the symbol $\emptyset$ is used to denote the empty set, so that $\emptyset = \{ \}$.

Definition 3: Equal sets

Two sets A and B are equal if and only if the sets contain the same elements. We write $A = B$. If A is not equal to B, we write $A \neq B$.

Definition 4: Subset of a set

A set A is called a subset of B if every element of A is an element of B. A is then said to be "included in" or "contained in" B, and this is denoted by $A \subset B$. If B contains at least one element that is not in A. then A is called a *proper subset* of B.

Since a null set has no elements, the definition of a subset is not violated when we assert as a logical truth that every element of a null set is contained in any set whatsoever, and hence we agree that a null set is a subset of every set. Consequently, there can only be one null set. It is meaningful to write $\emptyset \subset A$ regardless of what set A may represent.

Note that the definition of subset allows us to write $A \subset A$ so that every set may be considered as a subset of itself. Also the statement $A \subset B$ does not preclude the statement $B \subset A$. When both statements are true, then the sets must be equal. An alternative definition of equality of sets is: Two sets A and B are equal if and only if $A \subset B$ and $B \subset A$.

The two symbols $\in$ and $\subset$ denote two distinct notions—set membership and set inclusion, respectively. The difference between these ideas is dramatized by comparing the element x of set S and the subset $\{x\}$ of set S. Clearly, $x \in S$ and $\{x\} \subset S$. For example, let S be the players of a football squad that has but one captain. If x is the captain, then $x \in S$, while $\{x\}$ is the subset of all players on the squad that are captains, hence $\{x\} \subset S$.

It is natural to ask how sets can interact or interrelate with each other. Consider a system of sets that are subsets of some fixed, universal set U. The universal set U is fixed for a discussion where U changes as the discussion changes. The universal set U, for example, may be all the books in a library, or all the integers.

Definition 5: Complement

Let A be any subset of U. The set of all elements of U not in A is the complement of A with respect to U and is denoted by A'. In notation,

$$A' = \{x \mid x \in U, x \ not \ in \ A\}.$$

Definition 6: Union

Let A and B be any subsets of U. The set whose elements belong either to A or B (or both) is the union of A and B and is denoted by $A \cup B$. In notation,

$$A \cup B = \{x \mid x \in U, x \in A \ or \ x \in B\}.$$

Definition 7: Intersection

Let A and B be any subsets of U. The set whose elements belong to both A and B is the intersection of A and B and is denoted by $A \cap B$. In notation,

$$A \cap B = \{x \mid x \in U, x \in A \text{ and } x \in B\}.$$

Example 1

Let $U = \{0, 1, 2, 3, 4, 5, 6, 7, 8, 9\}$, $A = \{0, 2, 4, 6, 8\}$, $B = \{0, 3, 6, 9\}$, and $C = \{1, 2, 7\}$.

(a) $A' = \{1, 3, 5, 7, 9\}$.
(b) $A \cup B = \{0, 2, 3, 4, 6, 8, 9\}$.
(c) $A \cap B = \{0, 6\}$.
(d) $A \cap C = \{2\}$.
(e) $B \cap C = \varnothing$.   ▮

Two sets may not have common elements, as in part (e). If this is the case, the intersection is the null set and we say that the two sets are *disjoint*.

There are many similarities between set operations and the arithmetic operations of elementary algebra—so many, in fact, that set operations may be called algebraic operations, and the system consisting of the sets and their operations may be called an algebraic system. A list of some of the significant relations for set operations is given in Table 1.

These rules of operation are independent of the nature of the elements of the universal set. We call statements (1)–(17) *identities* because they do not depend on the choice of subsets. In (1)–(4), $\varnothing$ and U are called *identity elements* because they perform the same duties as 0 and 1 do in arithmetic.

Of course, these rules of operation require proofs. One way is to invoke the rule for equality, that is, to show that the left side is contained in the right side and that the right side is contained in the left side. For example let us prove identity (15). Let x be any element of $A \cap (B \cup C)$. This implies that x belongs to A and that x belongs to $B \cup C$. Therefore, x belongs to A and B or x belongs to A and C. But this means that x belongs to the set on the right side of identity (15), so the set on the left side is contained in the set on the right side.

To show that the right side of identity (15) is contained in the left, suppose y is an element of $(A \cap B) \cup (A \cap C)$. Therefore, y belongs to $A \cap B$ or y belongs to $A \cap C$. If y belongs to $A \cap B$, then y belongs to A and B, so y belongs to $B \cup C$ as well. If y belongs to $A \cap C$, then y belongs to A and $B \cup C$, so the right side of identity (15) is included in the left side. By definition, therefore, the two sets are equal.

Table 1

Identity laws

    (1) $A \cup \varnothing = A$ (2) $A \cap \varnothing = \varnothing$

    (3) $A \cup U = U$ (4) $A \cap U = A$

Idempotent laws

    (5) $A \cup A = A$ (6) $A \cap A = A$

Complement laws

    (7) $A \cup A' = U$ (8) $A \cap A' = \varnothing$

    (9) $(A')' = A$

Commutative laws

    (10) $A \cup B = B \cup A$ (11) $A \cap B = B \cap A$

Associative laws

    (12) $A \cup (B \cup C) = (A \cup B) \cup C$

    (13) $A \cap (B \cap C) = (A \cap B) \cap C$

Distributive laws

    (14) $A \cup (B \cap C) = (A \cup B) \cap (A \cup C)$

    (15) $A \cap (B \cup C) = (A \cap B) \cup (A \cap C)$

De Morgan's laws

    (16) $(A \cup B)' = A' \cap B'$

    (17) $(A \cap B)' = A' \cup B'$

To help clarify one's thinking on the complex relations of subsets, a simple pictorial method of representing sets and set operations known as *Venn diagrams* is useful (see Figure 1). In Venn diagrams the universal set U is pictured as all the points in a large geometric rectangle, and subsets are represented as regions of the rectangle. By shading appropriate regions the elementary set operations can be represented visually. In many cases these diagrams are an excellent way to demonstrate that two operations are equivalent.

Venn diagrams are not proofs. They help in establishing the plausibility of a relation, but a rigorous proof requires a mathematical argument such as the one used to establish identity (15).

Let us pose a question: How many subsets can be created from a finite set? More precisely, if $S = \{a_1, a_2, \ldots, a_N\}$, how many subsets are there?

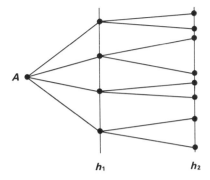

Figure 1

A proof that a finite set with N elements has 2^N subsets follows from a basic principle of counting. This statement is known as the basic principle of combinatorial analysis.

> **Basic Principle of Combinatorial Analysis.** Suppose there is a task T that can be split into component tasks T_1 and T_2, where T_2 will follow T_1. Suppose further that task T_1 can be performed in n_1 ways and task T_2 in n_2 ways. The principle asserts that task T can be performed as task T_1 followed by task T_2 in $n_1 n_2$ ways.

The validity of the principle can be checked using Figure 2. Suppose

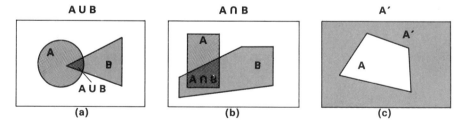

Figure 2

$n_1 = 4$ and $n_2 = 2$, and each line segment from point A to line h_1 represents one way of executing T_1. Each of the line segments from line h_1 to line h_2 is a way that task T_2 can be performed. The totality of ways in which T_1 and T_2 can be performed is simply the number of distinct routes from A to h_2. In this case $n_1 n_2 = 8$.

Example 2

In order to travel from a suburb of Chicago to New York, the first task is to get to the airport and the second task is to select a flight from Chicago to New York. There are 3 ways to get to the airport and 15 flights from Chicago to New York. Using the basic principle we see that the trip can be performed in 45 ways.   ∎

The extension of the principle to any finite set of tasks is clear. Let task

$$T_1 \text{ be performed in } n_1 \text{ ways,}$$
$$T_2 \text{ be performed in } n_2 \text{ ways,}$$
$$T_n \text{ be performed in } n_n \text{ ways.}$$

For future notation we can write this statement as "Tasks T_i, $i = 1, 2, \ldots, n$ can be performed n_i, $i = 1, 2, \ldots, n$ ways, respectively." The total task T, consisting of T_1, followed by T_2, and so on, can be performed in $n_1 n_2 \cdots n_n$ ways.

Example 3

A man has five suits, three pairs of shoes, and two hats. How many different combinations of this attire can he wear? The selection consists of three tasks: selection of suit, selection of a pair of shoes, and selection of a hat. By the basic principle of combinatorial analysis, there are $5 \cdot 3 \cdot 2 = 30$ combinations of attire.   ∎

To deduce that a finite set S of N elements has 2^N subsets, the following argument suffices. The overall task T to be performed is to select a subset of S. This task can be broken down into N tasks with the consideration of each element as a member of the subset a separate task that can be performed two ways. A subset is defined when these N tasks are performed. Thus there are as many different subsets as there are ways to perform N tasks, each with two alternatives, and this number is 2^N.

Problems

1. Which of the following sets are identical?
   (a) $\{1, 2, 3, 4\}$ (b) $\{2, 1, 4, 3\}$
   (c) $\{(1, 2), 3, 4\}$ (d) $\{1, 3, 2, 4\}$
   (e) $\{(1, 2, 3, 4)\}$

2. List the members of the following sets:
   (a) $\{x \mid x^2 - 1 = 0, x \text{ an integer}\}$
   (b) $\{x \mid x = a \text{ or } x = b\}$
   (c) $\{x \mid x^2 + 1 = 0, x \text{ an integer}\}$

3. Let $Z = \{x \mid x$ is an integer$\}$ and let $Z^+ = \{x \mid x$ is a positive integer$\}$.
   List the members of the following sets:
   (a) $\{n \mid n \in Z^+, n \leq 4\}$
   (b) $\{n \mid n \in Z, n^2 \leq 17, n < 0\}$
   (c) $\{n \mid n \in Z, n = 2m,$ where $m \in Z^+\}$

4. For each set, state whether the set is finite or infinite. Where
   feasible, list the members of the set.
   (a) $\{x \mid x$ is an industry, x made a profit in 1961$\}$
   (b) $\{x \mid x$ is a prime number$\}$
   (c) $\{x \mid x$ is a Ford dealer$\}$
   (d) $\{x \mid x$ is a letter in the word probability$\}$
   (e) $\{x \mid x$ is an integer divisible by 3$\}$
   (f) $\{x \mid x$ is a human who has played football$\}$

5. Construct all the subsets of the set $M = \{a, b, c\}$. Which of the
   subsets are proper subsets?

6. The National Baseball League at one time was a set of 10 baseball
   teams, each of which was a set of players; thus the members of the
   set are themselves sets. Give examples of sets whose members are
   themselves sets in
   (a) geography   (b) biology
   (c) mathematics   (d) economics

7. If $A = \{(1, 3), (2, 4), (5, 6)\}$, $B = \{(1, 2, 3), (4, 5, 6)\}$, and $C = \{1, 2, 3, 4, 5, 6\}$, are these sets the same or different?

8. Which of the following are true?
   (a) $\emptyset \subset \emptyset$ (b) $\emptyset \in \emptyset$
   (c) $\emptyset = \{\emptyset\}$ (d) $\emptyset \in \{\emptyset\}$
   (e) $\emptyset \in S$, where S is any set whatsoever.

9. Let
$$U = \{1, 2, 3, 4, 5, 6, 7, 8, 9, 10, 11, 12\},$$
$$A = \{1, 2, 3, 4, 5, 6\},$$
$$B = \{4, 5, 6, 7, 8, 9\}.$$

   Write out the following sets:
   (a) $A \cap B'$ (b) $(A \cap B') \cup (A' \cap B)$
   (c) $(A \cup B)'$ (d) $A' \cup B'$
   (e) $(A \cap B)'$ (f) $A' \cup B'$

10. Let U be the set of all products manufactured by a food processor
    with sales divisions in this country and overseas. Let
    $A = \{x \mid x$ is a product on which the profit margin is greater than
        10 percent$\}$
    $B = \{x \mid x$ is a product sold overseas$\}$
    Describe the following sets:
    (a) $A \cap B$ (b) $A \cup B$
    (c) $A' \cap B$ (d) $A' \cap B'$

11. Let U be the set of all undergraduate students at a university, and let A_1, A_2, A_3, and A_4 be, respectively, the subsets of freshman, sophomore, junior, and senior students. Further, let B be the subset of female students and let C be the subset of students owning cars. For each of the following sets, give an alternative verbal specification.
    (a) $A_1 \cup A_2$ (b) $C \cap (A_1 \cup A_2)$
    (c) $B \cap A_3$ (d) $(C \cup B)'$
    (e) $B \cap (A_3 \cup A_4)'$

12. Construct a Venn diagram for three sets represented as three circles that all intersect. Let the sets be denoted by A, B, and C, and identify the eight regions created by the three circles in terms of the sets A, B, and C.

13. The blood type of humans in a double classification depending on the presence of the three antigens, A, B, and Rh. The type AB, A, or B is assigned depending on the presence of either A, B, or both, and the type O is assigned if neither A nor B is present. The second classification is called Rh$^+$ if the Rh antigen is present, and Rh$^-$ if it is not. Draw a Venn diagram to represent these subsets and identify each of the eight types.

14. Consider the following subsets of Z^+, the set of positive integers:

$$A = \{x \mid x \in Z^+, x = 2y, \text{ for } y \in Z^+\}$$
$$B = \{x \mid x \in Z^+, x = 2y + 1, \text{ for } y \in Z^+\}$$
$$C = \{x \mid x \in Z^+, x = 3y, \text{ for } y \in Z^+\}$$

    (a) Describe the set $A \cap C$ in set notation.
    (b) Describe the set $B \cup C$ in set notation.
    (c) Verify that $A \cap (B \cup C) = (A \cap B) \cup (A \cap C)$.

15. Which of the following are true for *all* subsets of any universal set U?
    (a) $A \cap B \subset A$ (b) $A \cap B \subset B$
    (c) $A \cap B' \subset A$ (d) $A \cap B' \subset B$
    (e) $A' \cap B' \subset A'$ (f) $A' \cap B' \subset A$

16. (a) If $x \in A$ and $A \subset B$, is it necessarily the case that $x \in B$?
    (b) If $x \in A$ and $A \in B$, is it necessarily the case that $x \in B$?
    If your answer is in the negative, given an example.

17. Prove: If A, B, and C are sets such that $A \subset B$ and $B \subset C$, then $A \subset C$. (That is, the relation of being a subset is transitive.)

18. If A and B are nonempty subsets such that $A \cup B = B$, which of the following, if any, are true?
    (a) $A \subset B$ (b) $B \subset A$
    (c) $A \cap B = A$ (d) $A \cap B = B$

19. In planning a round trip from Chicago to Paris by way of New York, a traveler decides to travel between Chicago and New York by air and New York and Paris by sea. There are six airlines operating

between Chicago and New York, and four steamship lines operating between New York and Paris. In how many ways can the round trip be made without traveling over any line twice?

20. How many four-digit numbers can be formed from the digits 1, 2, 3, 4, and 5
    (a) if no digit may be repeated?
    (b) if the number must be odd without any repeated digit?

21. In how many ways can 6 students be seated in a classroom with 12 seats?

22. Two sets A and B are said to be *disjoint* if $A \cap B = \emptyset$. Which of the following statements are true of all subsets of a set S?
    (a) A and A' are disjoint.
    (b) If A and B are disjoint, then $B = A'$.
    (c) If A and B are disjoint, then $A \subset B'$.
    (d) If $A \subset B$, then A' and B are disjoint.
    (e) If $A \subset B$ and B and C are disjoint, then A and C are disjoint.

23. If A and B are subsets of a universal set U, the *difference* $A - B$ is defined to be the set

$$A - B = A \cap B'.$$

If $B \subset A$, then $A - B$ is the complement of B relative to A. Verify:
    (a) $B - (B - A) = A \cap B$
    (b) $(A \cap B) - C = A \cap (B - C)$
    (c) $A \cap (B - C) = (A \cap B) - (A \cap C)$
    (d) If $A \subset B$, then $B = (B - A) \cup A$.

24. If A and B are subsets of a universal set U, the *symmetric difference* $A \triangle B$ is defined to be the set

$$A \triangle B = (A \cap B') \cup (A' \cap B)$$

    (a) Draw the Venn diagram for $A \triangle B$.
    (b) Prove that $A \triangle A = \emptyset$.
    (c) Prove that $A \triangle \emptyset = A$.
    (d) Prove that $A \triangle U = A'$.
    (e) Prove that $A \triangle B = B \triangle A$.

25. One hundred households were surveyed with respect to ownership of electrical appliances with the following results: dishwashers, 28; can openers, 30; toasters, 42; dishwashers and can openers, 8; dishwashers and toasters, 10; can openers and toasters, 5; all three, 3.
    (a) How many households had none of the three?
    (b) How many households had only toasters?
    (c) How many households had can openers if and only if they had toasters?

26. In another survey of 100 households (see Problem 25) the numbers of the electrical appliances found were as follows: can openers only, 18; can openers but not dishwashers, 23; can openers and toasters, 8; can openers, 26; dishwashers, 48 ; dishwashers and toasters, 8; no appliances, 29.

     (a) How many households have toasters?

     (b) How many households have can openers and dishwashers but not toasters?

     (c) How many households have toasters if and only if they do not have dishwashers?

27. In still another survey of 100 households (see Problem 25) the report of the number of electrical appliances was as follows: all appliances, 10; dishwashers and toasters, 18; toasters and can openers, 13; can openers and dishwashers, 14; can openers, 30; dishwashers, 20; toasters, 50. As a result, the interviewer was dismissed. Why?

1.5 Events Again

Now let us apply the language and the notation of set theory to a sample space of outcomes of a random phenomenon. If the universal set U is a sample space S of some random phenomenon, then the events associated with the random phenomenon are subsets of S; thus the algebra of sets becomes the algebra of events.

Every thinkable outcome of the conceptual experiment is represented in S, so S will be known as the certain event. The null set $\emptyset$ is the *impossible* event, since there are no outcomes of S in $\emptyset$. Every event is a subset of outcomes of X, and an event A is said to occur whenever the result of the random experiment is an outcome that belongs to the subset A. The set-inclusion relation, $A \subset B$, means that every outcome in subset A is also an outcome in subset B. Thus the occurrence of event A implies the occurrence of event B. The complement of an event is the subset of all outcomes that

Table 1

Neither event A nor event B occurs	$(A \cup B)' = A' \cap B'$
A occurs but not B	$A \cap B'$ or $A - B$
At least one of A or B occurs	$A \cup B$
A or B occurs but not both	$(A \cap B') \cup (A' \cap B)$
At most one of A and B occurs	$(A \cap B)' = A' \cup B'$
Both A and B occur	$A \cap B$

are not in the subset of outcomes for event A, so in this case A' refers to the event that occurs if the event A does not occur. The union of two events $A \cup B$ is the event that contains the outcomes of either A or B, so the occurrence of $A \cup B$ is the occurrence of either A or B. Similarly, $A \cap B$ is the event that contains the outcomes of both A and B, and the occurrence of $A \cap B$ is the occurrence of A and B.

A list of some common statements for events and their equivalent form in set notation is given in Table 1.

Example 1

Toss a coin three times and denote every trial of three tosses with three letters. For example, (H, T, H) represents three tosses with a head on the first toss, a tail on the second toss, and a head on the third toss. The sample space for this random phenomenon is listed in Table 2. If

Table 2

(H, H, H)	(T, H, H)
(H, H, T)	(T, H, T)
(H, T, H)	(T, T, H)
(H, T, T)	(T, T, T)

the event A is "the first toss is H," it is represented by the subset {(H, H, H), (H, H, T), (H, T, H), (H, T, T)}. If the event B is "the third toss is H," it is represented by the subset {(H, H, H), (H, T, H), (T, H, H), (T, T, H)}. The event $A \cap B$ is the subset {(H, H, H), (H, T, H)}, which is the event that both A and B occur. The event $A' \cup B'$ is the subset {(T, H, H), (T, H, T), (T, T, H), (T, T, T), (H, T, T), (H, H, T)}, which is the event that at most one of A or B occurs. ▌

Example 2

Suppose you are given three urns, labeled 1, 2, and 3, and two balls, one red, and the other green. In some random fashion, select an urn for the red ball and put the red ball in it. Then select an urn for the green ball and put the green ball in it. Each pair of selections can be represented by a pair of numbers such as (2, 1), where the urn selected for the red ball is represented by the first number, and the urn selected for the green ball is represented by the second number. The sample space for this random phenomenon is listed in Table 3. If event A is that two urns are empty, it is represented by the subset {(1, 1), (2, 2), (3, 3)}. If event B is that the third urn is not empty, it is represented

Table 3

(1, 1)	(1, 2)	(1, 3)
(2, 1)	(2, 2)	(2, 3)
(3, 1)	(3, 2)	(3, 3)

by the subset $\{(3, 1), (3, 2), (3, 3), (1, 3), (2, 3)\}$. The event $A \cap B$ is the subset $\{(3, 3)\}$. The event that the sum of the numbers of the occupied urns is even is the subset $\{(1, 1), (3, 1), (2, 2), (1, 3), (3, 3)\}$.   ▍

Problems

1. The numbers, 1, 2, ..., 20 are printed on twenty identical chips. A chip is drawn at random and the number on the chip is noted. What is the event corresponding to each of the following?
   (a) The number is divisible by 5.
   (b) The number is odd and a prime.
   (c) The number has two digits and is even.
   (d) The number is prime and two more than another prime.
   (e) The number is even or divisible by 3.
   (f) The number is greater than 5 or divisible by 3 and even.

2. An experiment consists of selecting three radio tubes from a lot and testing them. If the tube is defective, assign the letter D to it. If the tube is good, assign the letter G to it. A selection is then described by three letters. For example, (D, G, G) denotes the outcome that the first tube selected is defective and the remaining two are good. Let A_1 be the event that the first tube selected is defective, A_2 the event that the second tube selected is defective, and A_3 the event that the third tube is defective. List the elementary events in the sample space, and describe the following events by listing their elements.
   (a) A_1 (b) A_2
   (c) A_3 (d) $A_1 \cup A_2$
   (e) $A_1 \cup A_3$ (f) $A_2 \cup A_3$
   (g) $A_1 \cup A_2 \cup A_3$ (h) $A_1 \cap A_2$
   (i) $A_1 \cap A_3$ (j) $A_2 \cap A_3$
   (k) $A_1 \cap A_2 \cap A_3$

3. A coin is tossed four times. List the 16 possible outcomes of the sample space and list the elements in each of the following events:
   (a) Exactly two heads occur.
   (b) Head on the first toss and exactly two tails occur.
   (c) Heads on the first two tosses or tails on the last two tosses.
   (d) At least three heads.
   (e) Neither three heads nor three tails.

4. Two dice, on black and one red, are tossed and the number of dots on their upper faces are noted. (Refer to Example 5 in Section 1.2.) List the members of the following events:
    (a) a total of more than 10
    (b) an even total and the red die even
    (c) black die shows more than 5
    (d) black die showing more than 5 and the red die showing less than 3
    (e) red die showing more than 3 or the black die showing an even number
    (f) red die showing an odd number or black die showing a number 2 more than the number the red die is showing

5. A city council has five major programs under consideration but can afford a cost analysis of only three. Let A1, A2, A3, A4, and A5 denote the programs. List the 10 possible triples of programs and describe the following events for the random experiment of choosing three programs to analyze:
    (a) Program A1 will be chosen.
    (b) Programs A1 and A2 will be chosen.
    (c) At least one of program A1 or program A2 will be chosen.
    (d) At most one of program A1 or program A2 will be chosen.

6. Assume that in an isolated physical system there are three molecules M_1, M_2, and M_3 each having 0, 1, or 2 units of energy, but the sum of their energies is 2 units. Describe a sample space for this phenomenon and list the elementary events in each of the following events:
    (a) M_1 will have energy 0.
    (b) M_1 and M_2 will have the same energy.
    (c) Neither M_2 nor M_3 will have 2 units of energy.
    (d) M_2 will not have 0 energy.
    (e) M_2 or M_3 will have 1 unit of energy.

7. Let A and B be any two events of a sample space S. For each of the following events, draw a Venn diagram and shade the region corresponding to the following events:
    (a) the event that exactly 0 of the events A and B occurs
    (b) the event that exactly 1 of the events A and B occurs
    (c) the event that exactly 2 of the events A and B occur
    (d) the event that at least 0 of the events A and B occurs
    (e) the event that at least 1 of the events A and B occurs
    (f) the event that at least 2 of the events A and B occur
    (g) the event that at most 0 of the events A and B occurs
    (h) the event that at most 1 of the events A and B occurs
    (i) the event that at most 2 of the events A and B occur

8. For the events stated in Problem 7, write expressions using intersection, union, complement, and A, B.

9. Let A, B, and C be any three events of a sample space S. For each of the following events, draw a Venn diagram and shade the region corresponding to the event.
   (a) At least one occurs.   (b) Only A occurs.
   (c) A and B occur, but not C.   (d) All three occur.
   (e) Exactly two occur.   (f) At most two occur.

10. Let A, B, and C be any three events of a sample space S. Using intersection, union, and complement, write expressions for the following:
    (a) At least one of A, B, or C occurs.
    (b) Only A occurs.
    (c) A and B occur, but not C.
    (d) All three of the events occur.
    (e) Exactly two of the three events occur.
    (f) At most two of the events occur.

1.6 Functions

The concept of *function* plays an extremely important part in the analysis of physical as well as abstract phenomena, and provides a useful way to study complicated ideas. Many physical laws are simple statements of a relationship that exists between two sets of phenomena, such as atmospheric pressures and heights above sea level, or the pitches of the note of a plucked string and the tensions of the string. Other examples are sales taxes and the amounts of purchase or growth rates of a plant and nutrients in the soil. Abstract ideas, too, such as a random variable and a sequence of numbers, can be conveniently defined in terms of function. A major task in scholarly endeavors—medicine, business, social science—is to discover how observable quantities are related, and then to express this relationship in some exact manner. The function concept provides a neat way in which the relationships can be stated. Economists, for example, usually preface the word function with a word that describes the relationship, such as "demand function" or "utility function." The mathematician usually omits the adjective because the same function can be used to describe many different phenomena. Instead, all functions that have the same form are grouped into one class.

Here are two examples that should help identify the idea of a function:

Example 1

For the set of all the people living in one city, the number representing the age of an individual to the nearest year is uniquely specified for each person. A person $27\frac{1}{2}$ years of age is put with the 27-year-olds. Thus there is a relationship between the set of people of the city and the set of nonnegative integers. ▋

Example 2

For the set of all words in a dictionary, the first letter of the word is uniquely specified for each word. Thus there is a correspondence between the set of all the words in a dictionary and the set of the 26 letters of the alphabet. ▋

Definition 1 : Function

A rule or a correspondence that assigns to every element of a set H a unique element of a set K is called a *function* from H into K. The set H is called the *domain* of the function, and the set K is called the *co-domain* of the function. The set of all elements of K that are related to the elements of H by the correspondence is called the *range* of the function.

By this definition we see that for the function from H to K to be well defined, every element of H must correspond to some element of K. This does not mean that every element of K will be assigned to an element of H; when this does occur the codomain and the range will coincide.

Traditionally, mathematicians represent the functional correspondence by letters such as f, g, F, or G in the following way. Let x represent an arbitrary, unspecified element of the set H; then it is customary to write $f(x)$ for the element of K that corresponds to x. The element $f(x)$ of K is called the value of the function f for the element x and it is read as "f of x" or "f at x." Another name for $f(x)$ is the *image* of x.

Figure 1 is a graphical representation of a function, with the sets H and

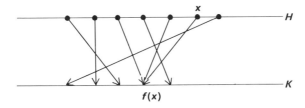

Figure 1

K pictured as sets of points and the arrow from the point x to the point $f(x)$ representing the functional correspondence.

Figure 2 is a "flow-diagram" interpretation of the same function with the element x of H considered as "input," the function f as some sort of operation performed on the input, and the value $f(x)$ of K as the "output."

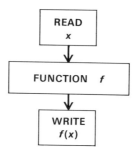

```
┌──────────┐
│   READ   │
│    x     │
└──────────┘
     │
     ▼
┌────────────────────┐
│   FUNCTION   f     │
└────────────────────┘
     │
     ▼
┌──────────┐
│  WRITE   │
│   f(x)   │
└──────────┘
```

Figure 2

Additional examples of the function concept follow:

Example 3

(a) H: the set of all circles in a plane
     K: the real numbers
     $f(x)$: the area of circle x

(b) H: the set of all salesmen for a company
     K: the nonnegative integers
     $g(x)$: the number of new customers obtained by salesman x in a particular year

(c) H: the set of all countries in a state
     K: the nonnegative integers
     $h(x)$: the number of unemployed in county x

(d) H the set of all words in a dictionary
     K: the nonnegative integers
     $F(x)$: the number of vowels in word x

(e) H: the set of all words in a dictionary
     K: the nonnegative integers
     $G(x)$: the number of letters in word x.   ▮

Notice that the functions in Example 3 parts (d) and (e), are different, even though the two domain sets are the same.

Definition 2: Identical functions

Two functions are identical if and only if they have the same domain, range, and correspondence.

Clearly, f and g are the same function if for every element $x \in H$, $f(x) = g(x)$.

In speaking of functions, we often say that f is defined on H and takes its values in K. The set H is sometimes referred to as the domain of definition, or the domain of the variable; the set K is often thought of as containing the set of values of the function. Sometimes, as noted, the entire set K will be values of the function; therefore, it is both the codomain and the range of f.

Do not confuse the value of a function with the function itself. The value of the function $f(x)$ is an element of the set K, while the function f is the relationship between the two sets and includes the two sets.

Functions may be described in several ways. In the preceding examples, the rule was given by a verbal statement. Another method is to list the corresponding pairs, such as a list of grades for the members of a class. If the domain and the range of a function are subsets of real numbers, the image of an element in the domain can sometimes be represented by an algebraic expression; e.g., the image of x is $3x^2 + 1$. Customarily, a function whose *domain* is the real numbers or a subset of the real numbers is called a function of a real variable; and, if the range of the function is also the real numbers or a subset of the real numbers, then the function is called a *real-valued function of a real variable*. A function whose domain consists of sets of elements, such as the subsets of a sample space, is called a *set function*.

A real-valued function of a real variable defines a set of ordered pairs of real numbers. An ordered pair of real numbers can be identified with a unique point in the coordinate plane (see Appendix C). Consequently, a diagram of this kind of function, in the form of a sketch or a graph, is readily obtainable. The collection of points corresponding to the ordered pairs is a description of the function. From time to time, techniques for the efficient determination of the graph of a numerical function will be topics for study. For now, one will have to be content with the calculation of several of the ordered pairs and guessing.

Example 4

The real-valued function f of a real variable x defined by

$$f(x) = \begin{cases} x, & x \leq 1, \\ 1, & x > 1 \end{cases}$$

can be described pictorially by the sketch in Figure 3. For all values of $x \leq 1$, the graph is part of the straight line through the points $(0, 0)$ and $(1, 1)$. For all values of $x > 1$, the graph is part of the line through $(1, 1)$ and $(2, 1)$. ∎

For the sample space associated with a random phenomenon, an event A is by definition a subset of the sample space. What is the probability that A occurs? Since we shall define probability to be a number, we shall

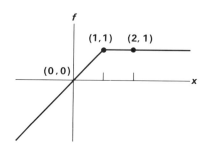

Figure 3

be constructing a function whose domain will be subsets of the sample space and whose codomain is the set of real numbers between 0 and 1 inclusive. The function will be called the probability of the event. We shall pursue these ideas in Chapter 2.

Problems

1. Let H be a set of authors and K the set of all books. If x is an author and y is a book, associate y to x if and only if x is the author of the book y. Does this rule establish a function from H to K?

2. Let S be the set of students who wrote an examination in history for which grades of A, B, C, D, and F were given. Let $M = \{A, B, C, D, F\}$. If x is a student and $y \in M$, associate y to x if and only if y is the grade student x received on the examination. Does this rule establish a function from S to M?

3. If $H = \{1, 2, 3, 4, 5\}$ and $K = \{a, b, c\}$, define two different functions from H to K by listing the pairs in the correspondence.

4. If $H = \{a, b, c\}$ and $K = \{0, 1\}$, write all the possible functions from H to K. (There is a total of eight.)

5. Which of the following correspondences between the set $\{1, 2, 3\}$ and the set $\{4, 5\}$ are functions?
   (a) $1 \rightarrow 4, 2 \rightarrow 4, 3 \rightarrow 5, 1 \rightarrow 5$
   (b) $1 \rightarrow 5, 2 \rightarrow 4, 3 \rightarrow 5$
   (c) $1 \rightarrow 4, 2 \rightarrow 4, 3 \rightarrow 4$

6. Let R denote the set of real numbers. Define two functions as follows:
   f: associate with each $x \in R$ its magnitude; that is, if $x \in R$,

   $$f(x) = \begin{cases} x & \text{for } x \text{ positive or zero,} \\ -x & \text{for } x \text{ negative.} \end{cases}$$

   f is called the *absolute value* of x and written $f(x) = |x|$.
   g: associate with each $x \in R$ the greatest integer less than or equal to the number x.

g is called the *bracket function* and written $g(x) = [x]$.

Compute:

(a) $f(-1.2), g(-1.2)$ (b) $f(4.5), g(4.5)$

(c) $f(\frac{3}{4}), g(\frac{3}{4})$ (d) $f(-\frac{3}{4}), g(-\frac{3}{4})$

7. Define two functions f and g with domain and codomain Z^+ (the set of positive integers) as follows:

    f: associate with $x \in Z^+$ the remainder obtained by dividing x by 2. Thus $f(7) = 1$ and $f(10) = 0$.

    g: associate with $x \in Z^+$ the remainder obtained by dividing x by 3. Thus $g(7) = 1$, $g(12) = 0$, and $g(26) = 2$.

    Compute:

    (a) $f(4), g(4)$ (b) $f(11), g(11)$

    (c) $f(27), g(27)$ (d) $f(15), g(15)$

8. Let f and g denote two functions with domain and codomain Z^+ such that

    f: associates with $x \in Z^+$ the number x itself, $f(x) = x$.

    g: associates with $x \in Z^+$ the number 2, $g(x) = 2$.

    Thus $f(6) = 6$ and $g(6) = 2$.

    Compute:

    (a) $f(4), g(4)$ (b) $f(23), g(23)$

    (c) $f(2), g(2)$ (d) $f(1), g(1)$

    The function f is called the *identity* function and g is an example of a *constant* function.

9. If F is a function that takes the elements of Z^+ into Z^+ such that $F(1) = 3$, and for each $n \in Z^+$, $F(n+1) = 2\ F(n)$, find $F(5)$.

10. If $f(x) = \sqrt{3 - x}$ is a rule that relates certain sets H and K of real numbers,

    (a) determine the largest subset H of real numbers for which the image of $f(x)$ is a real number.

    (b) find the value of $f(2), f(-3), f(-33)$, and $f(0)$.

11. If $f(x) = 1/(x^2 + 1)$ is the law of correspondence of a function defined over the set of real numbers,

    (a) determine the subset (proper or improper) of the real numbers for which the value of the function is a real number.

    (b) find the value of $f(0), f(2)$, and $f(-10)$.

    (c) prove that $f(-x) = f(x)$ for all x for which the function is defined.

12. A function need not be defined by a single expression or a single statement. If g is defined for the real numbers as

$$g(x) = \begin{cases} -1, & x < -1, \\ 0, & -1 \le x \le +1, \\ +1, & 1 < x, \end{cases}$$

(a) determine the domain and range of $g(x)$.
(b) find the value of $g(-13)$, $g(-1)$, and $g(1.3)$.
(c) find the set of values of x such that $-\frac{1}{4} < g(x) < \frac{3}{4}$.
(d) find the set of x such that $g(x) > 2$.

13. The function h is defined for the real numbers as follows:

$$h(x) = \begin{cases} 0, & x < -1, \\ \frac{1}{4}, & -1 \le x < 1, \\ \frac{1}{2}, & 1 \le x < 2, \\ \frac{2}{3}, & 2 \le x < 3, \\ 1, & x > 3. \end{cases}$$

(a) Determine the range of h.
(b) Find the value of $h(-3)$, $h(\frac{3}{4})$, and $h(\sqrt{2})$.
(c) Find the set of x such that $\frac{1}{2} \le h(x) < \frac{3}{4}$.

14. If $f(x) = 1/(2-x)$ relates two sets of real numbers,
(a) determine the subset (proper or improper) of the real numbers for which the value of the function is a real number.
(b) find the value of $f(0)$, $f(2.1)$, and $f(-1)$.

15. If $f(x) = 1/\sqrt{x}$,
(a) determine the subset (proper or improper) of the real numbers for which the value of the function is a real number.
(b) find the value of $f(25)$, $f(0.01)$, and $f(10^2)$.

16. Let H be any set and let A be any subset of H. Define a function on H as follows:

$$f(x) = \begin{cases} 0 & \text{for } x \in H, \, x \notin A, \\ 1 & \text{for } x \in H, \, x \in A. \end{cases}$$

This function is called the *characteristic* function of subset A.
Prove: If A and B are two subsets of H and if f and g are the characteristic functions of A and B, respectively, then the characteristic function h of $A \cap B$ is given by

$$h(x) = \text{lesser of } f(x) \text{ and } g(x) \text{ for every } x \text{ in } H.$$

17. If $f(x) = 2x^2 - 1$ and $g(x) = 1 - 3x$,
(a) find a subset of the real numbers for which the two functions are the same.
(b) find another subset of the real numbers for which the two functions are not the same.

18. The dimensions of a rectangle are 10 and b. Express the area of the rectangle as a function of b and describe the domain and range of the function.

19. The fare for a taxi is generally 35 cents for the first $\frac{1}{4}$ mile or any fraction thereof and 25 cents for each subsequent $\frac{1}{4}$ mile or fraction thereof. Let x be the distance traveled in a taxi and define the function $r(x)$ that is the fare (without tip) for a ride of x miles for $0 \le x \le 1$.

20. A factory worker is employed at the rate of $3 per hour for a normal 8-hour day. If he works overtime, he is paid at the hourly rate of $1\frac{1}{2}$ times the normal rate for the first 4 hours and at the hourly rate of two times the normal rate for the next 4 hours. Let t be the time in hours that the employee works in 1 day and express his total wages as a function $w(t)$ for $0 \leq t \leq 16$.

21. Let H be the words in a standard dictionary, K the positive integers, and f the function that associates with each word the number of distinct vowels in the word. What is the range of f? What is the codomain of f? What is $f(\text{Mississippi})$?

chapter two

Definition of Probability

2.1 Probability Space

For each random experiment there is a sample space, for each sample space there are events, and for each event there is the question of the occurrences of the event. Of greater interest perhaps is "will the event occur?" In this connection the terms chance, likelihood, and probability arise; that is, "What is the chance that it occurs?", "Is it likely to occur?", or "What is the probability that the event occurs?" In some sense these terms offer a measure of the occurrence of the event. Since measure usually connotes a number, in this chapter we shall define the probability that an event occurs by giving a precise and definite way of assigning numbers to events of a random experiment.

The first method for systematically assigning numbers to events is credited to Laplace; in it he considered only random experiments with a finite set of outcomes. The elementary outcomes were taken to be "equally likely" and the probability of the event was defined to be the ratio of the number of elementary outcomes favorable to the event to the total number of the elementary outcomes. Finding the probability of an event therefore involved, at times, sophisticated counting techniques. For this reason the science and art of counting without tedious enumeration, known as *combinatorics*, flourished and developed along with probability theory. Apparently, Laplace assumed that either all random experiments consisted of "equally likely" elementary outcomes or that the theory of probability did not apply otherwise. Anyway, attempts to extend the equally likely notion in a natural fashion were futile.   Perhaps some of the difficulty is hidden by the circular nature of the definition of probability, where one must know what "equally likely" is in order to define probability.   Certainly part of the difficulty in trying to extend this concept lies in specifying the term equally likely. Intuitively, two outcomes are equally likely if they have the same chance of occurring, but "the same chance" means equal probability.

In a second approach to probability, the number assigned to an event is related to how often the event occurs in a series of repetitions of an experiment. In n_1 trials of an experiment, if the event occurs in f_1 trials, the ratio f_1/n_1 is recorded, and this ratio is called the *relative frequency* of occurrence.   If the same experiment is repeated n_2 times and the same event occurs in f_2 of these trials, then the ratio f_2/n_2 is calculated.   Conceivably this process could be continued indefinitely, and in doing so we would generate many series of relative frequencies for the same event.   There is no reason to believe that all these ratios would be identical; but, hopefully, the collection would have some sort of regularity and would "stabilize" at or near some specific number, which we could call the limiting value of the sequence of relative frequencies.   This limiting value is a number that can be assigned to the event, and the relative frequencies can be considered as approximations of the limiting value.

Notice that the relative-frequency approach in no way requires that the outcomes be equally likely.   Life expectancy in the insurance business and the transfer of characteristics to offspring in genetics are examples of problems for which this approach to probability is very significant.   In mathematics, however, neither the relative-frequency method nor the equally likely method has been extended to a rigorous system that would suffice in wide application. Probabilists have therefore tried other ways of providing a general setting for probability theorems.

In 1933 Kolmogorov [10] proposed a set of axioms for the definition of probability, and it is this axiomatic approach that we shall follow.

The central idea in the definition of probability is the assignment of numbers to the events of a sample space so that the numbers can serve as a measure of the occurrence of these events, just as we assign a number as a measure of the length of a line segment.   Although the reader has undoubtedly calculated probabilities informally in games of chance, at this time we shall be assigning arbitrary numbers to events of a sample space so as to satisfy certain axioms.   Whether or not these numbers are realizable as a measure of the occurrence of an event is not relevant to our work.

Since we shall assign a unique real number as the probability of an event in a sample space, probability will be a function.   Not every conceivable function can serve as a probability function, so our formal definition of probability function will have its own distinguishing properties.

We shall begin by assuming that we have some elements, which we shall also call outcomes of a sample space.   We shall then establish explicitly the properties that the set of events and the probability function must satisfy. This will be done with two lists of axioms. The first list of axioms will specify properties for the set of events. The second list will specify properties for the probability function.

Definition 1: Probability space

A probability space $\mathscr{P}$ is a sample space S of outcomes, a set $\mathscr{E}$ of subsets of S called events, and a set function $P(A)$, where A is an event in the domain and the image $P(A)$ is a real number such that the following two sets of axioms are satisfied:

Axioms for the domain $\mathscr{E}$:

B1. The sample space S is in $\mathscr{E}$.
B2. The empty set $\varnothing$ is in $\mathscr{E}$.
B3. If a finite or countable number of events $A_1, A_2, \ldots$, belong to $\mathscr{E}$, then their union belongs to $\mathscr{E}$.
B4. If the event A belongs to $\mathscr{E}$, then the complement A' belongs to $\mathscr{E}$.
B5. If a finite or countable number of events $A_1, A_2, \ldots$, belong to $\mathscr{E}$, then their intersection belongs to $\mathscr{E}$.

Axioms for the set function P (for the codomain R of real numbers):

PA1. For every event A of $\mathscr{E}$, $0 \leq P(A) \leq 1$.
PA2. $P(S) = 1$.
PA3. The probability of the union of a finite or countable number of pairwise exclusive events is the sum of the probabilities of the events; that is, $P(A_1 \cup A_2 \cup A_3 \cdots) = P(A_1) + P(A_2) + P(A_3) + \cdots$ whenever $A_i \cap A_j = \varnothing$ for all i and j, $i \neq j$.

At this point it is convenient to introduce some shorthand notation. $A_1 \cup A_2 \cup A_3 \cup \cdots \cup A_n$ will be written $\bigcup_{i=1}^{n} A_i$, entirely analogous to the summation notion $\sum_{i=1}^{n} a_i$ that is used to denote $a_1 + a_2 + a_3 + \cdots + a_n$. (See Appendix B.) If we are dealing with the union of a countably infinite number of sets we denote $A_1 \cup A_2 \cup A_3 \cup \cdots$ by $\bigcup_{i=1}^{\infty} A_i$. The sum of a countably infinite sequence of numbers is analogously written as $\sum_{i=1}^{\infty} a_i$. While the union of an infinite number of sets is well defined ($\bigcup_{i=1}^{\infty} A_i$ consists of those elements belonging to at least one of the sets A_i), it is not clear what is meant by the sum $\sum_{i=1}^{\infty} a_i$. This is a problem in infinite series; see [18, Chap. 8]. For the present only the finite case is needed, although

detail in Chapter 8. For the present only the finite case is needed, although the theorems are true for the countable case as well. Observe that axiom PA3 can be written in the new notation as

$$P\left(\bigcup_{i=1}^{\infty} A_i\right) = \sum_{i=1}^{\infty} P(A_i).$$

The set $\mathscr{E}$ of subsets of the sample space S with the properties specified by axioms B1 through B5 is called a *Borel field* of sets of S. The statements for the probability function characterize what is known in probability theory as a *normed, nonnegative, countably additive measure* for the elements of the Borel field of sets $\mathscr{E}$ of the sample space S. It may not seem either reasonably or possible that we should need all this machinery, language, and abstraction in order to discuss rigorously such a simple thing as the flip of a coin, but we do.

Let us look at these axioms informally. The first two axioms for the set of events require that the two special sets, the entire sample space S and the null set $\varnothing$, belong to $\mathscr{E}$. The third axiom refers to the union of sets of $\mathscr{E}$ and asserts that if two events are in $\mathscr{E}$, then their union must be an event in $\mathscr{E}$, and so on with three events, four events, or any countable number of events in $\mathscr{E}$. The fourth axiom asserts that the complement of any event A (that is, the event A does not occur) must be a member of $\mathscr{E}$. (See Problem 4 for an equivalent formulation.) The fifth axiom provides the analogue of axiom B3 for the intersection of events; namely, the intersection of two events in $\mathscr{E}$ is in $\mathscr{E}$. In order to assert that a particular set of events is a Borel field of sets, we must show rigorously that the set of events has the five properties stated in the axioms.

Axioms PA1, PA2, and PA3 are general statements about the properties that a function must have to qualify as a probability function. PA1 asserts that the values of the function must lie between 0 and 1 inclusive, which is quite consistent with the relative-frequency notion of probability. The second axiom asserts that we must assign the number 1 to the entire sample space considered as an event. The third axiom is a little more complex. If A_1 and A_2 are two events for which $A_1 \cap A_2 = \varnothing$, then, according to this axiom, the probability of the union of the events A_1 and A_2 is the sum of the probabilities for A_1 and A_2,

$$P(A_1 \cup A_2) = P(A_1) + P(A_2), \qquad (A_1 \cap A_2 = \varnothing).$$

This statement relates an operation between events to an operation between probability values. Although it has been detailed here for only two events, axiom PA3 applies to any countable set of events such that the intersection of every pair of events in the set is empty.

Most of the sample spaces that have occurred so far in this book have been finite sets. When this is the case, the axioms can be made less complex.

For example, axioms B3, B5, and PA3 would need to be stated only for a finite number of events. However, both in theory and practice, infinite sample spaces do occur frequently, and the axioms as we have stated them will apply without modification to the most general case.

Although the sample space can be considered as a model of a random physical experiment, a probability space is a mathematical concept. The set of events must have the properties stated in the axioms, and the probability function must be specified in such a way that the remaining axioms are valid. Note that we have some freedom in selecting the probability function; that is, the axioms do not prescribe a unique probability function.

As a matter of notation we shall use the symbol $P(\cdot)$ to denote the probability function when the events are unspecified. When the probability function is to be applied to a specific event A, then the dot is replaced with A.

On the basis of these axioms, theorems about probability spaces and theorems that extend the properties of the probability function $P(\cdot)$ will be deduced in Section 2.4. In the next two sections the essentials of probability theory for finite sample spaces are developed.

Problems

1. Restate the axioms of Definition 1 for the case when the sample space is finite.

2. Show that the set $\mathscr{E} = \{\varnothing, S\}$, where S is any sample space, satisfies axioms B1–B5.

3. Show that if there are only a finite number of elementary events, so that every set of outcomes is an event (that is, every subset of S), then the axioms for the set of events $\mathscr{E}$ are satisfied.

4. Show that axiom B4 is equivalent to the following axiom, B4': If the events A_1 and A_2 belong to $\mathscr{E}$, then their difference $A_1 - A_2$ belongs to $\mathscr{E}$. (*Hint*: See Problem 23 in Section 1.4 for a definition of set difference.)

5. Suppose the sample space S has a finite number n of elements. The set of all subsets suffices for the set $\mathscr{E}$. Define $P(A)$ to be the number of elements in the subset A divided by n, and show that $P(\cdot)$ satisfies axioms PA1, PA2, and PA3.

2.2 Finite Probability Spaces

Although the definition of an abstract probability space in Section 2.1 is quite general, we shall restrict our discussions in the next five chapters to finite probability spaces; that is, the sample space on which a probability function is defined will be a finite set. In this section we show that for a finite set a probability function is completely defined if a set of values is assigned to the so-called elementary events.

Since the sample space is finite, we represent the elements with the symbols $e_1, e_2, \ldots, e_n$ and the sample space as $S = \{e_1, e_2, \ldots, e_n\}$, where n is a positive integer. The set of all subsets of S (2^n as we have seen) comprises the set $\mathscr{E}$ of events, and this set satisfies axioms B1–B5 of Definition 1 (see Problem 3, Section 2.1). For example, if $n = 4$, $S = \{e_1, e_2, e_3, e_4\}$ and $\mathscr{E}$ consists of the following 16 subsets (or events):

S	$\varnothing$	$\{e_1\}$	$\{e_2\}$
$\{e_3\}$	$\{e_4\}$	$\{e_1, e_2\}$	$\{e_1, e_3\}$
$\{e_1, e_4\}$	$\{e_2, e_3\}$	$\{e_2, e_4\}$	$\{e_3, e_4\}$
$\{e_1, e_2, e_3\}$	$\{e_1, e_2, e_4\}$	$\{e_1, e_3, e_4\}$	$\{e_2, e_3, e_4\}$

We wish next to define the probability function $P(\cdot)$ by assigning numbers to the 2^n subsets of S in such a way that axioms PA1, PA2, and PA3 hold. This assignment of numbers can be considerably simplified if we concentrate on those special events of S that have been defined previously as elementary events. Earlier, the sample space was defined as the set of outcomes of a random experiment; and in Definition 2 of Section 1.3 an elementary event was defined as a set consisting of a single outcome. Therefore, an outcome is an event.

Every event in a finite space is a finite subset of outcomes. Since each outcome can also be considered as an elementary event, every event can be expressed as the union of elementary events. If the event is $\{e_1, e_2\}$, then we can write $\{e_1, e_3\} = \{e_1\} \cup \{e_3\}$. Thus there is a distinction between e_2 and $\{e_2\}$. The former is an outcome in the sample space S; the latter is an event in the set of events $\mathscr{E}$. Therefore, it is appropriate to write $e_2 \in S$ and $\{e_2\} \in \mathscr{E}$. To simplify the exposition, let us denote the elementary event $\{e_1\}$ by E_1, so we can write such expressions as

$$\{e_1, e_2\} = E_1 \cup E_2 = \{e_1\} \cup \{e_2\}.$$

With these understandings we are now ready to define a probability function for a finite probability space. Suppose $S = \{e_1, e_2, \ldots, e_n\}$. The relationship between the arbitrary event A in $\mathscr{E}$ and the elementary events $E_1, E_2, \ldots, E_n$, along with axiom PA3, provides us with a direct way to define the probability function for the general case. An event A, which is an element of $\mathscr{E}$, is an elementary event or a finite union of elementary events. Moreover, the elementary events as elements of $\mathscr{E}$ are pairwise exclusive; therefore, the probability for A can be expressed as the sum of the probabilities for the elementary events in the union, by axiom PA3. The probability function is specified by assigning to the elementary events any set of n nonnegative numbers $p_1, p_2, \ldots, p_n$ such that $\sum_{i=1}^{n} p_i = 1$. We put $P(E_i) = p_i$ and verify easily that PA1 and PA2 are satisfied:

$$P(S) = \sum_{i=1}^{n} P(E_i) = \sum_{i=1}^{n} p_i = 1$$

$$0 \leq p_i \leq 1, \qquad i = 1, 2, \ldots, n.$$

For PA3 we observe that each event A is the union of at most a finite number of elementary events E_i. Hence $P(A) = \sum_{e_i \in A} P(E_i)$. For two events A_1 and A_2 such that $A_1 \cap A_2 = \emptyset$, we see that

$$P(A_1 \cup A_2) = \sum_{e_i \in A_1 \cup A_2} P(E_i),$$

where e_i is a member of $A_1 \cup A_2$. Since there is no elementary event in both A_1 and A_2,

$$P(A_1 \cup A_2) = \sum_{e_i \in A_1} P(E_i) + \sum_{e_i \in A_2} P(E_i)$$
$$= P(A_1) + P(A_2).$$

The general statement

$$P\left(\bigcup_{j=1}^{m} A_j \right) = \sum_{j=1}^{m} P(A_j)$$

follows directly from a mathematical-induction argument. (See Problem 20.)

This completes the verification of the fact that the assignment of probabilities to the elementary events of a finite sample space is sufficient to define a probability function for the set of events $\mathscr{E}$. This also completes the definition of a finite probability space for any finite sample space whatsoever. Observe that if there are n elementary events, then as soon as n nonnegative numbers whose sum is 1 have been selected, the probability function has been determined. For example, when $n = 2$, any two numbers p_1 and p_2 such that $p_1 \geq 0$, $p_2 \geq 0$, and $p_1 + p_2 = 1$ are acceptable. In fact, as soon as p_1 has been selected, $p_2 = 1 - p_1$ necessarily. Choosing a value for p_1 is generally not a mathematical question. In fact, it is at this point where the actual experiment may influence the choice of probability values.

Example 1

Two draws are made from a jar filled with an unknown number of red, white, and blue chips that are indistinguishable except for color. (This is an example of a sample of size 2.) Using R, W, and B to represent the colors, the elements of S can be written as

$$e_1 = (R, R), \qquad e_2 = (R, W), \qquad e_3 = (R, B), \qquad e_4 = (W, R),$$
$$e_5 = (W, W), \qquad e_6 = (W, B), \qquad e_7 = (B, R), \qquad e_8 = (B, W),$$
$$e_9 = (B, B),$$

where the first letter represents the result of the first draw and the second letter the result of the second draw. One assignment of probabilities is as follows:

$$P(E_1) = \tfrac{1}{36}, \qquad P(E_6) = \tfrac{1}{6},$$
$$P(E_2) = \tfrac{1}{18}, \qquad P(E_7) = \tfrac{1}{12},$$
$$P(E_3) = \tfrac{1}{12}, \qquad P(E_8) = \tfrac{1}{6},$$
$$P(E_4) = \tfrac{1}{18}, \qquad P(E_9) = \tfrac{1}{4}.$$
$$P(E_5) = \tfrac{1}{9},$$

If A is the event of a red chip on either the first or second draw, then $A = \{(R, R), (R, W), (R, B), (W, R), (B, R)\}$ and

$$P(A) = \tfrac{1}{36} + \tfrac{1}{18} + \tfrac{1}{12} + \tfrac{1}{18} + \tfrac{1}{12} = \tfrac{11}{13}.$$

Any other assignment of numbers for probabilities would in general yield a different value for $P(A)$. ∎

Example 2

Given two identical balls, place these balls in three similar cells, which are identified by the letters A, B, and C. If this is done in a "random fashion," such as by an honest, blindfolded person, then we can consider this a random experiment. The totality of possible outcomes can be represented by triples of integers, where the first integer is the number of balls in cell A, the second integer the number in cell B, and the third integer the number in cell C.

$$e_1 = (2, 0, 0), \qquad e_4 = (1, 1, 0),$$
$$e_2 = (0, 2, 0), \qquad e_5 = (1, 0, 1),$$
$$e_3 = (0, 0, 2), \qquad e_6 = (0, 1, 1).$$

A rather natural assignment of probabilities to the elementary events would be to assign the value $\tfrac{1}{6}$ to each; $P(E_i) = \tfrac{1}{6}, i = 1, 2, 3, 4, 5, 6$. If this is so, and E is the event that cell B is occupied, then $E = \{e_2, e_4, e_6\}$ and $P(E) = \tfrac{1}{2}$. Another assignment of probabilities is as follows: $P(E_1) = P(E_2) = P(E_3) = \tfrac{1}{9}$, while $P(E_4) = P(E_5) = P(E_6) = \tfrac{2}{9}$. In this case $P(E) = \tfrac{5}{9}$. ∎

Example 3

A group of five students is asked to rate a new course as either "favorable" or "unfavorable." The outcomes for this poll can be represented by quintuples of 0's and 1's, where the first integer is the response of the first person, so that $(1, 0, 1, 1, 0)$ would be the outcome that the first, third, and fourth person expressed a judgment of "favorable." Since there are two possibilities for each of five tasks, there are $2^5 = 32$ outcomes in the sample space S. With no more information than is given, a reasonable assignment of probabilities would be to set $P(E_i) = \frac{1}{32}$ for $i = 1, 2, \ldots, 32$. If A is the event "at least four students respond favorably," then

$$A = \{(1, 1, 1, 1, 0), (1, 1, 1, 0, 1), (1, 1, 0, 1, 1), (1, 0, 1, 1, 1),$$
$$\times (0, 1, 1, 1, 1), (1, 1, 1, 1, 1)\}$$

and $P(a) = \frac{3}{16}$. ∎

Example 4

A research organization has assembled three potential projects for study in the next year. Three groups in the organization independently select one project as their choice for the study.

The outcomes of the selection process can be represented by triples of a's, b's, and c's. where the letters a, b, and c denote the three projects. The first letter in the triple represents the selection of the first group, the second letter that of the second group, and the third letter that of the third group. Thus (b, a, b) would mean that project b was the choice of the first and third group, and project a was the choice of the second group. To count the number of outcomes, take the three selections as three tasks with three possibilities for each task. There are altogether $3^3 = 27$ triples. If one is willing to assume that the rational, discriminating judgement of each group is, in fact, random, then the assignment of a probability $\frac{1}{27}$ to each outcome is appropriate. If B is the event "project b is selected by at least two groups," then

$$B = \{(b, b, a), (b, b, c), (b, a, b), (b, c, b), (a, b, b), (c, b, b), (b, b, b)\}$$

and $P(B) = \frac{7}{27}$. ∎

Problems

1. Suppose a coin is tossed four times, and the number of heads is recorded. Determine the outcomes and arbitrarily assign a probability function for this random experiment.

2. If a die has three sides marked with three dots, two sides with two dots, and one side with one dot, then a probability function for the elementary events E_i, $i = 1, 2, 3$, where E_i represents i dots on the uppermost face is $P(E_1) = \frac{1}{6}$, $P(E_2) = \frac{1}{3}$, $P(E_3) = \frac{1}{2}$. Determine the probability for all events in $\mathscr{E}$.

3. Three people are asked to rate a movie as "poor," "ordinary," or "excellent." Determine the elements of the sample space, and, assuming the three people were randomly selected, assign a probability function to complete the definition of the probability space.

4. A coin is tossed two times one after another and the outcome, head or tail, is recorded using H and T, respectively. An elementary event is any such double toss. Determine all the elementary events and list the subsets (that is, events for the sample space).

5. A city council consisting of five members is presented with a petition to hire a city recreation director. The petition needs a simple majority to pass. How many voting outcomes are possible? Assign a probability function to the same space.

6. Which of the following are probability functions on the events of $S = \{e_1, e_2, e_3\}$?

	P_1	P_2	P_3	P_4
$\varnothing$	0	0	0	0
$\{e_1\}$	$\frac{1}{4}$	$\frac{1}{4}$	$\frac{1}{3}$	$\frac{1}{2}$
$\{e_2\}$	$\frac{1}{4}$	$\frac{1}{2}$	$\frac{1}{3}$	0
$\{e_3\}$	$\frac{1}{2}$	$\frac{1}{4}$	$\frac{1}{3}$	$\frac{1}{2}$
$\{e_1, e_2\}$	$\frac{3}{8}$	$\frac{3}{4}$	$\frac{2}{3}$	$\frac{1}{2}$
$\{e_2, e_3\}$	$\frac{1}{2}$	$\frac{3}{4}$	$\frac{2}{3}$	$\frac{1}{2}$
$\{e_1, e_3\}$	$\frac{3}{4}$	$\frac{1}{2}$	$\frac{2}{3}$	1
S	1	1	1	1

7. Determine the probability of each event of $S = \{e_1, e_2, e_3, e_4\}$ if $P(E_1) = \frac{1}{8}$, $P(E_2) = \frac{2}{8}$, $P(E_3) = \frac{3}{8}$, $P(E_4) = \frac{2}{8}$.

8. One card is selected from a deck of 52 ordinary playing cards. Describe a sample space for the experiment. What is the probability of drawing a heart? An ace? A black card?

9. A certain paragraph contains 10 words of one letter, 10 words of two letters, 15 words of three letters, 15 words of four letters, and 50 words of five or more letters. A word is chosen at random. What is the probability that the word will have
   (a) fewer than three letters?
   (b) at least three letters?
   (c) at most three letters?
   (d) more than three letters?

10. Urn I contains one red and three white balls, urn II contains two white and two red balls. A coin is flipped, and, if a head comes up, a ball is drawn from urn II, otherwise a ball is drawn from urn I. An elementary event then is a pair, where the first symbol H or T de-

notes the outcome of the coin toss and R_i or W_i denotes the outcome of the draw. (R_1, W_1, W_2, and W_3 represent the balls in urn I and R_2, R_3, W_4, and W_5 represent the balls in urn II.) Describe the sample space for the experiment by listing all the outcomes. What is the probability that a white ball will be drawn?

11. People passing a street corner are interviewed with regard to their marital status and ownership of a car. If it is assumed that the probability that a person is neither married nor owns a car is $\frac{1}{4}$, and the probability that a person either owns a car or is married, but not both, as $\frac{1}{3}$, what is the probability that the next person interviewed is married and owns a car? From the information given, is it possible to find the probability that a person interviewed is married?

12. There are four candidates in a school board election. The first and second candidates seem to have the same chance of winning, and the first candidate seems to be twice as likely to win as the third candidate and four times as likely to win as the fourth candidate. What is the probability that the first or fourth candidate will win?

13. Three identical balls are placed in three distinguishable cells marked A, B, and C in random fashion, and the number of balls in each cell is recorded, making an element of the sample space a triple of numbers. List all the elementary events and assign a probability function to complete the definition of the probability space for this experiment.

14. An animal may turn either right or left in a T-maze. Suppose a rat runs a T-maze five times, with the behavior recorded for each run. Describe a sample space for this experiment by listing all the outcomes and assign a probability function. What is the probability the rat ran to the left once?

15. Suppose that a die with faces marked 1 to 6 is "loaded" in such a manner that for $k = 1, 2, \ldots, 6$ the probability of the face marked k turning up when the die is tossed is proportional to k. Find the probability of the event that the outcome of a toss of a die will be an even number.

16. There are two urns of marbles. Urn I has two red marbles and one white marble, and urn II has one red marble and two white marbles. A marble is selected at random from urn I and placed in urn II, and then a marble is selected at random from urn II. Let R_1, R_2, and W_1 denote the marbles in urn I, and R_3, W_2, and W_3 those in urn II. An elementary event is a pair of these symbols where the first of the pair denotes the marble drawn from urn I and the second that drawn from urn II. Set up a sample space and a probability function for this experiment.

17. A fair coin is tossed repeatedly until either a head appears or the coin has been tossed three times. Determine the elementary events and assign a probability function to complete the definition of the probability space. What is the probability that the coin will be tossed three times?

18. Refer to Problem 17. Determine a probability space if the coin is tossed repeatedly until a head appears or the coin has been tossed four times.

19. Refer to Problems 17 and 18. Determine a probability space if the coin is tossed repeatedly until either a head appears or the coin has been tossed n times. (n is a positive integer.)

20. Prove the statement

$$P(A \cup B) \le P(A) + P(B),$$

where A and B are two events of a finite probability space $\mathscr{E}$.

2.3 Finite Probability Spaces with Equally Likely Outcomes

In Section 2.2 we saw that the probability function for the finite sample space can be specified for all events by defining $P(\cdot)$ for the elementary events as a set of n nonnegative numbers whose sum is 1. Now let us add one additional restriction on the set of nonnegative numbers, namely, that they are all equal; that is, $P(E_i) = P(E_j)$ for all values of i and j from 1 to n. Consequently,

$$\sum_{i=1}^{n} P(E_i) = \sum_{i=1}^{n} p_i = n \cdot p_1 = 1,$$

so that $P(E_i) = 1/n$ for $i = 1, 2, \ldots, n$. This is equivalent to assuming that all outcomes are equally likely.

The computation of the probability of an arbitrary event E of a probability space with equal probabilities for the elementary events can be simplified in the following way. By axiom PA3 the sum of the probabilities of the elementary events that make up E is $P(E)$; but since these probabilities are equal it is only necessary to count the number of elementary events for E and divide this by the total number of elementary events in S. If we use the term "size of E" to refer to the number of elementary events in E, then we can write

$$P(E) = \frac{\text{size of } E}{\text{size of } S}.$$

Example 1

Given two balls, one red and one white, indistinguishable except for color, place these balls in three similar cells that are identified by the letters A, B, and C. All the possible ways can be recorded as follows:

$$
\begin{array}{ccc}
(R\ \ W\,|\,\_\,|\,\_\,) & (W\,|\ \ R\ \ |\,\_\,) & (W\,|\,\_\,|\ \ R\ \ ) \\
(\ \ R\ \ |\,W\,|\,\_\,) & (\_\,|\,R\ \ W\,|\,\_\,) & (\_\,|\,W\,|\ \ R\ \ ) \\
(\ \ R\ \ |\,\_\,|\,W\,) & (\_\,|\ \ R\ \ |\,W\,) & (\_\,|\,\_\,|\,R\ \ W\,)
\end{array}
$$

Each of the parentheses represents an outcome of the sample space. If we assume that the placement of the balls was done in a random manner, then we can assign the probability of $\frac{1}{9}$ to each elementary event. The compound event E described by "cell B is occupied" has size 5, and $P(E) = \frac{5}{9}$. ∎

Example 2

Two dice are thrown, one red and one white. Since each die has 6 possible outcomes, $6 \cdot 6 = 36$ different pairs of integers are required to record the outcomes. The pair (2, 3) represents the outcome of a 2 on the first or red die and a 3 on the second or white die. If it is assumed that the dice are perfect, the assignment of $\frac{1}{36}$ as the probability of each elementary event follows. The event E described by "the sum of the spots on the two dice is 4 or less" has size 6, because

$$E = \{(1, 1), (2, 1), (1, 2), (3, 1), (2, 2), (1, 3)\},$$

and

$$P(E) = \frac{6}{36} = \frac{1}{6}. \quad ∎$$

Although we arrived at the equally likely probability space as a specialized case, it played a central role in the development of probability theory (as briefly noted on page 34). The classical definition grew out of the analysis of games of chance, where the thinkable outcomes were called cases and the probability of any event was the ratio of the number of favorable cases to the total number of cases, assuming that all cases were equally likely. This definition appears here as a natural consequence of the axioms, so the classical approach is not rejected by the axiomatic approach.

The equally likely character of the outcomes can also be described by saying that the probability function is a *uniform probability function*. The numerical measure assigned to each elementary event is not as important in the calculation of the probability of an event as is the number of elementary events in the subset defined by the event. Our next concern therefore will be to study techniques for counting the number of elements in a set.

A word of caution: When the sample space is finite, a common assumption is that the outcomes are equally likely. For example, if a jar has one white chip and one red chip, indistinguishable except for color, and one chip is drawn, there are two outcomes, each one as likely as the other. However, if a jar has one white chip and ten red chips, indistinguishable except for color, and one chip is drawn, there are still two outcomes, but these outcomes are no longer equally likely.

Problems

1. An employer wishes to fill a position from a group of 13 employees, 5 male and 8 female. If he selects one without knowing the identity of the employee, what is the probability that the person selected is a male?

2. If two coins are tossed, what is the probability that both show heads? What is the probability there is one head and one tail?

3. Two students are asked to predict whether or not a record will be classified a "hit." Assuming that their judgment is unbiased and random, what is the probability that they disagree?

4. In a family with three children, assuming that boys and girls have an equal chance of being born, what is the probability that there are two boys and one girl?

5. A single letter is selected at random from the word "probability." What is the probability that it is a vowel?

6. A letter is chosen at random from the word "Mississippi." What is the probability that the letter is a vowel? An "s"?

7. Suppose a die is thrown twice with 36 outcomes considered equally likely. Find the probability that the
    (a) second die is odd.
    (b) sum is at least 4.
    (c) first die is larger than the second die.
    (d) sum is exactly 8.

8. Six locations, L1, L2, L3, L4, L5, and L6, are under consideration as potential sites for two new manufacturing plants. If the two are selected at random, find the probability that L1 or L2 will be selected.

9. A certain class has 10 students whose first names have fewer than three letters, 10 students whose first names have three letters, 15 students whose first names have four letters, 15 students whose first names have five letters, and 20 students whose first names have six or more letters. A student is chosen at random. What is the probability that

(a) the student's name will have fewer than four letters?
(b) the student's name will have at least four letters?
(c) the student's name will have exactly five letters?
(d) the student's name will have three or five letters?

10. Of 1000 families in a certain town, 500 have two cars, and 950 have television sets. Of the 500 with two cars, 25 have unemployed heads of household. Of the 500 without two cars, 75 have unemployed heads of household. A family is chosen at random in this town. What is the probability that

(a) the family does not have two cars?
(b) the family has a television set?
(c) the family has an unemployed head of household?
(d) the family has an employed head of household and two cars?

11. In a lot of 500 bolts, 10 percent are usually too large and 5 percent too small. The remainder are usually sufficiently correct to be acceptable. A prospective buyer selects a bolt at random. What is the probability that it will be acceptable?

12. A firm employs 1000 people; 650 are men, 800 own a car, and 555 men own a car. A prize drawing is held with all employees eligible. What is the probability that the winner will be

(a) a male employee?
(b) an employee who owns a car?
(c) a female employee who does not own a car?

13. A boy has four coins in his pocket: a penny, a nickel, a dime, and a quarter. He takes two coins out, one after another. What is the probability that he has more than 11 cents?

14. An urn contains three red and two white balls. You plan to draw two simultaneously. What is the probability that they will be of the same color?

15. Three men, Adams, Brown, and Carlson, each toss a fair coin to determine who pays for the coffee. The person whose coin does not match the other two pays. If all three faces match, they throw again until an odd face occurs.

(a) Construct a probability space appropriate to this experiment exhibiting the elementary events and their probabilities.
(b) What is the probability that Adams will be the odd man?

16. Urn I contains one red and two white balls, and urn II contains two red and one white ball. You plan to transfer one ball from urn I to urn II and then to draw one from urn II. What is the probability that the ball drawn from urn II will be white?

17. Two students and two professors are arranged in a row for a panel discussion. What is the probability that the students and professors alternate in the row?

18. A clinic waiting room has six seats in a row. Suppose three people enter the room and select their seats at random. What is the probability that they sit with no empty seats between them? What is the probability that there is exactly one empty seat between any two of them?

19. Two cards are drawn at random from an ordinary deck of 52 playing cards. If the first card is replaced before the second is drawn, what is the probability of drawing an ace followed by the king of spades?

20. The specifications for the assembly of an electronic device require two different tubes. If both these tubes are defective, the device will not function; however, if just one of these tubes is defective, it will function. The first tube is selected from a box of five such tubes, two of which are defective. The second tube is selected from a box of four tubes, one of which is defective. What is the probability that the device will fail to function?

21. A penny is tossed twice and a nickel is tossed three times. What is the probability that the penny showed heads exactly the same number of times that the nickel did?

2.4 Properties of the Probability Function

Using as a starting point the axioms stated in Section 2.1 for the probability function $P(\cdot)$, we shall derive some general formulas and theorems for $P(\cdot)$ that are valid for an arbitrary probability space. In what follows, letters E and F denote any two events of the set $\mathscr{E}$ of subsets of the sample space S on which a probability function $P(\cdot)$ has been defined.

Theorem 1

The probability of the impossible event is zero; that is, $P(\varnothing) = 0$.

PROOF: The certain event S and the impossible event $\varnothing$ are mutually exclusive, so by using PA3 we obtain

$$P(S \cup \varnothing) = P(S) + P(\varnothing).$$

On the other hand, $S \cup \varnothing = S$, so

$$P(S \cup \varnothing) = P(S).$$

But then $P(S) + P(\varnothing) = P(S)$; from this it follows that $P(\varnothing)$ must be zero. ∎

Theorem 2

The probability of the difference of two events is given by
$$P(F - E) = P(F) - P(E \cap F).$$

PROOF: The events $F - E$ and $F - E'$ are mutually exclusive (why?), so by using PA3, we obtain
$$P((F - E) \cup (F - E')) = P(F - E) + P(F - E').$$
Now
$$\begin{aligned}(F - E) \cup (F - E') &= (F \cap E') \cup (F \cap E) \\ &= F \cap (E' \cup E) \\ &= F \cap S \\ &= F.\end{aligned}$$
Therefore, $P[(F - E) \cup (F - E')] = P(F)$, so
$$P(F) = P(F - E) + P(F - E').$$
But $F - E' = F \cap E$, so
$$P(F - E) = P(F) - P(F \cap E). \quad \blacksquare$$

Theorem 3

The probability of the complement of the event E is $1 - P(E)$; that is,
$$P(E') = 1 - P(E).$$

PROOF: For any event E, $E' = S - E$. Let S and E be the two events in Theorem 2; then
$$P(S - E) = P(S) - P(E \cap S),$$
$$P(E') = 1 - P(E). \quad \blacksquare$$

Theorem 4

The probability of the union of two events E and F is given by
$$P(E \cup F) = P(E) + P(F) - P(E \cap F).$$

PROOF: Since E and F are any two events on S, we can write $E \cup F = E \cup (F - E)$, where the intersection of E and $F - E$ is the empty set.

PA3 and Theorem 2 can be applied to this union:

$$P(E \cup F) = P(E \cup (F - E))$$
$$= P(E) + P(F - E)$$
$$= P(E) + P(F) - P(E \cap F). \quad \blacksquare$$

Please notice that whenever $E \cap F = \emptyset$, this result is the same as PA3. Theorem 4 extends PA3 to the union of a finite number of *arbitrary* events, that is, events that are not necessarily pairwise exclusive.

Theorem 5

The probability of a subevent E of an event F is less than or equal to the probability of event F; that is, if $E \subset F$, then $P(E) \leq P(F)$.

PROOF: Event E is a subevent of F so that $E \subset F$ and $E \cap F = E$. If E and F are the two events in Theorem 2,

$$P(F - E) = P(F) - P(E \cap F),$$
$$P(F - E) = P(F) - P(E).$$

But $P(F - E) \geq 0$, by PA1, so $P(F) - P(E) \geq 0$, or $P(F) \geq P(E)$. $\blacksquare$

Theorems 1–5 are statements about the probability of certain combinations of events in terms of the probabilities of the events that make up the combinations. Theorem 1 is the companion for axiom PA2. In Theorem 2, if we consider the sample space to be finite, then the probability of the difference of two events is simply the difference between the probability of all the elementary events in event F and those elementary events in E that are also in F. In Theorem 3, either the event E or the event E' occurs, so the sum of the probabilities for E and E' must be 1. In Theorem 4, $P(F)$ is the sum of the probabilities of the elementary events in F, and $P(E)$ represents the probability of the elementary events in E. If any elementary events are in both E and F, then their probabilities would have been counted twice, once for $P(E)$ and once for $P(F)$. Hence by subtracting $P(E \cap F)$ we obtain the desired equality. In Theorem 5, the inequality holds, since there may be elementary events in F that are not in E. The following examples show how these formulas are used in computing probabilities.

Example 1

An integer is selected at random from the set of integers from 1 to 1000 inclusive. What is the probability that the integer so selected is divisible by 6 or by 8?

Let the sample space be the 1000 integers with the uniform probability function; that is $P(\text{integer } k) = \frac{1}{1000}$, $k = 1, 2, \ldots, 1000$. If A is the event "integers divisible by 6" and B the event "integers divisible by 8," then the required probability is $P(A \cup B)$. However, A and B are not mutually exclusive events, since the integer 48, for example, lies in both A and B. Thus we must use Theorem 4. The number of integers in A is the number of multiples of 6 not exceeding 1000. There are 166 of them. Similarly, the number of integers in B is 125. An integer in both A and B is divisible by both 6 and 8 if and only if it is divisible by the least common multiple of 6 and 8, which is 24. There are 41 integers divisible by 24 in the first 1000 integers. Thus

$$P(A \cup B) = P(A) + P(B) - P(A \cap B)$$
$$= \tfrac{166}{1000} + \tfrac{125}{1000} - \tfrac{41}{100}$$
$$= \tfrac{1}{4}. \quad \blacksquare$$

Example 2

Five fair coins are tossed simultaneously. What is the probability of at least one head?

The sample space appropriate to this experiment will have 32 elementary events, corresponding to the 2^5 possible combinations of heads and tails. We consider the elementary events that are particular combinations (for example, H H T T H) as equally likely, and therefore use the uniform probability function $P(E_i) = \frac{1}{32}$ for each elementary outcome E_i. Now let E be the event of at least one head; then the complement E' is the event of no head at all, which is equivalent to the elementary event of all five coins with a tail. Therefore, $P(E') = \frac{1}{32}$, so by Theorem 2, $P(E) = \frac{31}{32}$. $\quad \blacksquare$

Some comments on the contents of this chapter:

1. In PA2, the probability of the certain event is postulated to be 1 and in Theorem 1 the probability of the impossible event is asserted to be 0. The converse statements are not true statements. There are probability spaces where the probability of an event is 1 and yet the event is not the certain event. The following example will make this clear. The sample space for the toss of a die is $S = \{1, 2, 3, 4, 5, 6\}$. Let the probabilities assigned to the elementary events be as follows:

$$P(\{1\}) = P(\{2\}) = P(\{3\}) = \tfrac{1}{3},$$
$$P(\{4\}) = P(\{5\}) = P(\{6\}) = 0.$$

If A is the event $= \{1, 2, 3\}$, $P(A) = 1$, yet A is anything but S.

2. The phrase "at random" has been used rather freely in the preceding material, and the adjective "random" has occurred with the expression

"random phenomena" (or random experiment). Unfortunately, the intention in both cases is not quite the same. When an object is selected "at random" from a set of objects, each object available for selection is considered to have an equal probability of being selected. This is equivalent to the selection of the equal-likely probability space with the uniform probability function for the random experiment.

3. In applications of probability theory, the student should be aware that many probability functions are possible for the same random experiment and that there is usually no mathematical reason to prefer one to another. For the simple experiment to tossing a coin, the sample space $S = \{H, T\}$, with $P(\{H\}) = \frac{1}{3}$, $P(\{T\}) = \frac{2}{3}$, or $P(\{H\}) = \frac{5}{6}$, $P(\{T\}) = \frac{1}{6}$, are as acceptable and reasonable as the usual assignment of $\frac{1}{2}$ to each elementary event. In fact, we can take $P(\{H\}) = p$, where p is any number between 0 and 1, and $P(\{T\}) = 1 - p$ for an acceptable probability-function assignment. To ascertain which value of p is best, we need to investigate the physical experiment for which the sample space is a model.

4. Finally, we remark that by starting with undefined terms and elements, defining relations between elements, listing axioms for the relations and the elements, and deducing theorems from the axioms, we obtain a mathematical system.

Problems

1. Let A and B be events in a sample space S such that $P(A) = 0.4$, $P(B) = 0.3$, and $P(A \cap B) = 0.2$. Find the probabilities of

   (a) $A \cup B$ (b) A'
   (c) $A' \cap B$ (d) $A' \cap B'$

2. Let A and B be events in a sample space S such that $P(A) = \frac{1}{2}$, $P(B) = \frac{1}{8}$, and $P(A \cap B) = 0$. Find the probabilities of

   (a) A' (b) $A' \cap B$
   (c) $A \cup B$ (d) $A' \cap B'$

3. A ball is drawn from a bag containing thirty balls numbered from 1 through 30. The number of the ball is recorded and the ball is replaced. A second ball is drawn, the number noted, and the ball replaced. What is the probability that the two balls did not have the same number?

4. The probability that Bill is the first in the class to debug his computer program is 0.1 and that Joe is first is 0.2. What is the probability that either Bill or Joe is first?

5. One card is drawn from each of two ordinary decks of 52 playing cards. What is the probability that at least one of the cards is the ace of hearts?

6. A programmer applies for employment at two companies, A and B. His estimate of the probability that he will be offered a job by company A is 0.7, and his estimate of the probability that he will be rejected by company B is 0.5. He feels that the probability of being rejected by at least one company is 0.6. What is the probability he is offered a job by at least one company?

7. Determine whether these statements are true or false and give reasons.

    (a) The probability that common-stock prices rise on a particular day is 0.4 and the probability that bond prices fall on the same day is 0.2, so the probability that either common-stock prices rise or bond prices fall is 0.6.

    (b) The probability of a business closing during the first year is 0.1 and during the second year is 0.05, so the probability of a business closing during the first 2 years is 0.15.

8. A real estate dealer estimates that within the next month the probability that he will sell at least five houses is 0.8, and the probability that he will sell less than eight houses is 0.4. On the basis of these assumptions, what is the probability that he will sell five, six, or seven houses?

9. A contractor estimates that the probability that he can finish a certain job within the next x days is $x/10$, for $x = 1, 2, 3, \ldots, 10$.

    (a) Describe an appropriate probability space.

    (b) What is the probability that it takes exactly five days to finish the job?

    (c) What is the probability that it takes more than seven days to finish the job?

    (d) What is the probability that it will take an odd number of days to finish the job?

10. A biologist believes that the probability of his finding at least five members of a species of bat in a cave in Texas is 0.9, and the probability of his finding fewer than eight is 0.6. What is the probability that he will find at least five but fewer than eight members?

11. Suppose that the probability of an event A is the square of the probability of an event B'. The probability that the event A does not occur is the square of twice the probability that B does occur. Find the probability of A.

12. A literature class was given an objective exam and an essay exam on Shakespeare's *Hamlet*. For a student chosen by lot, it was found that the probability of a score below 80 on the objective exam was $\frac{1}{2}$. The probability of a score of at least 80 on the objective exam and a score below 80 on the essay exam was $\frac{1}{3}$. What is the probability that a student chosen at random had a score of at least 80 on both exams?

13. Prove: If A is a subevent of B and $P(B) = 0$, then $P(A) = 0$.

14. Suppose that A and B are two events in a sample space S with $P(A) = 0$. Show that

    (a) $P(A \cap B) = 0$.   (b) $P(A \cup B) = P(B)$.

15. Suppose that A and B are two events in a sample space S with $P(A) = 1$. Show that

    (a) $P(A \cup B) = 1$.   (b) $P(A \cap B) = P(B)$.

16. Prove that the following relations hold for any two events A and B in a probability space:

$$P(A \cap B) \leq P(A) \leq P(A \cup B) \leq P(A) + P(B).$$

17. Suppose that A and B are two events in a sample space S with $P(A) = \frac{3}{4}$ and $P(B) = \frac{5}{8}$. Show that

    (a) $P(A \cup B) \geq \frac{3}{4}$.   (b) $\frac{3}{8} \leq P(A \cap B) \leq \frac{5}{8}$.

18. Show that for any two events A and B on a probability space, $P((A \cap B') \cup (A' \cap B)) = P(A) + P(B) - 2P(A \cap B)$.   [Notice that $(A \cap B') \cup (A' \cap B)$ is the event that exactly one of A and B occur.]

19. Prove Boole's inequality: If $A_1, A_2, \ldots, A_n$ are n events,

$$P(A_1 \cup A_2 \cup \cdots \cup A_n) \leq P(A_1) + \cdots + P(A_n).$$

(*Hint*: Use mathematical induction.)

20. Show that for any three events, A, B, and C on a probability space,

$$P(A \cup B \cup C) = P(A) + P(B) + P(C) - P(A \cap C) - P(B \cap C)$$
$$- P(A \cap B) + P(A \cap B \cap C).$$

21. Let A and B be two events on a probability space. In terms of $P(A)$, $P(B)$, and $P(A \cap B)$, express for $k = 0, 1, 2$,

    (a) $P(\text{exactly } k \text{ of the events } A \text{ and } B \text{ occur})$.
    (b) $P(\text{at least } k \text{ of the events } A \text{ and } B \text{ occur})$.
    (c) $P(\text{at most } k \text{ of the events } A \text{ and } B \text{ occur})$.

22. Evaluate the probabilities asked for in Problem 21 in the case that

    (a) $P(A) = P(B) = \frac{1}{3}, P(A \cap B) = \frac{1}{6}$.
    (b) $P(A) = P(B) = \frac{1}{3}, P(A \cap B) = \frac{1}{9}$.
    (c) $P(A) = P(B) = \frac{1}{3}, P(A \cap B) = 0$.

23. Let A, B, and C be three events on a probability space. In terms of $P(A)$, $P(B)$, $P(C)$, $P(A \cap B)$, $P(A \cap C)$, $P(B \cap C)$, and $P(A \cap B \cap C)$, express:

    (a) $P(\text{exactly 2 of the events, } A, B, \text{ and } C \text{ occur})$.
    (b) $P(\text{at least 2 of the events } A, B, \text{ and } C \text{ occur})$.
    (c) $P(\text{at most 2 of the events } A, B, \text{ and } C \text{ occur})$.

chapter three

Sampling

3.1 Sampling and Counting

There are many instances in the application of probability theory where it is desirable and necessary to count the outcomes in the sample space and the outcomes in an event. In Chapter 2 we saw that in the special instance of a finite probability space with a uniform probability function, the probability of an event is known when the number of outcomes that comprises the event is known, that is, as soon as the number of outcomes in the subset that defines the event is known. Practical examples are given by the student who, on considering the possibility of guessing his way through a 10-question true–false test, ponders the number of ways six correct answers (a passing score) can be selected, or by the biologist who contemplates the number of different ways in which 5 mice can be chosen from a pen of 100 mice.

A typical and important application of counting techniques arises in choosing a sample from all possible samples. Consider the following situation. One hundred families, each with the same number of children, have volunteered for a consumer expenditure study. Each week three of the families will be selected at random; that is, they will be chosen in such a way that each family is as likely to be selected as any other, for a detailed analysis of their food costs for the week. How many different triples of families can be chosen to be visited each week?

Enumeration would be one way to determine the answer to the question, but by applying the basic principle of combinatorial analysis we can obtain the result with less effort. For the first task, choosing the first family, there are 100 choices; for the second task, choosing the second family, there are 99 possible choices; and for the third, 98, so that a total of $100 \cdot 99 \cdot 98 = 970{,}200$ different options exists. This solution takes into consideration the order in which the families are visited and shows how many elements there are in the collection of all possible triples. In statistical jargon, this collection comprises all possible samples, and the selection of a sample is the selection of one of the 970,200 outcomes.

The next example, although it may be impractical, illustrates how knowledge of counting can be helpful in finding a new probability space.

A fishbowl is filled with pennies that have been minted in the last 25 years. A coin is drawn, the date on the coin is recorded, and the coin is returned to the fishbowl. This procedure is repeated two more times. How many different triples of dates can be recorded?

Because each of the 3 coins has one of 25 possible dates, the number of triples is $25^3 = 15,625$. For by applying the basic principle of combinatorial analysis again, we can get any one of the 25 choices for the first entry of the triple, and 25 for the second, and 25 for the third. Hence 25^3 is the correct number.

Observe that in this case, unlike the preceding case of the "family triples," the number of triples of dates is independent of the number of coins in the bowl. The original set of 25 dates gives rise to the set of triples of dates, which are therefore a new set of outcomes; and selecting three coins is to select one of the 15,625 outcomes in the new set.

The statistician may have less frivolous uses for developing sound sampling techniques. A typical sampling problem is to estimate the number of defective items in a mass-produced lot.

Example 1

A purchasing agent for a large institution purchases mass-produced light bulbs in lots of 1000 bulbs. Although theoretically every light bulb is good if not perfect, in all likelihood this is not the case. However, to test each and every bulb before accepting delivery would probably be more expensive than the cost of all the defective bulbs that were turned up. Hence the purchasing agent should consider some sort of inspection plan that will economically estimate the number of defectives. He decides to choose a random sample of 20 bulbs and test each of them. This innocently unleashes a host of questions:

1. How many samples of 20 bulbs are possible? The numerical answer to this question will have to wait for the moment, even though the solution depends only on the basic principle of combinatorial analysis.

2. Suppose that all 20 in the sample are good light bulbs. What information, if any, is then known about the remaining 980?

3. Suppose that two of the tested light bulbs are defective. Does this mean that $\frac{1}{10}$ (or 98) of the remaining 980 is defective?

4. Was a sample of 20 the right number to choose to obtain the desired estimate?

We must defer the answers to these questions for now. ∎

Sampling is an example of an experiment that fits the description of a random phenomenon. Whether one selects every fourth item or whether names are pulled out of a hat, the outcome for each selection is not deterministic. In most cases, each element in the sample space of all possible samples will be considered to have the same chance of being selected. This is equivalent to assuming the uniform probability function for the sample space. Consequently, our primary interest now is to count the number of possible samples. In Section 3.4 we shall again consider these and other sampling problems.

All the sets that appear from now on in this chapter will have two definite features: (1) each set will contain a finite number of members, and (2) each member can be distinguished from any other; that is, the objects will be considered distinguishable. For example, if the set contains five white chips and three red chips, where the chips are identical except for color, we shall consider the white chips to be numbered 1, 2, 3, 4, and 5, and the red chips 6, 7, and 8. Sometimes the elements might be better distinguished by assigning names. If there are five people in the Jones family and three in the Smith family, we would naturally use the names of the individuals to distinguish the Joneses and the Smiths.

Problems

1. A true–false examination has 10 questions. Assuming that every question is answered, how many different possible sets of 10 answers are there?

2. A die is tossed six time in succession, with the number of spots on the uppermost face recorded after each toss. How many elements are there in the sample space for this experiment?

3. A jar contains a large number of candy mints of five colors: red, white, green, yellow, and blue. Two mints are selected one at a time. How many different combinations of colors of mints are possible?

4. A die is rolled, a card selected from an ordinary deck of 52 cards, and a coin is tossed. An outcome of this experiment is a triple. How many triples are there in the sample space?

5. An examination consists of four sections: I has four problems, II has three problems, III has four problems, and IV has two problems. If a student must answer one problem from each section, how many different examinations of four problems are possible?

6. A well-known ice cream dealer markets 28 different flavors. Four ice cream cones are ordered, one after another. How many different sequences of four cones are possible?

7. A biologist would like to run a series of experiments on some mice using different diets, water supply, and temperature. If there are three diets, four categories of water supply, three temperature levels, and each experiment runs for 4 weeks, how many weeks are required for the entire experiment, if the experiments are run consecutively?

8. An economist classifies 100 industries into three categories with 20 in category *A*, 50 in category *B*, and 30 in category *C*. He plans to study three of these, with one industry selected at random from each category. How many triples are possible?

9. A stock enthusiast plans to chart three securities each day of a 5-day week with regard to an increase in value, a decrease, or no change. How many different charts are possible?

10. A man is told to purchase a tube of toothpaste and a jar of shampoo. There are 8 brands of toothpaste, with each brand in four different sizes, and 12 brands of shampoo, each in two different sizes. How many different pairs of toothpaste and shampoo can be purchased?

11. A psychology professor wants to perform an experiment in learning nonsense symbols, and he needs 3 subjects from his class of 34 students. How many different collections are possible?

12. There are 1000 recruits in a camp with 600 from New York and 400 from California. How many samples of 5 men, with 3 men from New York and 2 from California, are possible?

3.2   Samples and Permutations

The reason for taking a sample often influences the manner in which it is carried out. The following two situations are clearly different: (1) From a group of people choose two people to serve as president and secretary of the group; (2) choose two people from the same group to move a piano. If it is agreed by everyone involved that the first person selected will be the president, then the selection (Jones, Smith) gives a different set of officers than the selection (Smith, Jones). Hence the order of the selection is important. A sample is *ordered* if for any two choices in the sample a_i and a_j, the listing $(\ldots, a_i, \ldots, a_j, \ldots)$ is different from the listing $(\ldots, a_j, \ldots, a_i, \ldots)$. In situation (2) the sample is *unordered*.

Another problem in sampling procedures is to decide whether or not an object should be returned to the collection before the next selection is made. If the process is defined so that an object once selected or identified is removed from further consideration, then we say that the elements have been selected *without replacement*. In this case the elements selected are necessarily distinct. On the other hand, if the process requires us to return each object to the set before the next selection is made, the elements should be selected *with replacement*. If one were selecting a key (in order to open a door) from a key chain with 10 keys, the selection would be most efficiently accomplished if each key that failed to open the door were held apart from the rest. That is, the selection would be without replacement. If five people were selecting a flavor of ice cream from a list of five flavors, the same flavor could be selected more than once. That is, the selection would be with replacement.

We make the following definitions:

Definition 1 : *n*-set and *r*-subset

A set S consisting of n elements will be called an *n-set*. A subset of S consisting of r elements is called an *r-subset* of the *n*-set S.

From an *n*-set S of distinguishable elements a sample of size r is drawn in sequence. Each element is replaced after its characteristics have been recorded and before the next item is chosen. In other words, draw an ordered sample with replacement and record the result of the draw by $(a_1, a_2, \ldots, a_r)$, where the subscripts indicate the order in which the elements were selected. The first element drawn was a_1, the second a_2, and so on. If $r = 2$, the result of the draw would be an ordered pair (a_1, a_2). If $r = 3$, the result would be an ordered triple (a_1, a_2, a_3). The term used to describe these ordered configurations is *r-tuple*, so the ordered pairs are called 2-tuples and the ordered triples 3-tuples.

Definition 2: *r*-permutation

An *r*-tuple of elements selected from an *n*-set S of distinguishable objects without replacement is an *r*-permutation, denoted by $(a_1, a_2, \ldots, a_r)$, where the a_j are elements of the set S.

Other verbalizations for the *r*-permutation symbol are the *permutation of size r*, or the *permutation of n things taken r at a time*. The elements of an *r*-permutation are necessarily distinct. Therefore, r must be less than or equal to n. Two *r*-permutations are equal if and only if the corresponding *r*-tuples are identical.

Definition 3: *r*-permutation with replacement

An *r*-tuple of elements selected with replacement from an *n*-set S of distinguishable objects is an *r*-permutation with replacement.

Because the selection is with replacement, the elements $a_1, a_2, \ldots, a_r$ of the r-tuple are not necessarily distinct. There is no restriction on the magnitude r relative to n; that is, it is meaningful to talk about a 100-permutation with replacement from a 4-set. Two r-permutations with replacement are equal if the corresponding r-tuples are identical.

Example 1

Let us take four objects (a 4-set), indistinguishable except by the numbers 1, 2, 3, and 4. These objects could be four balls in an urn, four people, or four brands of toothpaste. The set of all possible 2-permutations with replacement is

$$
\begin{array}{llll}
(1, 1) & (2, 1) & (3, 1) & (4, 1) \\
(1, 2) & (2, 2) & (3, 2) & (4, 2) \\
(1, 3) & (2, 3) & (3, 3) & (4, 3) \\
(1, 4) & (2, 4) & (3, 4) & (4, 4)
\end{array}
$$

while the possible 2-permutations (that is, without replacement) are

$$
\begin{array}{llll}
(1, 2) & (2, 1) & (3, 1) & (4, 1) \\
(1, 3) & (2, 3) & (3, 2) & (4, 2) \\
(1, 4) & (2, 4) & (3, 4) & (4, 3) \quad \blacksquare
\end{array}
$$

Example 2

A symposium committee has arranged for four speakers to appear on a program. There is no restriction on the order in which they can appear, but each speaks only once. What are the possible arrangements?

If the letters a, b, c, and d designate the four people, we can list precisely 24 arrangements in the set of all possible 4-permutations.

$$
\begin{array}{llllllll}
a & b & c & d & \quad & c & a & b & d \\
a & b & d & c & \quad & c & a & d & b \\
a & c & b & d & \quad & c & b & a & d \\
a & c & d & b & \quad & c & b & d & a \\
a & d & b & c & \quad & c & d & a & b \\
a & d & c & b & \quad & c & d & b & a \\
\\
b & a & c & d & \quad & d & a & b & c \\
b & a & d & c & \quad & d & a & c & b \\
b & c & a & d & \quad & d & b & a & c \\
b & c & d & a & \quad & d & b & c & a \\
b & d & a & c & \quad & d & c & a & b \\
b & d & c & a & \quad & d & c & b & a. \quad \blacksquare
\end{array}
$$

In both of the previous examples, the possible permutations were enumerated, but unless the list is small, it is tedious. By using the basic principle of combinatorial analysis intelligently, we can easily develop formulas for the number of r-permutations of an n-set, and this we do next.

Theorem 1

The number of r-permutations with replacement of an n-set is n^r.

PROOF: Let S be an n-set. Using the basic principle of combinatorial analysis, we see that the selection of an r-permutation with replacement is simply r tasks with n choices for each task. Thus the total is $n \cdot n \cdots n$ with n repeated r times. ∎

Theorem 2

The number of r-permutations of an n-set is $n(n-1) \cdots (n-r+1)$.

PROOF: Again, refer to the basic principle of combinatorial analysis. This time task T_1 has n options, task T_2 has $(n-1)$ options because one element has been removed, and so on to task T_r with $n-(r-1)$ options because $(r-1)$ elements have been removed. The product $n \cdot (n-1) \cdots (n-r+1)$ is the desired result. ∎

In the case of the n-permutation from an n-set, that is, $r = n$ (Example 2), every element appears in the n-tuple and the number of permutations is $n(n-1) \cdots (n-n+1) = n(n-1) \cdots 2 \cdot 1$. This product of the first n integers is read n-*factorial* and is denoted by the symbol $n!$. The number of n-permutations of an n-set is $n!$. Five books with different titles can be arranged on a shelf in $5! = 5 \cdot 4 \cdot 3 \cdot 2 \cdot 1 = 120$ ways.

Several notations are used to write the product $n(n-1) \cdots (n-r+1)$. One way is to write

$$(n)_r = n(n-1) \cdots (n-r+1),$$

where n and r are both positive and $r \leq n$. Another way uses the factorial notation:

$$(n)_r = \frac{n!}{(n-r)!}.$$

In this form the right side is meaningless if $r = n$, for then

$$(n)_n = \frac{n!}{0!}.$$

However, $(n)_n = n(n-1) \cdots (n-n+1) = n!$ is legitimate. Thus, to be consistent, $0!$ is by definition set equal to 1.

Example 3

With eight different flags, how many signals of three flags can be displayed if it is assumed that only one flag of each type is available?

Since we want the number of 3-permutations from an 8-set, the result is $(8)_3 = 8 \cdot 7 \cdot 6 = 8!/5! = 336.$ ∎

Example 4

Suppose that there are three plumbers in a town and six different people call a plumber. What is the total number of possible calls made by these six people?

We let a 6-tuple represent the six people who call and number the plumbers, 1, 2, and 3. Then each place in the 6-tuple stands for a person, and the entry represents a call to plumber 1, 2, or 3. The three plumbers form a 3-set and the number of 6-permutations with replacement is desired, since a plumber may be called by more than one person. The result is $3^6 = 729.$ The 6-tuple $(2, 1, 3, 2, 2, 2)$ is one of the 729 possibilities. ∎

Example 5

A restaurant menu lists 3 soups, 10 meats, 5 desserts, and 3 beverages. How many ways can a meal of soup, meat dish, dessert, and beverage be ordered?

A meal is a 4-tuple with the first component a soup, the second a meat dish, the third a dessert, and the fourth a beverage. Direct application of the basic principle of combinatorial analysis yields the result of $3 \cdot 10 \cdot 5 \cdot 3 = 450$ different meals. Notice that in this example the collection of 450 4-tuples of meals was *not* generated as r-permutations of some n-set. We must understand the problem before we choose a method of counting. ∎

Problems

1. Find the value of

    (a) $(5)_3$ (b) $(5)^3$ (c) $5!$

2. For the five objects a, b, c, d, and e, list the 2-permutations without replacement.

3. Six men compete in a race. In how many ways can the first two places be taken?

4. How many license plates can be made using two letters followed by a three-digit number?

5. If there are four flags of different colors, how many different signals can be made by arranging them on a flagpole, if at least two flags must be used for each signal?

6. A man has five coins in his pocket. He agrees to give one coin to his daughter and one to his son. How many ways can this be done?

7. How many three-letter words, where a word is defined as any three letters, can be formed in the English language?

8. Five people enter an elevator in a 50-story building. Each person calls out the floor at which he will get off. In how many ways can the five floors be designated? If no two of the people get off at the same floor, how many ways can this be done?

9. An inspector visits six different machines during the day. In order to prevent operators from knowing when he will inspect, he varies the order of his visits. In how many ways can this be done?

10. A multiple-choice test has 100 questions with six possible answers for each question. How many different sets of 100 answers are possible?

11. A product is assembled in three stages. At the first stage there are five assembly lines, at the second stage there are four assembly lines, and at the third stage there are six assembly lines. In how many different ways may the product be routed through the assembly process?

12. A certain chemical substance is made by mixing five separate liquids. It is proposed to pour one liquid into a tank, and then to add the other liquids in turn. All sequences must be tested to determine which gives the best yield. How many tests must be performed?

13. From the digits 1, 2, 3, 4, and 5, how many four-digit numbers can be formed without any repeated digits if the number must be odd?

14. In dialing a telephone number, one has to select seven slots, three for the exchange and four to identify the telephone in that exchange. The telephone dial contains 10 slots, one each for 0, 1, 2, ..., 9. If the first digit in the exchange cannot be a zero, how many different telephone numbers are possible?

15. How many three-digit even integers can be formed from the digits 1, 5, 6, and 8, with (a) no digit repeated, and (b) with repeated digits allowed?

16. How many ways can three coins fall? How many ways can m coins fall, where m is a positive integer?

17. Given an alphabet of n symbols, in how many ways can one form "words" consisting of exactly k symbols?

18. Two cards are drawn without replacement (one after the other) from a standard deck of 52 cards. In how many ways can one draw

    (a) first a spade, then a heart?
    (b) first a spade, then a heart or diamond?
    (c) first a spade, then another spade?

19. Repeat Problem 18, assuming that the first card is put back before the second is drawn.

20. What is the smallest value of n such that $n!$ exceeds

    (a) a hundred?   (b) a thousand?   (c) a million?

21. Let n be a positive integer.   Simplify

    (a) $(n+1)! - n!$ (b) $\dfrac{n!}{(n-1)!}$

    (c) $\dfrac{(n+1)!}{(n-1)!}$

22. A city council has nine items of unfinished business pending. Two items are policy matters and seven are administrative questions. Four items will be selected and arranged in order as the agenda for the next meeting. The bylaws prescribe that, at any meeting, all unfinished policy matters for discussion must precede any unfinished administrative question on the agenda. Find the number of possible agendas.

23. A survey of tariff experts is conducted to determine which industries are most deserving of continued protection.   Each expert is asked to rank four industries from a list of eight.   How many different-ranked ballots are possible from each expert?

24. Each permutation of the digits 1, 2, 3, 4, 5, and 6 determines a six-digit number. If the six-digit numbers corresponding to all possible permutations are listed in order of increasing magnitude, which is the 417th?

25. How many distinct k-tuples can be formed from the integers 1, 2, $\ldots$, n

    (a) without repetition?   (b) with repetition?

26. A class consists of 20 students. How many distinct arrangements of these 20 into four rows of five students each are possible? (The arrangement within the row is relevant.)

3.3 Samples and Combinations

In Section 3.2 the sampling was ordered; that is, the elements of the sample were considered to be selected one at a time. But sampling need not be done this way. Consider a set of red and white chips in a jar. If order is not important, we could simply draw a handful of chips.

Suppose that we have an n-set S of distinguishable objects and we wish to draw a sample of size r from the set S without regard for the order and, at first, without replacement. This can be done either one at a time or by the handful. Because the objects selected are distinct elements of S, the sample is actually an r-subset of S. The result of the draw can be recorded as the r-subset $\{a_i, \ldots, a_r\}$, where the a_i, $i = 1, 2, \ldots, r$ denote those r elements of S that have been selected.

Definition 1 : r-combination

An unordered collection of r elements selected from an n-set S of distinguishable objects without replacement is an r-combination.

Because an r-combination is the result of a draw performed without replacement, the selected elements are necessarily distinct. An r-combination is actually an r-subset of S, so two r-combinations are equal if the corresponding subsets are equal. (*Note*: Order is of no consequence with r-subsets and hence r-combinations.)

When the unordered selection is made with replacement, another mathematical object arises.

Definition 2 : r-combination with replacement

An unordered collection of r elements selected with replacement from an n-set S of distinguishable objects is an r-combination with replacement.

Because an r-combination with replacement is the result of a draw performed with replacement, the same element can be selected several times; thus the selected elements need not be distinct. Two r-combinations with replacement are equal if the elements of each are the same. There is no restriction on the magnitude of r relative to the magnitude of n.

Example 1

Take four objects (refer to Example 1 of Section 3.2) indistinguishable except by the numbers 1, 2, 3, and 4, and list the 2-combinations without and with replacement.

There are six 2-combinations without replacement:

$$\{1, 2\}, \{1, 3\}, \{1, 4\}, \{2, 3\}, \{2, 4\}, \{3, 4\},$$

and there are ten 2-combinations with replacement:

$$\{1, 1\}, \{1, 2\}, \{1, 3\}, \{1, 4\}, \{2, 2\},$$
$$\{2, 3\}, \{2, 4\}, \{3, 3\}, \{3, 4\}, \{4, 4\}. \quad \blacksquare$$

Example 2

A symposium committee has drawn up a list of six experts who would be available to speak. However, owing to budgetary limitations, only three can be invited. What are the possible collections of three speakers?

If the letters a, b, c, d, e, and f denote the six people, we can list 20 elements in the set of all possible 3-combinations.

$\{a, b, c\}$	$\{b, c, d\}$
$\{a, b, d\}$	$\{b, c, e\}$
$\{a, b, e\}$	$\{b, c, f\}$
$\{a, b, f\}$	$\{b, d, e\}$
$\{a, c, d\}$	$\{b, d, f\}$
$\{a, c, e\}$	$\{b, e, f\}$
$\{a, c, f\}$	$\{c, d, e\}$
$\{a, d, e\}$	$\{c, d, f\}$
$\{a, d, f\}$	$\{c, e, f\}$
$\{a, e, f\}$	$\{d, e, f\}$ $\quad \blacksquare$

The theorems concerning the possible number of r-combinations in an n-set, corresponding to the theorems for r-permutations, are as follows:

Theorem 1

The number of r-combinations of an n-set is $(n)_r/r!$.

PROOF: Let S be an n-set and let x denote the number of r-combinations from S. For each r-combination we can write $r!$ r-permutations. Therefore, the totality of r-combinations gives rise to $x \cdot r!$ r-permutations. But this exhausts the r-permutations, because every

r-permutation is some r-subset of S. Since there are $(n)_r$ permutations (Section 3.2, Theorem 1),

$$x \cdot r! = (n)_r,$$

$$x = \frac{(n)_r}{r!}. \quad \blacksquare$$

The symbol $\binom{n}{r}$ is the usual notation for the number of r-combinations of an n-set, and we write

(1) $$\binom{n}{r} = \frac{(n)_r}{r!} = \frac{n(n-1) \cdots (n-r+1)}{r!}.$$

These numbers are also known as *binomial coefficients* and can be written in the form

(2) $$\binom{n}{r} = \frac{n!}{r!(n-r)!},$$

which has been obtained by multiplying and dividing the right side of expression (1) by $(n-r)!$ and realizing that $n! = n(n-1) \cdots (n-r+1) \cdot (n-r)!$. Because every r-combination is also an r-subset, the number of r-subsets of an n-set is $\binom{n}{r}$. These numbers, in addition to being interesting in themselves, are also quite important to later material and we shall study them in some detail.

Theorem 2

The number of r-combinations with replacement of an n-set is

$$\binom{n+r-1}{r}.$$

This theorem is similar to the preceding three theorems in its statement but very different in the kind of argument that is required to justify the assertion. The proof is somewhat complicated but at the same time an excellent example of a mathematically elegant argument. We shall give the proof at the end of the section, although the proof can be omitted without affecting the continuity of our discussion.

Example 3

From a group of eight people, how many different committees of three people can be chosen?

The committee is considered to be an unordered group with no person selected more than once. The problem then is to count the number of 3-combinations of an 8-set, and this is

$$\binom{8}{3} = \frac{8!}{3!\,5!} = 56. \quad \blacksquare$$

Example 4

Throw five dice simultaneously. Because each die represents a 1-combination from a 6-set, the five dice are a 5-combination with replacement from a 6-set. Hence the total number of different 5-combinations with replacement is

$$\binom{6+5-1}{5} = \binom{10}{5} = 252. \quad \blacksquare$$

Example 5

A bridge hand consists of the selection of 13 cards from a deck of 52 cards. Since the order in a hand is of no concern, this is the problem of the number of 13-combinations of a 52-set; that is,

$$\binom{52}{13} = 635{,}013{,}559{,}600. \quad \blacksquare$$

Example 6

From a collection of 150 rats, 3 are taken for a special maze experiment. How many outcomes are there in the sample space of all possible samples? Since the order is of no concern, the number of outcomes is the number of 3-combinations of a 150-set, which is

$$\binom{150}{3} = 551{,}300. \quad \blacksquare$$

The assertions in the preceding two theorems and the corresponding theorems of Section 3.2 are summarized in Table 1. In each case it is assumed that the selection is made from an n-set of distinguishable objects.

At this point we digress to examine some of the properties of the numbers $\binom{n}{r}$. The binomial coefficients are defined for the positive integers n and r, $0 < r \le n$, and are themselves positive integers. We can extend the definition to $r = 0$ by defining $\binom{n}{0} = 1$.

Two of the more important properties of these numbers are

$$(3) \qquad \binom{n-1}{r-1} + \binom{n-1}{r} = \binom{n}{r},$$

$$(4) \qquad \binom{n}{r} = \binom{n}{n-r}.$$

Table 1

	Without replacement	With replacement
Ordered	r-permutation $(n)_r$	r-permutation with repetition n^r
Unordered	r-combination $\binom{n}{r}$	r-combination with repetition $\binom{r+n-1}{r}$

(Handwritten annotations: $\frac{n!}{(n-r)!}$ and $\frac{n!}{r!\,(n-r)!}$)

These statements are easily verified using the definition of the symbols in (1) or (2). However, a set-theoretic argument based on the interpretation of $\binom{n}{r}$ as the number of r-subsets (where the 0-subset is the empty set) is also possible as follows. With regard to (3), consider all the r-subsets of an n-set S that do not contain some specified element of S. Actually, these r-subsets would have been selected from the remaining $(n-1)$ elements and so there are $\binom{n-1}{r}$ such subsets. Every other r-subset of the n-set S must contain the specified element. These r-subsets can be conceived as $(r-1)$-subsets with the specified element added for the rth member, so the total number is $\binom{n-1}{r-1}$. Because this exhausts the r-subsets, the sum of the two must be the number of r-subsets of the n-set S.

The validity of (4) can be established by observing that every selection of an r-subset of an n-set means that the elements not selected will be an $(n-r)$-subset of the same n-set. Thus there are as many $(n-r)$-subsets as r-subsets of an n-set.

So far we have established that the binomial coefficients (or binomial numbers) are a collection of integers defined for any positive integer n. We have not discussed why they are called coefficients or why the term binomial is used. We rectify this deficiency by stating one of the important theorems of mathematics—the formula for the expansion of a binomial expression (an expression with two terms).

Theorem 3

If n is a positive integer, then

$$(x+y)^n = x^n + \binom{n}{1}x^{n-1}y + \cdots$$
$$+ \binom{n}{r}x^{n-r}y^r + \cdots + y^n.$$

Using the summation symbol, we can write

$$(x+y)^n = \sum_{r=0}^{n} \binom{n}{r} x^{n-r} y^r.$$

Another way of stating the theorem is to say that the coefficients in the algebraic expansion of the binomial expression $(x+y)^n$ are the integers $\binom{n}{r}$, which are called binomial coefficients. Theorem 3 is a most remarkable formula because it gives us an algebraic identity in x and y, where x and y do not even have to be numbers. Of course, the sum and product of the symbols must have meaning. Because of property (3) a proof by mathematical induction is not very difficult.

■ **PROOF:** The assertion to be verified is the identity

(5) $$(x+y)^n = \sum_{r=0}^{n} \binom{n}{r} x^{n-r} y^r.$$

When $n=1$,

$$(x+y) = \sum_{r=0}^{1} \binom{1}{r} x^{1-r} y^r$$

$$= \binom{1}{0} x^1 y^0 + \binom{1}{1} x^0 y^1$$

$$= x+y.$$

Assume that the identity is true for n; that is, assume that (5) is true, and on the basis of this hypothesis we must show that

(6) $$(x+y)^{n+1} = \sum_{r=0}^{n+1} \binom{n+1}{r} x^{n+1-r} y^r.$$

Multiply both members of (5) by $(x+y)$; then

$$(x+y)^{n+1} = x \sum_{r=0}^{n} \binom{n}{r} x^{n-r} y^r + y \sum_{r=0}^{n} \binom{n}{r} x^{n-r} y^r.$$

The factors x and y can be included in the sum as

(7) $$(x+y)^{n+1} = \sum_{r=0}^{n} \binom{n}{r} x^{n+1-r} y^r + \sum_{r=0}^{n} \binom{n}{r} x^{n-r} y^{r+1}.$$

To complete the induction argument, we must show that the right members of (6) and (7) are the same. A careful look at the two sums in (7) reveals that there are terms with the same exponents on x and y. For example, the second term of the first sum is $\binom{n}{1} x^n y$, while the first term of the second sum is $\binom{n}{0}x^n y$.   When $r = m + 1$ in the first sum, the term is

$$\binom{n}{m+1}x^{(n+1)-(m+1)}y^{m+1} = \binom{n}{m+1}x^{n-m}y^{m+1}.$$

The same exponents occur in the second sum when $r = m$, namely, $\binom{n}{m}$ $\times x^{n-m}y^{m+1}$.   Add these two terms to obtain

$$\binom{n}{m+1}x^{(n+1)-(m+1)}y^{m+1} + \binom{n}{m}x^{n-m}y^{m+1},$$

which can be written as

$$\left[\binom{n}{m+1} + \binom{n}{m}\right]x^{(n+1)-(m+1)}y^{m+1}.$$

By property (3) we can write the two terms as

$$\binom{n+1}{m+1}x^{(n+1)-(m+1)}y^{m+1},$$

which has the same form as a term on the right side of (6) when $r = m + 1$.   This argument is valid for all values of $r = 1$ to n in the first sum of (7), and $r = 0$ to $n - 1$ in the second sum of (7).   Hence all terms in (6) for $r = 1$ to n are accounted for. Since the term for $r = 0$ in the first sum of (7) and the term for $r = n$ in the second sum of (7) are left over, the right side of (7) can be rewritten as

$$\binom{n}{0}x^{n+1} + \sum_{m+1=1}^{m+1=n}\binom{n+1}{m+1}x^{(n+1)-(m+1)}y^{m+1} + \binom{n}{n}y^{n+1}.$$

However,

$$\binom{n}{0} = \binom{n+1}{0} \qquad \text{and} \qquad \binom{n}{n} = \binom{n+1}{n+1},$$

so if we make these substitutions and replace $m + 1$ by r, we have the result desired—that (6) and (7) are identical.   ∎

The proof of Theorem 2 follows:

■**PROOF** (suggested by [20], pp. 9–10): Let S be a set of size n. Select a collection of r elements of S without regard to order and with replacement.   Because S is a finite set of n elements, the set of integers $S_1 = \{1, 2, \ldots, n\}$ can be identified with S.   That is, no matter what the elements in S may be, each element can be "named" by using a different positive integer as a name.   (If the set S is {Paul, Mike, Ralph, Carl}, then S_1 might be {3, 4, 2, 1}, where Paul is named with 3, Mike with 4, Ralph with 2, and Carl with 1.) Corresponding to any r-combination with replacement of the elements of S, there is an r-combination with replacement of the elements of S_1, which can be written {$a_1, a_2, \ldots, a_r$}, where $a_i, i = 1, 2, \ldots, r$ are integers from 1 to n inclusive. We write the integers a_i in ascending order and suppose that $a_1 \leq a_2 \leq \cdots \leq a_r$. Because the selection is with replacement, some (or all) of the a_i may be equal. Let S^* be the set of integers {1, 2, ..., $r + n - 1$}. Take any r-combination with replacement {$a_1, a_2, \ldots, a_r$} of S_1 and construct a new set of integers {$b_1, b_2, \ldots, b_r$} of S^* in the following fashion:

$$b_1 = a_1 + 0,$$
$$b_2 = a_2 + 1,$$
$$b_3 = a_3 + 2,$$
$$\vdots$$
$$b_r = a_r + r - 1.$$

Although the $a_i, i = 1, 2, \ldots, r$, are not necessarily distinct integers, the $b_i, i = 1, 2, \ldots, r$ are, and all the b_i are between 1 and $r + n - 1$ inclusive. Thus {$b_1, b_2, \ldots, b_r$} is an r-combination or r-subset of S^*. Every r-combination with replacement of S leads to an r-combination of S^*; and because two r-combinations with replacement are equal when the elements are the same, each r-combination with replacement of S leads to a distinct r-subset of S^*. Hence the number of r-combinations with repetition in S is less than or equal to the number of r-combinations in S^*. The relationship is reversible, however. Let {$b_1, b_2, \ldots, b_r$} be an r-combination (or r-subset) of the integers of S^*, written with $b_1 < b_2 < \cdots < b_r$. The integers $a_1, a_2, \ldots, a_r$ defined by

$$a_1 = b_1 - 0,$$
$$a_2 = b_2 - 1,$$
$$a_3 = b_3 - 2,$$
$$\vdots$$
$$a_r = b_r - (r - 1)$$

are an r-combination with replacement of S_1 with $a_1 \leq a_2 \leq \cdots \leq a_r$, and thus are equivalent to an r-combination with replacement of S. Once again every distinct r-combination of S^* will produce a distinct r-combination with replacement of S. This says that the number of r-combinations with repetition in S is at least as large as the number of r-combinations in S^*. Consequently, the number of r-combinations with repetition of S must be exactly the number of r-combinations of S^*, and this is $\binom{r + n - 1}{r}$. ∎

Problems

1. A coach must pick a team of 5 from 10 players. In how many ways can he do this

   (a) if there are no restrictions?
   (b) if two specified boys must be included?

2. An electronic circuit may fail at 15 stages. If it fails at exactly 3 stages, in how many ways may this happen?

3. There are six candidates in an election for two at-large positions on the city council, and a voter may mark his ballot either for one or for two candidates. In how many ways can he cast his vote?

4. A committee of three is to be chosen from four married couples. If all are equally eligible, in how many ways can the committee be selected? If a husband and wife cannot both serve on the committee, in how many ways can the three people be selected?

5. A traveler has his choice of six different airlines for a nonstop flight between London and Chicago. Why can each of the following be considered as correct answers to the question: How many ways can one make a flight from Chicago to London and return?

   (a) $(6)_2$ (b) 6^2 (c) $\binom{6}{2}$ (d) $\binom{7}{2}$

6. A tourist in London plans to devote a morning and an afternoon touring two art galleries, spending the morning in one and the afternoon in another. If there are four art galleries on his list, in how many ways can he spend his day?

7. From a panel of 10 seniors and 5 juniors, a committee of 5 is selected at random. How many committees would have a majority of seniors?

8. A poker hand is a subset of 5 cards of a standard deck of 52 cards. What is the number of poker hands that contain

   (a) exactly one ace?   (b) a pair of aces?

9. A manufacturer receives a lot of 20 motors, 6 of which are defective. Five motors are selected at random and if more than two of the motors are defective, the entire lot is rejected. How many different batches of five motors are possible, and how many of these will lead to rejection of the entire lot?

10. A book publisher is assigned the task of buying rights to English novels. He has narrowed his choice to nine books; yet he is authorized to purchase only four. In how many ways can he select four books from the nine? If four of the books were paperback

and five clothbound, how many ways could the publisher choose the four books if two were to be paperbacks and two were to be cloth?

11. An ice cream manufacturer sells 10 flavors. Four customers request one flavor each. How many different requests are possible, and how many of these requests would not include the flavor vanilla?

12. How many different committees of size 4 may be formed from a group of 25 people if one committeeman is designated as chairman?

13. A committee of 6 is drawn from a group consisting of 25 men and 15 women. How many different committees are there on which the tallest man is a member?

14. A reader selects 3 books from a set of 10 of which 4 are classified as nonfiction and 6 as fiction. How many ways can this be done with the requirement that the reader selects at least one nonfiction book?

15. One straight line is determined by two points in a plane. Three lines are determined by three noncollinear points. How many lines are determined by n points, no three of which are collinear?

16. Write an expression for the number of subsets containing at least one member that can be formed from a set of 100 elements.

17. Eleven people are to travel in two cars, six in one, five in the other. How many ways can this be done?

18. Fifteen people are to travel in three cars: six in one, five in a second, and four in a third. How many ways can this be done?

19. A group of 50 people consists of 20 men and 30 women. A committee consisting of 3 men and 2 women is to be selected with one of the committee members designated as chairman. How many distinct committees are possible?

20. There are three offices available for seven people. The first office can accommodate two people, the second office, three people, and the third office, two people. How many different assignments of people to offices are possible?

21. How many different five-digit numbers can be formed from three 1's and two 2's?

22. The eleven digits, 1, 2, 2, 2, 3, 3, 3, 3, 4, 4, and 4 are arranged in all distinguishable ways. How many permutations begin with 22? How many begin with 343?

23. A quarterback has a set of 10 plays, 5 of which are passing plays, 2 wide running plays, and 3 power running plays. In a 10-play sequence with no play repeated,

   (a) how many sequences can begin with a passing play?
   (b) how many sequences can have a passing play every other play?

24. How many ways can you answer a 10-question true–false exam, making the same number of answers true as you do false? How many if there are to be no two consecutive answers the same?

25. In how many ways can six nickels and four dimes be arranged in a circle?

26. A college has a schedule of eight football games in a season. How many ways can the season end in four wins, three losses, and a tie?

27. How many ways can three men and six women be assigned to three cars if there is to be one man and two women in each car?

28. Six curriculum proposals are under consideration by a foundation for grants. Funds are available for only two, but it is possible that none of the requests will qualify. How many different ways may the grants be made?

29. Show that at least three people in Philadelphia have the same set of three initials.

30. Show that the sum of the binomial coefficients of $(x+y)^n$ is 2^n. This is another demonstration that the number of subsets of an n-set is 2^n.

31. Write the binomial coefficients in rows according to increasing values of n, with ascending values of r in each row for $n = 1, 2, 3, 4, 5,$ and 6. The triangle arrangement is known as the *Pascal triangle*.

    $n = 1$:   1 1

    $n = 2$:   1 2 1

    $n = 3$:   1 3 3 1

    Use property (3) to find a scheme for writing the coefficients of row $k+1$ from the coefficients in row k.

32. Using the definition of the binomial numbers in terms of factorials, prove that
$$\binom{n}{r} = \binom{n}{n-r}.$$

33. Using the definition of the binomial numbers in terms of factorials, prove that
$$\binom{n-1}{r-1} + \binom{n-1}{r} = \binom{n}{r}.$$

34. By writing the binomial coefficients in terms of factorials, derive the formula
$$\binom{n}{r+1} = \frac{n-r}{r+1}\binom{n}{r}.$$

Use this to deduce that $\binom{n}{r}$ is smaller than $\binom{n}{r+1}$ if $2r < n-1$,

and is greater than $\binom{n}{r+1}$ if $2r > n-1$.

35. If $\binom{n}{7} = \binom{n}{11}$, find n.   If $\binom{18}{r-2} = \binom{18}{r}$, find r.

36. Expand the following using the binomial theorem.

(a) $(x + 2y)^4$ (b) $(x^3 - 1)^5$

(c) $(1 + y^2)^7$ (d) $\left(\dfrac{2}{x} + \dfrac{1}{y}\right)^4$.

37. Establish the identity

$$\sum_{r=0}^{n} (-1)^r \binom{n}{r} = 0.$$

38. Establish the identity

$$\sum_{0}^{n/2} \binom{n}{2k} = 2^{n-1} \qquad \text{if } n \text{ is even.}$$

39. Prove: The product of any r successive positive integers is divisible by $r!$.

3.4 Probability Problems

In this section we shall use our new theorems of combinatorial analysis to determine probabilities of events that arise from different types of probability problems.   We shall do this by presenting some typical applications of the theory of probability, so that the reader might gain some insight and feeling for the processes of probability as well as the way in which the fundamental concepts can be applied to practical problems.

The procedure for solving these problems will usually involve four steps:

1. Select a sample space for the random phenomenon.
2. Assign a probability function $P(\cdot)$ to the events of the sample space.

In this section we shall always assume that the uniform probability function applies.   Therefore, we can complete the assignment of the probability function by counting the number n of elementary events in the sample space and assigning the value $1/n$ to each elementary event.

3. Define the event in question as a subset of the sample space.
4. Determine the value of the probability function for this event.

In order to carry out the above schedule, we should normally prove theorems to justify each step. We shall not state the theorems formally, but the solutions to the problems will use the main idea of the proofs of the theorems.

One additional bit of notation. We shall denote by $N(S)$ the size or number of elements in the set S.

Example 1

Winning tickets for first, second, and third prizes are drawn from a bowl containing 450 tickets numbered from 1 to 450 inclusive. What is the probability that ticket 1 won a prize?

The 450 tickets are a set from which 3-permutations are formed, with the 3-tuple of numbers the prize winners. Thus the sample space consists of $(450)_3$ outcomes, so $N(S) = (450)_3$. The event A that ticket 1 wins a prize is the subset of outcomes with a 1 in one of the three spots. If 1 is the first component, there are $(449)_2$ choices for the second and third components. If 1 wins second prize, there are also $(449)_2$ choices for first and third prizes. Since the same reasoning applies if ticket 1 wins third prize, there are $3 \cdot (449)_2$ outcomes in event A, so $N(A) = 3 \cdot (449)_2$. Consequently,

$$P(A) = \frac{N(A)}{N(S)} = \frac{3(449)_2}{(450)_3} = \frac{3}{450} = \frac{1}{150}.$$

Observe that event A could be written as the union of the three events A_1, A_2, A_3, where A_i is the event "ticket 1 won prize i," $i = 1, 2, 3$. Moreover, ticket 1 can win only one prize, so the three events are mutually exclusive and

$$P(A) = P(A_1) + P(A_2) + P(A_3). \quad \blacksquare$$

Example 2

Six people in a class of 40 volunteer to participate in an experiment. What is the probability that the shortest person in the class has been selected?

The sample space is the total number of 6-combinations from a 40-set, so $N(S) = \binom{40}{6}$. The size of the event B—that the shortest person is in the volunteer group—is the number of 6-combinations that include a particular person, or the number of 5-combinations that does not include a particular person. In this case, then, $N(B) = \binom{39}{5}$ and

$$P(B) = \frac{N(B)}{N(S)} = \frac{\binom{39}{5}}{\binom{40}{6}} = \frac{3}{20}. \quad \blacksquare$$

Example 3

A committee of 5 is to be selected at random from a group of 10 men and 8 women. What is the probability that there will be exactly 3 men on the committee?

The 10 men and 8 women constitute an 18-set with 10 elements of one kind and 8 of another kind. The committee can be regarded as a 5-combination of the 18-set, so the sample space S is the set of all such combinations, $\binom{18}{5}$ in number. The event A is the subset of S with 3 men and 2 women in the selection. Now this selection can be broken into two parts with the selection of the men one task and the selection of the women a second task. The 3 men represent a 3-combination of a 10-set and there are $\binom{10}{3}$ such objects. The 2 women are but one of the $\binom{8}{2}$ 2-combinations of an 8-set. Thus there are $\binom{10}{3} \cdot \binom{8}{2}$ 5-combinations in event A and

$$P(A) = \frac{\binom{10}{3}\binom{8}{2}}{\binom{18}{5}} = \frac{20}{51}. \quad \blacksquare$$

Example 4

A rat runs through a T-maze six times. What is the probability that the rat went left more times than it went right?

The experiment may be regarded as six successive operations, but a sample space can be constructed as the totality of 6-permutations with replacement from the 2-set $\{L, R\}$. Thus $N(S) = 2^6$. The event A, "more L's than R's," consists of all 6-permutations of L and R with 4, 5, or 6 L's. There is exactly one permutation with 6 L's. For 5 L's there is 1 R and the position for the R can be selected in 6 ways. For 4 L's two positions are selected for the R's, and this can be done in $\binom{6}{2}$ ways. Thus

$$N(A) = \binom{6}{2} + \binom{6}{1} + 1 = 23$$

and

$$P(A) = \frac{N(A)}{N(S)} = \frac{23}{64} = 0.359. \quad \blacksquare$$

Example 5

Two balls are drawn at random from an urn containing three red balls and two green balls, with the balls indistinguishable except for color. (1) What is the probability that the first ball drawn is red? (2) What is the probability that the second ball drawn is green?

We answer each question for ordered selections with and without replacement. We distinguish the balls by numbers from 1 to 5 with the red balls numbered 1, 2, and 3 and the green balls numbered 4 and 5.

Question 1: For a selection with replacement, the sample space S could be the set of 2-tuples (z_1, z_2), where z_1 or z_2 could have the values 1, 2, 3, 4, and 5. Symbolically,

$$S = \{(z_1, z_2) \mid z_1, z_2 = 1, 2, 3, 4, 5\}.$$

Because S is the set of 2-permutations with replacement of a 5-set, there are 5^2 elementary events and $N(S) = 25$. Now let A be the event that the first ball is red. The subset A of S is all 2-tuples with the first component 1, 2, or 3; since there are 5 choices for the second component, $N(A) = 3 \cdot 5$. The probability of A is the ratio $N(A)/N(S)$; that is, $P(A) = 3 \cdot 5/5^2 = \frac{3}{5}$.

If the selection is without replacement, S will be the 2-permutations of a 5-set and $N(S) = (5)_2 = 5 \cdot 4$. The size of A is now $3 \cdot 4$ because there are only four choices for the second component. However, $P(A) = 3 \cdot 4/5 \cdot 4 = \frac{3}{5}$.

Question 2: Let B be the event that the second ball drawn is green. The subset B of S is all 2-tuples with the second component 4 or 5 and the first component 1, 2, 3, 4, or 5. Consequently, $N(B) = 5 \cdot 2$ and $P(B) = 5 \cdot 2/5^2 = \frac{2}{5}$.

If the selection is without replacement, the event B has size $4 \cdot 2$, because there are only 4 choices left for the first component after the second component has been designated, and $P(B) = 4 \cdot 2/5 \cdot 4 = \frac{2}{5}$. ∎

■ Example 6

(Suggested by [16], pp. 55–56.) Suppose that we want to inspect a lot of articles of size n to see that they conform to certain minimum standards. If n is at all large, it is quite unreasonable to expect the lot to be perfect; so let us assume that there are n_1 defectives and thus $n_2 = n - n_1$ nondefectives in the lot. Rather than inspect the entire lot, draw a sample of size r without replacement (r-permutation or r-combination) and test each of the articles in the sample. The probability that there will be k defectives in the sample is simply an application of Example 2. However, this requires us to assume that there are n_1 defectives among the n articles in the lot, where n_1 is not known.

We can approach the problem in a different manner. Suppose that in a sample of size r there are two defectives. Should the lot be accepted or rejected? It is not easy to give a simple yes or no answer. If we expect that the proportion of defectives is high, then the probability of two defectives is not necessarily small. However, if the proportion of defectives is low, the probability of two defectives is small. Consequently, the occurrence of two defectives when the proportion is supposed to be low may result in a rejection of the lot, because this may lead us to believe that there are more defectives in the lot than we expected.

Hence we want a sampling procedure that makes accepting a good lot (or rejecting a bad lot) probable. Consider the following plan. Decide on a number such as 1, to be called the acceptance number. The lot will be accepted if there are no more than this number of defectives in the sample. Assume that the proportion of defectives is p, so that the number of defectives may be taken as $n_1 = np$, where n is the size of the lot. We compute the probability that the lot will be accepted as a function of p. Thus $P(p)$ is the probability that an r-combination of an n-set with $n_1 = np$ objects of one kind and $n_2 = n - n_1 = n - np = n(1 - p)$ objects of another kind has 1 or fewer objects of the first kind:

$$P(p) = \frac{\binom{n_1}{0}\binom{n_2}{r}}{\binom{n}{r}} + \frac{\binom{n_1}{1}\binom{n_2}{r-1}}{\binom{n}{r}}$$

$$= \frac{\binom{n(1-p)}{r} + \binom{np}{1}\binom{n(1-p)}{r-1}}{\binom{n}{r}}$$

$$= \frac{\binom{n(1-p)}{r} + np\binom{n(1-p)}{r-1}}{\binom{n}{r}}.$$

If the lot size is 100 and the sample size 10, then

$$P(p) = \frac{\binom{100(1-p)}{10} + 100p\binom{100(1-p)}{9}}{\binom{100}{10}}.$$

The probability of accepting a lot containing a proportion of 0.05 defectives is 0.92, while the probability of acceptance when the proportion of defectives is 0.15 is 0.48. The graph of the function $P(p)$ as a function of p for specific values of n and r is called the *operating characteristic curve* of the acceptance sampling plan. The value of the function $P(p)$ is the probability of accepting the lot if in fact it contains the fraction p defective.

The same methods can be used for inspection sampling in order to maintain certain standards of quality in production. In general, the values of n and r must be adjusted for inspection or acceptance sampling as well as to accommodate the requirements of the producer and the consumer. The single sampling plan is but one of many methods of sampling inspection, and it is often used when it is desired to minimize the amount of sampling and yet keep a desired objective.   ▌

Lest the student think that pulling balls from urns is an unrealistic and useless endeavor, we present the following simplified version of counting the fish in a lake as a practical urn problem. Counting bees in a hive or the size of a flock of birds can be done similarly.

■ Example 7

(Suggested by [5], p. 44, and [8], p. 192.) How do you take a census when it is impossible to count the whole population, in this case the fish in a pond? The procedure is this:

1. Catch a fixed predetermined number of fish and mark them. Say that 500 fish are caught and marked.

2. Return the fish to the lake and allow them to get mixed up in the original population.

The pond is now like an urn that contains an unknown number of balls N, 500 of which are red (the marked fish) and $N - 500$ black balls (the unmarked fish).

3. Catch 500 fish again and count the number of marked fish in the second catch. Say that 100 marked fish are caught.

We assume that the second catch is a random sample and that the population does not change between the two catches.

4. Compute the probability of catching 100 marked fish in a catch of 500. This is the same as picking 100 red balls in a sample of 500 balls that come from an unknown population N with 500 red balls and $N - 500$ black balls. The probability is given by

$$P_{100}(N) = \binom{N - 500}{500 - 100}\binom{500}{100} \Big/ \binom{N}{500}$$

where the denominator is the number of ways of picking 500 balls from N without replacement, $\binom{500}{100}$ is the number of ways of picking exactly

100 red balls from the 500 red ones, and $\begin{pmatrix} N - 500 \\ 500 - 100 \end{pmatrix}$ is the number of ways of finding 400 black balls from the $N - 500$ black ones.

We are now in a position to estimate N, which is done by choosing a value of N so that $P_{100}(N)$ is as large as possible. Unfortunately, this is not a trivial matter and is accomplished by finding a condition so that $P_{100}(N)/P_{100}(N-1)$ changes from less than 1 to more than 1. It turns out that the largest value of $P_{100}(N)$ occurs when

$$N = [500 \cdot \tfrac{500}{100}],$$

where [ ] is the bracket function.

The general formula is given by $N = [nM/k]$, where M is the number of marked fish, n the number of fish in the second catch, and k the number of marked fish in the second catch. The estimate of N is called the *maximum likelihood estimate* of N. ∎

The reader might like to try a practical experiment using this method of estimating the population. Clean out the used IBM cards from the computing room's wastepaper basket at the end of a busy day, and estimate the number using the above method. Then verify the result by actually counting the cards.

Problems

1. A sample of size 3 is drawn without replacement from the digits 0, 1, . . ., 9 with the numbers drawn arranged in a row. What is the probability that there is a nine in either the first position or the third position but not both positions?

2. A committee of 5 is selected at random from a group of 45. What is the probability that the oldest of the group and the youngest of the group are on the committee?

3. One hundred white mice have been infected with a virus and 78 have contracted the disease. A sample of size 3 is drawn from the group and tested. What is the probability that none of these have the disease?

4. A group of 40 men contains 30 men over 5 feet 9 inches in height and 10 men under that height. Six men volunteer for a mission. What is the probability that exactly four are over 5 feet 9 inches?

5. A class is graded on a pass or fail basis. In general 85 percent of the students enrolled in the class receive a passing grade. If there are 1000 students in the class and a sample of size 2 is selected, find the probability that at least one of the students selected received a failing grade.

6. Two balls are drawn with replacement from an urn containing eight balls of which five are red and three are white. Find the probability that

   (a) both balls will be red.
   (b) both balls will be the same color. (Explain why two different answers are possible.)

7. An urn contains three red balls, four white balls, and five blue balls. Another urn contains five red balls, six white balls, and seven blue balls. One ball is selected from each urn. What is the probability that

   (a) both will be white?   (b) both will be the same color?

8. A sample of size 3 is drawn with replacement (without replacement) from the digits 0, 1, ..., 9. By placing the numbers in a row in the order in which they are drawn, an integer 0 to 999 is formed. What is the probability that the number formed is divisible by 39? (Regard 0 as divisible by 39.)

9. A firm buys three lots of material a month, choosing at random from four local and five out-of-state suppliers. What is the probability that the firm buys from two local and one out-of-state supplier?

10. The personnel of a company totals 100 and is made up of $\frac{4}{5}$ wage workers and $\frac{1}{5}$ salaried workers, $\frac{1}{2}$ of each category being male. What is the probability that two randomly selected individuals would be female salaried workers?

11. An instructor assigns two term papers to each of his students in a class of 25. Each paper may be on any one of six topics. If all combinations of topics are equally likely, what is the probability that all students do their papers on the same topics?

12. An urn contains eight red and five white balls. Three balls are drawn without replacement. Find the probability that at least one of the balls is white.

13. From a group of four freshmen, three sophomores, two juniors, and three seniors, a committee of three is to be selected in such a way that each combination of three students is equally likely to be selected. What is the probability that

   (a) all committee members are freshmen?
   (b) no committee members are seniors?

14. The key to a locked door is one of 12 keys in a cabinet. If a person selects 2 keys at random from the cabinet and takes them to the door with him, what is the probability that he can open the door without returning for another key?

15. A firm manufactures four different boats and each is available in three different colors. What is the probability that an order for *two* specific boats will be for the same boat? Consider two cases.

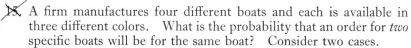 

16. Four mathematicians make plans to meet at the Holiday Inn in Philadelphia, but there are four motels in the city with that name. What is the probability that each mathematician will choose a different motel?

17. Three razor blades are selected from a box containing 20 blades, 5 of which are known to be used. What is the probability that those selected are unused?

18. A game warden inspects a catch of 10 fish by examining 2 which he selects at random. What is the probability that a fisherman will be arrested with a catch of 10 that includes 3 undersized fish?

19. A group of 100 people previewed a television show in a studio and their reaction was recorded with 95 percent favorable. At the conclusion 20 were selected at random for additional questions. What is the probability that all of those selected liked the show?

RECIPROCAL

20. In a music experiment, two subjects are asked to rank three different sounds according to their intensity. What is the probability that the rankings are identical if the intensity of the sounds is actually the same?

21. Six students place their textbooks for the same course on a table. If Dick and Alice are two of the six, and each of the six selects a textbook randomly from those on the table,

    (a) what is the probability that Alice will select her own book?
    (b) what is the probability that both Dick and Alice will select their own books?
    (c) what is the probability that at least one of the two, Dick and Alice, will select his own book?

22. An individual who claims to have extrasensory perception agrees to the following experiment. Six cards, three red and three black, are shuffled and placed face down on a table. The individual in question sits in another room and, taking the cards in order one through six, designates the numbers of those cards that he believes are red. If the individual does not really have extrasensory perception, what is the probability that he will guess correctly two of the three red cards?

23. Six residents of a town each call a plumber one morning, with the plumber called selected at random from a list of three in the telephone directory. What is the probability that the first plumber is called three times, the second plumber two times, and the third plumber one time?

$\frac{20}{243}$

24. Five digits are picked at random with replacement from $\{0, 1, \ldots, 9\}$. What is the probability that a digit does not appear more than once?

25. A committee of four is to be selected at random from a group consisting of eight faculty members and four students. What is the probability that the committee will have an equal number of faculty and students? What is the probability that the leader of the student group and the leader of the faculty group will be on the committee?

26. A true–false test consists of 10 questions, with 7 correct answers necessary for a passing score. If a student guesses at random at each question, what is the probability that he passes the examination?

27. There are 52 chips in an urn, numbered 1 to 52, with the numbers 1, 14, 27, and 40 considered "special." A sample of size 13 is drawn from the urn with replacement (without replacement). What is the probability that the sample will contain

    (a) exactly one "special" number?
    (b) at least one "special" number?

28. A man tosses a fair coin 10 times. Find the probability that he will have

    (a) heads on the first 5 tosses, tails on the second 5 tosses.
    (b) 5 heads and 5 tails.
    (c) at least 5 heads.

29. Given a group of four people, find the probability that at least two among them will have

    (a) the same birthday   (b) the same birth month

30. Ten persons in a room are wearing badges marked 1 through 10. Three persons are chosen at random and asked to leave the room simultaneously. Their badge numbers are noted.

    (a) What is the probability that the smallest badge number is 5?
    (b) What is the probability that the largest badge number is 5?

31. A mathematician turned magician claims that if a subject will concentrate on two numbers from the set $\{1, 2, 3, 4, 5\}$, he will "read the subject's mind" and name the numbers selected. Suppose that the mathematician is merely guessing. What is the probability that he will name the two numbers?

32. A manufacturer of transistors agrees to supply a purchaser with the following acceptance sampling plan: Out of each lot of 100 items, 2 will be selected at random and tested. If both items tested are good, the lot is accepted; otherwise the lot is rejected. Obtain the operating characteristic curve for this plan.

33. From six positive and eight negative numbers, four numbers are chosen at random without replacement and multiplied. What is the probability that the product is a positive number?

34. A box contains tags marked $1, 2, \ldots, n$. Two tags are chosen at random. Find the probability that the numbers on the tags will be consecutive integers if the tags are chosen

    (a) with replacement, (b) without replacement.

35. A group of 10 men and 10 women is divided into two groups of 10 each. Find the probability that each group contains as many men as women.

36. Prove: The probability that a specified element is included in an r-permutation with replacement from an n-set is $1 - (1 - 1/n)^r$.

37. Prove: The probability that a specified element is included in an r-permutation from an n-set is r/n.

38. Prove: For an n-set of n_1 distinguishable objects of one kind and $n_2 = n - n_1$ distinguishable objects of another kind, the probability of selecting exactly k objects of the first kind in an r-permutation of the n-set is

$$\frac{\binom{r}{k}(n_1)_k(n_2)_{r-k}}{(n)_r}.$$

39. Prove: For an n-set with n_1 distinguishable objects of one kind and $n_2 = n - n_1$ distinguishable objects of another kind, the probability of selecting exactly k objects of the first kind in an r-combination of the n-set is

$$\frac{\binom{n_1}{k}\binom{n_2}{r-k}}{\binom{n}{r}}.$$

40. Prove: For an n-set with n_1 distinguishable objects of one kind and $n_2 = n - n_1$ distinguishable objects of another kind, the probability of selecting exactly k objects of the first kind in an r-permutation with repetition of the n-set is

$$\frac{\binom{r}{k}n_1^k n_2^{r-k}}{n^r}.$$

41. Show that the probability expression in the theorem of Problem 38 is equivalent to the probability expression in the theorem of Problem 39.

42. Let $p = n_1/n$. Show that the probability expression of Problem 40 can be written

$$\binom{r}{k} p^k (1-p)^{r-k}.$$

43. As a consequence of Problem 39, one might expect the following identity to be true:

$$\sum_{k=0}^{n} \binom{n}{k}\binom{n}{n-k} = \binom{2n}{n}.$$

Establish this identity as well as the identity

$$\sum_{k=0}^{n} \binom{n}{k}^2 = \binom{2n}{n}.$$

44. A game warden in a wildlife preserve corralled 50 deer, banded them on the ear, and then released them. A week later he corralled 25 deer and found that 10 of them had bands. What was his estimate of the deer population?

45. Instead of the population N being taken as unknown, suppose that the size of the marked population is unknown. Everything else is the same as in Example 7. (This model also provides a way of estimating the number of defectives in a known lot of articles.)

(a) Show that the probability of obtaining k marked ones from an unknown number of marked ones in a sample size n is given by

$$P_k(M) = \binom{M}{k}\binom{N-M}{n-k} \Big/ \binom{N}{n}.$$

An alternative verbalization is that $P_k(M)$ is the probability of obtaining k defectives in a sample of size n from a population N that contains M defectives.

(b) By considering $P_k(M)/P_k(M-1)$, it is possible to obtain that a maximum likelihood estimate is $[(N+1)k/n]$. Use this result to find an estimate of the number of diseased rabbits in a flock of 2000, if 2 are ill in a sample of 25.

46. Using Problem 45, help a pearl diver estimate the number of pearls he might find in a catch of 200 oysters if he finds one pearl in a sample of size 50.

chapter four

Dependent and Independent Events

4.1 Conditional Probability

In this chapter we shall see how the occurrence of one event influences the probability that an event occurs, and how the probability of the intersection of two events is related to the probabilities of the individual events. To restate the problem, if E and F are two events of a general probability space (which need not have the uniform probability function defined on it), and if it is assumed (or known) that event F has occurred, what effect does the information about F have on the occurrence of E? Some examples will clarify the situation.

Example 1

A fair coin is tossed three times. We learn that the result of at least one toss is a head. What is the probability that there is exactly one tail? Let the sample space be the usual set of eight 3-tuples of H and T corresponding to the three tosses of the coin. Let F denote the event "at least one head," and E the event "exactly one tail."

$$F = \{(H, H, H), (H, H, T), (H, T, H), (T, H, H),$$
$$(H, T, T), (T, H, T), (T, T, H)\},$$
$$E = \{(H, H, T), (H, T, H), (T, H, H)\}.$$

Relative to the original sample space which includes (T, T, T), we see that $P(E) = \frac{3}{8}$; but how does the knowledge that F has occurred affect this probability? Every elementary event in E is also in F, so it makes sense to ask about the probability of E as an event in the space F. Let us consider F as a complete sample space with the uniform probability function and E as an event of F. Thus the probability that E occurs based on the knowledge that F has occurred is $\frac{3}{7}$, because there are seven equally likely elementary events in F and $E \subset F$. This new probability for the event E is usually written $P(E \mid F)$. Knowing that F has occurred gives us a different probability and indeed increases the probability that E occurs. ∎

Example 2

Suppose that two draws are made without replacement from an urn with four red and two green balls. If the first draw is red, what is the probability that the second draw is green? Noting that six balls are available for the first draw and five for the second draw, we represent one sample space S for this random experiment as $6 \cdot 5 = 30$ ordered pairs $\{(x_1, x_2) \mid x_1, x_2 = 1, 2, 3, 4, 5, 6, x_1 \neq x_2\}$, where 1, 2, 3, and 4 distinguish the red balls and 5 and 6 distinguish the green. The event F, first draw is red, is the subset of $4 \cdot 5 = 20$ ordered pairs listed in the first 4 columns of Table 1.

Table 1

(1, 2)	(2, 1)	(3, 1)	(4, 1)	(5, 1)	(6, 1)
(1, 3)	(2, 3)	(3, 2)	(4, 2)	(5, 2)	(6, 2)
(1, 4)	(2, 4)	(3, 4)	(4, 3)	(5, 3)	(6, 3)
(1, 5)	(2, 5)	(3, 5)	(4, 5)	(5, 4)	(6, 4)
(1, 6)	(2, 6)	(3, 6)	(4, 6)	(5, 6)	(6, 5)

Let E be the event that the second draw is green. The $5 \cdot 2 = 10$ ordered pairs in E are listed in Table 2.

Table 2

(1, 5)	(2, 5)	(3, 5)	(4, 5)	(6, 5)
(1, 6)	(2, 6)	(3, 6)	(4, 6)	(5, 6)

Because we know that event F has occurred, the set of elementary events has been reduced to the 20 in F; and because we ultimately want the probability that E occurs, $E \cap F$ is the only part of E that is important to us. The members of event $E \cap F$, that is, those elementary events of E that are also in F, are listed in Table 3.

Table 3

(1, 5)	(2, 5)	(3, 5)	(4, 5)
(1, 6)	(2, 6)	(3, 6)	(4, 6)

Thus there are 8 elementary events from the original sample space in $E \cap F$ and 20 elementary events from the original sample space in F. Consequently, using F as a new sample space with the uniform probability function, we find that the probability of a green ball on the second draw, knowing that a red ball was selected on the first draw, is $\frac{8}{20}$. Symbolically, we write $P(E \mid F) = \frac{2}{5}$. Observe that relative to the original sample space $P(E) = \frac{1}{3}$, so that $P(E \mid F) \neq P(E)$ in this case. ∎

Knowing that F has occurred forces an adjustment in $P(E)$, because the subset F replaces the sample space S as the reference space for the computation of the probabilities. Now, the probability of any event in the sample space is given by the probability function $P(\cdot)$, but what we want is the probability of any event of S, knowing that F has occurred, in terms of the same $P(\cdot)$. In other words, we seek a probability function for F, as the sample space, in terms of the probability function $P(\cdot)$ of S. If E is any event in S and if $E \cap F = \emptyset$, then $P(E \cap F) = 0$. However, when $E \cap F \neq \emptyset$, $P(F)$ provides the adjustment for $P(E \cap F)$ in terms of $P(\cdot)$ on S, as we indicate in the definition for conditional probability.

Definition 1: Conditional probability for two events

If E and F are two events of a probability space S with probability function $P(\cdot)$ and $P(F) > 0$, then the conditional probability of event E on the occurrence of event F, denoted by $P(E \mid F)$, is given by

(1) $$P(E \mid F) = \frac{P(E \cap F)}{P(F)}.$$

If $P(F) = 0$, then $P(E \mid F)$ is not defined.

Intuitively, we have already been reading the symbol $P(E \mid F)$ as "the probability that event E will occur knowing that event F has occurred."

In short form, the symbol may be read "probability of event E given event F," or "probability of event E on the hypothesis of event F." Conditional probability should not be considered as a new idea but only a formal way of adjusting the probabilities of some events when more information is available. Thus the new notation $P(E|F)$ merely shows what the information is that affects $P(E)$.

Let us solve another example.

Example 3

Six equally desirable sites a, b, c, d, e, and f are under consideration for the location of a new ice cream stand. Three sites are to be selected at random. What is the probability that sites a and b were selected if it is known that site f was not selected?

The sample space of outcomes are all possible 3-combinations from the 6-set $\{a, b, c, d, e, f\}$, so there are $\binom{6}{3}$ elementary events. The uniform probability function is also assumed. The two events in question are "a and b selected" and "f not selected." If E denotes the first event and F the latter, then the desired probability is $P(E|F)$. From equation (1), as soon as $P(F)$ and $P(E \cap F)$ are computed, we are done.

Event F is the subset of three sites that omits f, and there are $\binom{5}{3}$ of these. Therefore,

$$P(F) = \binom{5}{3} \bigg/ \binom{6}{3} = \frac{5!}{3!2!} \frac{3!3!}{6!} = \frac{1}{2}.$$

The event $E \cap F$ consists of those triples of sites that include a and b but not f. Since there are only three sites from which one is chosen, there are $\binom{3}{1}$ outcomes in $E \cap F$. Therefore, $P(E \cap F) = \binom{3}{1} \bigg/ \binom{6}{3} = \frac{3}{20}$.

Consequently,

$$P(E|F) = \frac{\frac{3}{20}}{\frac{1}{2}} = \frac{3}{10}. \quad \blacksquare$$

Example 4

Of 50 students in an economics statistics class, 20 had some calculus while the others have not had any. Of those with no calculus, 11 did not pass the final examination in statistics, and 3 of those with calculus failed. If a student, selected at random, passed the examination, what is the (conditional) probability that the student had calculus?

The sample space for this problem is the 50 people in the class, with each person as likely to be selected as another. Let A denote the event "the student passed the examination" and B the event "the student had calculus." The desired probability is the conditional probability

$$P(B \mid A) = \frac{P(A \cap B)}{P(A)}.$$

Since there are 17 who had calculus and passed the examination and 19 who did not have calculus but passed, $P(A) = \frac{36}{50}$. The size of event $A \cap B$ is the number who both had calculus and passed the examination. Thus $P(A \cap B) = \frac{17}{50}$, and so

$$P(B \mid A) = \frac{\frac{17}{50}}{\frac{36}{50}} = \frac{17}{36}.$$

The data for this problem can be summarized as shown in Table 4.

Table 4

	Passed	Failed
Students with calculus	17	3
Students without calculus	19	11

The conditional probability for two events in a sample space S has been defined in terms of the probability function $P(\cdot)$ for S. It is reasonable to assume that all axioms and theorems for general probability functions will hold for the conditional probability function. They do, and a few of them are listed below. In these statements S denotes a sample space for which $P(S) = 1$ and A, B, E, and F are arbitrary events in S with $P(F) > 0$.

1. $0 \leq P(E \mid F) \leq 1$.
2. $P(S \mid F) = 1$.
3. $P(A \cup B \mid F) = P(A \mid F) + P(B \mid F)$, if $A \cap B = \varnothing$.
4. $P(A \cup B \mid F) = P(A \mid F) + P(B \mid F) - P(A \cap B) \mid F)$.
5. $P(E' \mid F) = 1 - P(E \mid F)$.

The verifications of these statements can be made using direct application of the definition of conditional probability, and we shall leave them as an exercise for the reader. We can now assert that conditional probability is itself a probability function.

The relationship between $P(E \mid F)$ and $P(E \cap F)$ that is given by the definition of conditional probability is valuable for computational as well as theoretical purposes. Because $P(F) > 0$, we can rewrite the relationship given by the definition,

$$P(E \mid F) = \frac{P(E \cap F)}{P(F)},$$

as $P(E \cap F) = P(E|F)P(F)$ for any event E. Hence, if we know the conditional probability $P(E|F)$ and $P(F)$, we know $P(E \cap F)$. Often it is easy to compute $P(E|F)$ directly, as the next example shows.

Example 5

(Refer to Example 2.) Suppose that two draws are made without replacement from an urn with four red and two green balls. What is the probability that the first draw is red and the second draw is green (that is, the intersection of the two events)? Referring to the sample space described in Example 2, let F be the event "first draw is red" and let E be the event "second draw is green." We want $P(E \cap F)$. From Example 2,

$$P(F) = \frac{4 \cdot 5}{6 \cdot 5} = \frac{2}{3} \quad \text{and} \quad P(E|F) = \frac{8}{20} = \frac{2}{5}.$$

Hence $P(E \cap F) = \frac{2}{5} \cdot \frac{2}{3} = \frac{4}{15}$.   ∎

The probability of the intersection of two events, expressed as $P(E \cap F) = P(E|F)P(F)$, can be used to generate an expression for the probability of the intersection of three events.

$$\begin{aligned} P(A_1 \cap A_2 \cap A_3) &= P(A_3|A_1 \cap A_2)P(A_1 \cap A_2) \\ &= P(A_3|A_1 \cap A_2)P(A_2|A_1)P(A_1) \\ &= P(A_1)P(A_2|A_1)P(A_3|A_1 \cap A_2). \end{aligned}$$

The extension to four or more events follows this same pattern. We refer the reader to [16, p. 92] for the general theorem.

In Example 2 we saw that $P(E) \neq P(E|F)$. Unfortunately, $P(E)$ is frequently used when $P(E|F)$ should be used, so the distinction between these probabilities must be emphasized. Look at the following example.

Example 6

(Refer to Example 5.) Suppose that two draws are made without replacement from an urn with four red and two green balls. Let F be the event "first draw is red," and let G be the event "second draw is red." The probability that both draws are red is $P(G \cap F) = P(G|F)P(F) = \frac{3}{5} \cdot \frac{4}{6} = \frac{2}{5}$. But what is the probability of the event G? G is the subset of all pairs that have 1, 2, 3, or 4 in the second spot, and any of the other five numbers in the first spot. Thus

$$P(G) = \frac{5 \cdot 4}{6 \cdot 5} = \frac{20}{30} = \frac{2}{3}.$$

Clearly, $P(E) \neq P(E|F)$. The probability that the second draw is red is different from the probability that the second draw is red on the hypothesis that the first draw is red.   ∎

Problems

1. Consider a coin tossed three times. If there are exactly two heads, what is the probability that the first toss was a head?

2. Two draws are made without replacement from an urn with six red and four black balls. If the first draw is red, what is the probability that

   (a) the second draw is black? (b) the second draw is red?

3. A man tosses a coin four times in succession. What is the conditional probability of

   (a) three heads if the first toss is a head?
   (b) two heads and two tails if the first toss is a tail?

4. A player throws two dice, one red and one black. After the dice are thrown, an observer can see that the red die has turned up 2, but he cannot see the black die. What is the probability that the player has thrown a total of 7?

5. A male and a female rat run a T-maze in successive trials with the male first. The outcomes are considered to be equally likely. What is the probability that the female rat went left if the male went left?

6. The probability that a male has tuberculosis is 0.05, and the probability that a female has tuberculosis is 0.10. From a group of 350 men and 350 women, one is chosen at random. Given that the person selected has tuberculosis, find the probability that he is male.

7. The probability that a woman is a college graduate is 0.2. The probability that a woman is married, given that she is a college graduate, is 0.43; and the probability that a woman is married is 0.35. If a woman selected at random is married, what is the probability that she is a college graduate?

8. A bridge hand of 13 cards is selected without replacement from an ordinary deck of 52 cards. If it is known that the hand contains the ace of spades, what is the conditional probability that the hand will contain all four aces?

9. In a study of IQ and college graduates it was found that half of the people have an IQ over 110. Of those with an IQ over 110, 60 percent are college graduates, while only 20 percent of those with an IQ under 110 are college graduates. Find

   (a) $P(\text{college graduate} \mid \text{IQ over 110})$.
   (b) $P(\text{IQ over 110} \mid \text{college graduate})$.

10. For a family with two children, assume that each child is as likely to be a boy as it is to be a girl. What is the conditional probability that both children are boys, given that at least one of the children is a boy?

11. A card is selected at random from a deck of 52 playing cards. What is the conditional probability that the card is the jack of spades, given that it is a spade?

12. Students in a special summer program took two courses, mathematics and literature. Four percent failed mathematics, 3 percent failed literature, and 1 percent failed both. Among those who failed mathematics, what percentage also failed literature? What percentage passed mathematics and failed literature?

13. A student survey on the coeducational use of their dormitory rooms produced the following results:

	Never	Sometimes	Often
Women	11	28	7
Men	4	24	18

If a person is selected at random, find

    (a) P(student is a women | use is often).
    (b) P(student is a man | use is never).
    (c) P(use is sometimes | student is a man).

14. A red die and a green die are thrown. What is the conditional probability that

    (a) the red die is odd, given that the sum of the faces is 7?
    (b) the sum of the faces is 7, given that one die is odd and the other even?

15. Four balls are placed in an urn by the following scheme: A fair coin is tossed, and if it falls heads, a white ball is placed in the urn; but if it falls tails, a red ball is placed in the urn. What is the probability that

    (a) the urn will contain exactly three white balls?
    (b) the urn will contain exactly three white balls, given that the first ball placed in the urn is white?

16. The probability that a man who earns more than $15,000 annually owns two or more cars is 0.8, while the probability that a man earns more than $15,000 is 0.3. The probability that a man owns two or more cars is 0.25. If it is known that a man owns three cars, what is the probability that his income exceeds $15,000?

17. A survey of truck fleets of size 1, 2, and 3 with regard to the owner-
    ship of the trucks in the fleet produced the following data:

No. of trucks in fleet	No. of fleets owned	No. of fleets rented
One	4501	112
Two	562	89
Three	333	210

A truck fleet is chosen by lot. What is the probability that there is
one truck in the fleet? Given that there are two trucks in the
fleet, what is the probability that the fleet is owned? Given that
there are two or three trucks in the fleet, what is the probability
that the fleet is rented?

18. A drivers' education examination consists of six tasks, of which four
    must be completed successfully to pass the examination. If an
    examinee fails the first two tasks, what is the probability of failing
    the examination if we assume that all outcomes are equally likely?

19. A die is thrown three times in succession. What is the conditional
    probability of

    (a) a sum on the three tosses of at least 10, given that the first toss
        is a 2?
    (b) three tosses the same, given that the first toss is a 2?

20. An urn contains seven balls of which three are white. Three balls
    are drawn and laid aside (not replaced in the urn). Another ball
    is drawn. What is the probability that it will be white?

21. There are three candidates in an election, A, B, and C, with proba-
    bilities 0.5 that A will win, 0.3 that B will win, and 0.2 that C will
    win. Just before the election C withdraws. What are the new
    chances for the other two candidates?

22. A die is weighted so that the probability that a 6 occurs is 0.4, the
    probability that a 1 occurs is 0.3, and the probability of all other
    spots is 0.3. What is the probability that a person gets a six, given
    that a 1 does not occur?

23. Three cards are drawn from a deck of 52 playing cards.

    (a) If it is known that the hand contains at least two aces, what is
        the probability that it contains three aces?
    (b) If it is known that the hand contains the two red aces, what is
        the probability that it contains three aces?

24. Three draws are made with replacement from an 8-set of five red and three green balls in a bag. If the last draw is a red ball, what is the probability that there are exactly two red balls in the sample?

25. A sample of size 4 is drawn without replacement from an urn containing 12 balls, of which 8 are white. Find the conditional probability that the last ball drawn is white, given that the sample contains 3 white balls.

26. If $P(\cdot)$ is a probability function on a sample space S, and E and F are any events with $P(F) > 0$, prove that

(a) $P(S \mid F) = 1$.
(b) $P(E' \mid F) = 1 - P(E \mid F)$.

27. If $P(\cdot)$ is a probability function on a sample space S, and A, B, and F are three events with $A \cap B = \varnothing$ and $P(F) > 0$, prove that

$$P(A \cup B \mid F) = P(A \mid F) + P(B \mid F).$$

28. If $P(\cdot)$ is a probability function on a sample space S, and A, B, and F are three events with $P(F) > 0$, show that

$$P(A \cup B \mid F) = P(A \mid F) + P(B \mid F) - P(A \cap B \mid F).$$

29. If $P(\cdot)$ is a probability function on a sample space S, and A and B are events with $P(B) > 0$, prove that

(a) $A \subset B$ implies $P(A \mid B) = P(A)/P(B)$.
(b) $B \subset A$ implies $P(A \mid B) = 1$.

30. If $P(\cdot)$ is a probability function on a sample space S, and A, B, and C are three events with $P(A) > 0$, $P(C) > 0$, and $B \subset A$, $C \subset A$, prove that

$$\frac{P(B \mid A)}{P(C \mid A)} = \frac{P(B)}{P(C)}$$

31. A fair coin is tossed n successive times.

(a) Find the probability that the nth toss results in heads.
(b) Find the conditional probability that the nth toss results in heads, given that all preceding tosses resulted in heads.

4.2 Conditional Probability Applications

Let A and B be two nonempty events in a sample space S. The two events may or may not have elementary events in common. But in terms of A and its complement A', we can always write

(1) $B = (B \cap A) \cup (B \cap A').$

Since $A \cap A' = \varnothing$, we can show that $(B \cap A) \cap (B \cap A') = \varnothing$ (Exercise 21). By applying axiom PA3 to the right side of (1) we find that

$$P(B) = P(B \cap A) + P(B \cap A').$$

If $1 > P(A) > 0$, it follows that $P(A') > 0$; hence the definition of conditional probability yields

$$P(B) = P(B \mid A)P(A) + P(B \mid A')P(A').$$

We have thereby proved

Theorem 1

If A and B are two events with $1 > P(A) > 0$, then $P(B) = P(B \mid A) \cdot P(A) + P(B \mid A')P(A')$.

Example 1 shows how nicely the technique we used to prove Theorem 1 helps to solve certain problems.

Example 1

Suppose we have two urns. Urn I contains two black balls and three white balls; urn II contains two black balls and one white ball. An urn is chosen at random (for example, flip a fair coin, choose I if heads and II if tails) and a ball picked from this urn. What is the probability that a white ball is chosen?

Since there are four white and four black balls altogether, one is tempted to guess $\frac{1}{2}$. This is a wrong guess, as we now see. Let A be the event "selection of urn I." Then A' is "selection of urn II." Also $P(A) = P(A') = \frac{1}{2}$. If B is the event "picking a white ball," we obtain $P(B \mid A) = \frac{3}{5}$ and $P(B \mid A') = \frac{1}{3}$. Consequently, Theorem 1 gives us

$$P(B) = \tfrac{3}{5} \cdot \tfrac{1}{2} + \tfrac{1}{3} \cdot \tfrac{1}{2} = \tfrac{7}{15}. \quad \blacksquare$$

Example 2

A bond issue for the construction of a new public library is before the voters. A poll showed that 85 percent of those with a college education favored the construction of a new library, but only 20 percent of those not having a college education did so. Suppose that 90 percent of the voting population does not have a college education. What is the probability that a voter selected at random who favors the bond issue will be one with a college education?

The underlying sample space for this problem is the set of all voters. Let A denote the event "voter has a college education," and let B denote the event "voter favors the library construction." For a voter selected at random, then,

$$P(A) = 0.1 \quad \text{and} \quad P(A') = 0.9.$$

For the event B, we are given that

$$P(B \mid A) = 0.85 \qquad \text{and} \qquad P(B \mid A') = 0.2.$$

We are to find $P(A \mid B)$.

From the definition of conditional probability,

$$P(A \mid B) = P(A \cap B)/P(B),$$

and we can write $P(A \cap B) = P(B \mid A)P(A)$. Now we are given $P(A)$, $P(B \mid A)$, and we can find $P(B)$ using Theorem 1 as follows:

$$\begin{aligned}
P(B) &= P(B \mid A)P(A) + P(B \mid A')P(A') \\
&= (0.85)(0.1) + (0.2)(0.9) \\
&= 0.265.
\end{aligned}$$

Since $P(A \cap B) = (0.85)(0.1) = 0.085$, we have that

$$P(A \mid B) = \frac{0.085}{0.265} = 0.32. \qquad \blacksquare$$

The two events A and A' form a partition of the underlying sample space S, and Theorem 1 gives the probability of an arbitrary event in terms of this partition. Theorem 1 can be generalized, of course, to an arbitrary partition of the sample space. But first we define the term "partition."

Definition 1: Partition of a set

A partition of a set A is a set of nonempty subsets $\{B_1, B_2, \ldots, B_m\}$ of A such that

(1) $B_i \cap B_j = \varnothing, \qquad i, j = 1, 2, \ldots, m; i \neq j.$

(2) $\bigcup_{j=1}^{m} B_j = A.$

In other words, a partition of a set A is a set of mutually disjoint nonempty subsets of A whose union is the set A.

Theorem 2

If $B_1, B_2, \ldots, B_m$ is a partition of the sample space S, where $P(B_j) > 0$, $j = 1, 2, \ldots, m$, and if A is any event of S, then

$$P(A) = \sum_{j=1}^{m} P(A \mid B_j)P(B_j).$$

The proof follows the same argument that we used to prove Theorem 1. In fact, when $m = 2$, Theorem 2 becomes Theorem 1.

PROOF: Since $S = \bigcup_{j=1}^{m} B_j$, we have

$$A = A \cap S,$$

or

$$A = A \cap \left(\bigcup_{j=1}^{m} B_j \right),$$

so that

$$A = \bigcup_{j=1}^{m} (A \cap B_j),$$

where we have used the generalized distributive law for union and intersection. Because the sets B_j, $j = 1, \ldots, m$ are disjoint, the sets $A \cap B_i$ and $A \cap B_j$ are also disjoint. Therefore, by axiom PA3 we obtain

$$P(A) = \sum_{j=1}^{m} P(A \cap B_j).$$

Since $P(B_j) > 0$, $j = 1, 2, \ldots, m$, Definition 1 of Section 4.1 applied to each term in the sum gives us the required formula. ∎

A partition of a sample space stratifies the elementary events into disjoint sets, so Theorem 2 applies readily to any random experiment for which there is a natural classification of the elementary events.

Example 3 will clarify the procedure.

Example 3

In a small city there are four voting subdivisions, or precincts, identified as Ward 1, Ward 2, Ward 3, and Ward 4. The political composition of the wards with regard to the proportion showing a preference for the Democratic party is as follows:

Ward 1	0.45,	Ward 2	0.35,
Ward 3	0.45,	Ward 4	0.55.

A voter is selected from the total voting population by first selecting a ward and then choosing a voter, both selections being random. What is the probability that a voter with Democratic preferences is selected?

The sample space is the set of all voters in the city. Let B_i, $i = 1, 2, 3, 4$ be the event "a voter lives in Ward i," and let A be the event "a voter is a Democrat." We want to find $P(A)$. It is given that $P(B_i) = \frac{1}{4}$, $i = 1, 2, 3, 4$, and that

$$P(A \mid B_1) = 0.45, \qquad P(A \mid B_2) = 0.35,$$
$$P(A \mid B_3) = 0.45, \qquad P(A \mid B_4) = 0.55.$$

Since a voter does not live (legally) in two wards, we can assume that all voters in the city belong to the set S given by

$$S = B_1 \cup B_2 \cup B_3 \cup B_4,$$

where $B_i \cap B_j = \emptyset$ for $i, j = 1, 2, 3, 4$ and $i \neq j$. Thus $\{B_1, B_2, B_3, B_4\}$ is a partition of S, and by Theorem 2 we may write

$$P(A) = \sum_{j=1}^{4} P(A \mid B_j)P(B_j).$$

Substituting the values given earlier for the probabilities in the expression on the right side, we find

$$P(A) = (0.45 + 0.35 + 0.45 + 0.55)\tfrac{1}{4} = 0.45.$$

That is, the probability is 0.45 that a voter selected by this process will be a Democrat. ▮

Example 4

What is the probability that the ace of spades is next to the king of spades in a well-shuffled deck of 52 playing cards?

The 52 cards are a 52-set and any shuffling is, in effect, an arrangement of the 52-set. A sample space is the set of 52-tuples of cards, and the probability function is the uniform probability function. If the ace of spades is the first card, then the king of spades must be the next card for success. If the ace of spades is a card in the middle somewhere, the king of spades can be on either side of the ace of spades. If the ace of spades is the last card, then the king of spades must be the next-to-last card. So let us separate the elementary events into three cases: B_1 is the event that the ace of spades is the first card, B_2 is the event that the ace of spades is the last card, and B_3 is the collection of the remaining elementary events. Thus B_1, B_2, B_3 is a partition of the sample space. Now $P(B_1) = P(B_2) = \tfrac{1}{52}$, and $P(B_3) = \tfrac{50}{52} = \tfrac{25}{26}$. If A is the event that the king of spades is next to the ace of spades, then $P(A \mid B_1) = P(A \mid B_2) = \tfrac{1}{51}$, while $P(A \mid B_3) = \tfrac{2}{51}$. Using Theorem 2 with $n = 3$ we get

$$P(A) = (\tfrac{1}{51})(\tfrac{1}{52}) + (\tfrac{1}{51})(\tfrac{1}{52}) + (\tfrac{2}{51})(\tfrac{25}{26}) = \tfrac{1}{26}. \quad ▮$$

Conditional probability concepts are useful in the computation of probabilities of random phenomena known as *compound* experiments. Whenever we can describe a random experiment as a succession of separate random phenomena, such as tossing a coin and then rolling a die, the experiment is a compound experiment. Usually the separate experiments are executed in a definite order, such as first selecting a classroom and then picking a student from the class, as we see in Example 5.

Example 5

There are three urns containing red and blue chips with r_i red chips and b_i blue chips in urn i, $i = 1, 2, 3$. Two fair coins are tossed and one chip is drawn from the urn corresponding to the number of heads showing plus one. What is the probability that the chip is red?

Instead of describing a sample space for this experiment, we can go directly to the partition defined by the events B_1, B_2, and B_3, where B_i, $i = 1, 2, 3$ is the event that urn i is selected. If we let A be the event that a red chip is drawn, then

$$P(B_1) = \tfrac{1}{4}, \qquad P(B_2) = \tfrac{1}{2}, \qquad P(B_3) = \tfrac{1}{4},$$

and

$$P(A \mid B_i) = (r_i)/(r_i + b_i), \qquad i = 1, 2, 3.$$

Using the formula of Theorem 2, with $n = 3$, we get

$$P(A) = (r_1)/(r_1 + b_1)\tfrac{1}{4} + (r_2)/(r_2 + b_2)\tfrac{1}{2} + (r_3)/(r_3 + b_3)\tfrac{1}{4}. \quad \blacksquare$$

Another application of conditional probability is the famous formula known as *Bayes' theorem*. The formula is a simple and easy consequence of Theorem 2 and was first used by Thomas Bayes in a paper published posthumously in 1763.

Theorem 3

Let B_1, B_2, ..., B_m be a partition of a sample space. If the unconditional probabilities $P(B_i)$, $i = 1, 2, \ldots, m$, and the conditional probabilities $P(A \mid B_i)$, $i = 1, 2, \ldots, m$, are known, then

$$P(B_j \mid A) = \frac{P(A \mid B_j)P(B_j)}{\sum_{i=1}^{m} P(A \mid B_i)P(B_i)}.$$

PROOF: From the definition of conditional probability we can write $P(B_j \cap A)$ in two ways if both $P(B_j)$ and $P(A)$ are positive:

(a) $P(B_j \cap A) = P(B_j \mid A)P(A)$,

(b) $P(B_j \cap A) = P(A \mid B_j)P(B_j)$,

so that $P(B_j \mid A)P(A) = P(A \mid B_j)P(B_j)$. Solving for $P(B_j \mid A)$, we see that

(2)
$$P(B_j \mid A) = \frac{P(A \mid B_j)P(B_j)}{P(A)}.$$

Since $\{B_1, B_2, \ldots, B_m\}$ is a partition of S, we can apply Theorem 2 to $P(A)$ and obtain

$$P(A) = \sum_{j=1}^{m} P(A \mid B_j)P(B_j).$$

The theorem is proved after we substitute this expression for $P(A)$ in (2). $\quad \blacksquare$

Bayes' theorem has led to considerable controversy in theory and in application. Often the $P(B_i)$ are not known beforehand, so one must make an assumption about them. In some cases the assumption can lead to unusual conclusions. Care must be exercised in the application of Bayes' theorem, especially with regard to the hypotheses. (See Parzen [16, pp. 119–124] for some examples.)

Example 6 is an application of the theorem.

Example 6

(Refer to Example 5.) Let urn 1 have three red and four blue chips; urn 2, four red and three blue chips; and urn 3, one red and six blue chips. An urn is selected by tossing two fair coins and choosing the urn corresponding to the number of heads plus one. A chip is selected and it is red. What is the probability that the chip from came urn 2? Let A be the event of a red chip being selected and B_i, $i = 1, 2, 3$, the event of urn i being selected. As before, B_1, B_2, B_3 is a partition of the sample space and we know the following probabilities:

$$P(B_1) = \tfrac{1}{4}, \qquad P(B_2) = \tfrac{1}{2}, \qquad P(B_3) = \tfrac{1}{4},$$

$$P(A \mid B_1) = \tfrac{3}{7}, \qquad P(A \mid B_2) = \tfrac{4}{7}, \qquad P(A \mid B_3) = \tfrac{1}{7}.$$

Therefore,

$$P(B_2 \mid A) = \frac{(\tfrac{4}{7})(\tfrac{1}{2})}{(\tfrac{3}{7})(\tfrac{1}{4}) + (\tfrac{4}{7})(\tfrac{1}{2}) + (\tfrac{1}{7})(\tfrac{1}{4})} = \frac{2}{3}.$$

Similarly,

$$P(B_1 \mid A) = \frac{(\tfrac{3}{7})(\tfrac{1}{4})}{(\tfrac{3}{7})(\tfrac{1}{4}) + (\tfrac{4}{7})(\tfrac{1}{2}) + (\tfrac{1}{7})(\tfrac{1}{4})} = \frac{1}{4}$$

and

$$P(B_3 \mid A) = \frac{(\tfrac{1}{7})(\tfrac{1}{4})}{(\tfrac{3}{7})(\tfrac{1}{4}) + (\tfrac{4}{7})(\tfrac{1}{2}) + (\tfrac{1}{7})(\tfrac{1}{4})} = \frac{1}{12}. \qquad \blacksquare$$

Example 7

A subject is presented with four ways of solving a problem, a_1, a_2, a_3, and a_4, and is allowed to choose one at random. If we know that the probability of solving the problem in 3 minutes by each method is given by 0.6, 0.3, 0.2, and 0.1, respectively, and that the subject does solve the problem in 3 minutes, what is the probability that the subject selected the first method?

Let B_1, B_2, B_3, or B_4 be the events that "method a_1, a_2, a_3, or a_4" is selected and let A denote the event "the problem is solved." From the statement of the problem the following probabilities are known:

$$P(B_1) = P(B_2) = P(B_3) = P(B_4) = \tfrac{1}{4},$$

$$P(A \mid B_1) = 0.6, \qquad P(A \mid B_3) = 0.2,$$

$$P(A \mid B_2) = 0.3, \qquad P(A \mid B_4) = 0.1.$$

It is also given that A has occurred. Since we must find $P(B_1 \mid A)$ and the events B_1, B_3, B_2, and B_4 are a partition of the sample space, we may use Theorem 3:

$$P(B_1 \mid A) = \frac{P(A \mid B_1)P(B_1)}{\sum_{i=1}^{4} P(A \mid B_i)P(B_i)}.$$

Substituting the values given above for the probabilities on the right side, we find that

$$P(B_1 \mid A) = \frac{(0.6)(0.25)}{(0.25)(0.6 + 0.3 + 0.2 + 0.1)}$$

$$= 0.5;$$

that is, the probability is 0.5 that the first method was selected, given that a solution was obtained.　▌

Bayes' theorem is often used with a terminology different from Example 7. The set B_i in the partition of the sample space are referred to as *hypotheses*. The assertion $P(B_2 \mid A)$ of the theorem is a statement about the probability of a hypothesis, given the occurrence of an event A. The probability $P(B_2)$ is called *a priori* (prior) probability about the hypothesis B_2, and the probability $P(B_2 \mid A)$ is called *a posteriori* (posterior) probability about the same hypothesis. Often the prior probabilities are guesses for which the posterior probabilities are interpreted as corrections. The hypothesis interpretation of Bayes' theorem is used in Example 8.

Example 8

A sales manager is considering an expansion of his sales force. Because of the competitive nature of the personnel market, he reviews the educational requirements for new salesmen. The present salesmen can be divided into three educational categories: $H_1 = \{2$ years of college$\}$, $H_2 = \{$bachelor's degree$\}$, and $H_3 = \{1$ year beyond a bachelor's degree$\}$. For the sample space of all salesmen, assuming the uniform probability function, the manager has

$$P(H_1) = 0.1,$$
$$P(H_2) = 0.6,$$
$$P(H_3) = 0.3.$$

Looking at the sales records for these categories, he concludes that the event $A = \{$successful salesman$\}$, occurred in the three categories as follows:

$$P(A \mid H_1) = 0.25,$$
$$P(A \mid H_2) = 0.75,$$
$$P(A \mid H_3) = 0.90.$$

On this evidence he would be justified in requiring a high educational background. Now, using Baye's theorem, he forms the posterior probabilities:

$$P(H_1 \mid A) = \frac{P(H_1)P(A \mid H_1)}{P(H_1)P(A \mid H_1) + P(H_2)P(A \mid H_2) + P(H_3)P(A \mid H_3)}$$

$$= \frac{(0.1)(0.25)}{(0.1)(0.25) + (0.6)(0.75) + (0.3)(0.90)}$$

$$= \frac{0.025}{0.725}$$

$$= 0.04,$$

$$P(H_2 \mid A) = \frac{(0.6)(0.75)}{(0.1)(0.25) + (0.6)(0.75) + (0.3)(0.90)}$$

$$= \frac{0.43}{0.725} = 0.59,$$

$$P(H_3 \mid A) = \frac{(0.3)(0.9)}{(0.1)(0.25) + (0.6)(0.75) + (0.3)(0.90)}$$

$$= \frac{0.27}{0.725} = 0.37.$$

Strictly interpreted this means that the probability that a randomly selected successful salesman came from the category H_1 is 0.04, from the category H_2 is 0.59, and from the category H_3 is 0.37. Since the posterior probability for H_1 is less than the prior probability, while the posterior probability for H_3 is greater than the prior probability the sales manager might well insist on a higher level of educational experience for new salesmen. ∎

Problems

1. Bolts are produced by two different machines, with the probability of $\frac{1}{100}$ that a defective bolt is produced by the first machine and a probability of $\frac{2}{100}$ that a defective bolt is produced by the second

machine. A machine is selected at random and a bolt is picked. What is the probability that the bolt is defective?

2. If we choose two light bulbs at random, and without replacement, from a lot of 20 defectives and 80 nondefectives, what is the probability that

   (a) the first bulb chosen is defective?
   (b) the second bulb chosen is defective?

3. Three urns contain colored balls as follows: urn I, three red, four white, and one blue; urn II, one red, two white, and three blue; urn III, four red, three white, and two blue. If an urn is selected at random and a ball drawn, what is the probability that the ball is red?

4. Refer to the urns in Problem 3. If an urn is chosen at random and a red ball is drawn, what is the probability that it came from urn II?

5. A student depends on an alarm clock for which there is a probability of 0.2 that it will not work. If it does operate, there is a probability of 0.8 that he will waken. If it does not operate, there is a probability of 0.3 that he will waken anyway. What is the probability that he will get to his 10 o'clock class?

6. There are 20 boys and 30 girls in a kindergarten class. One quarter of the boys and $\frac{1}{10}$ of the girls wear glasses. A child is selected at random. What is the probability that the child does not wear glasses?

7. A class in economics is composed of 15 sophomores, 10 juniors, and 5 seniors. Three of the sophomores, 3 of the juniors, and 2 seniors received an A grade in the course. If a student is selected at random in the class, what is the probability that he has received an A grade?

8. Of the freshmen in a certain college, 20 percent attended private secondary schools and 80 percent attended public schools. Thirty percent of those freshmen who attended public schools and 25 percent of those freshmen who attended private schools made the honors list. If a freshman student selected at random is found to be on the honors list, what is the probability that the student attended public school?

9. Suppose 75 percent of the student body ride bicycles. Five percent of those who ride bicycles have had accidents, while 8 percent of those who do not ride bicycles have had accidents. If a randomly selected student has had an accident, what is the probability that he rides a bicycle?

10. Of 100 boxes of transistors, 10 per box, 20 boxes were produced at plant I, 30 at plant II, and 50 at plant III. The probability that a defective transistor is produced at plant I is 0.05, at plant II it is

0.04, and at plant III it is 0.02. A box is selected at random and a transistor in it is found to be defective. What is the probability it was produced at plant II?

11. The probability that a beginner at golf gets a good shot is $\frac{1}{3}$ if he selects the proper club, but it is only $\frac{1}{4}$ if he selects any other club. There are five clubs from which he may choose. What is the probability that he makes a good shot?

12. Urn I contains 10 white and 3 red balls. Urn II contains 3 white and 5 red balls. Two balls are to be transferred from I to II, and one ball is then to be drawn from II. What are the probabilities that the two balls transferred from I will be

(a) both white?   (b) one white and one red?
(c) both red?

What is the probability that a white ball will be drawn from II?

13. Jones chooses at random one of the integers 1, 2, and 3, then throws as many dice as are indicated by the number he chose. What is the probability that he will throw a total of 5 points?

14. Suppose you separate two aces and a jack from a deck of 52 playing cards, place them with their backs uppermost, mix them, choose one at random, and replace it in the deck. You then shuffle the deck and draw a card. What is the probability that you will draw an ace?

15. Three machines produce nails, with the weekly output of machine II $\frac{1}{2}$ that of machine I and the same as that of machine III. Approximately 3 percent of the product of machines I and II are defective, while 5 percent of the product of machine III is defective. The nails produced by these machines are mixed in a single storage bin. What is the probability that a single nail selected from the bin is defective?

16. The effect of television news specials on a class of high school seniors was the subject of a study. In a class with 30 males and 20 females 25 percent of the men and 45 percent of the women said that they watched at least one news special in the previous week. If a student is selected by lot and has seen a recent news special, what is the probability that the student is a male?

17. A trick coin with two tails is put in a jar with three normal coins. One coin is selected and tossed three times in succession, with tails the result each time. What is the probability that the trick coin was selected?

18. (Bertrand's Box Paradox.) There are three identical boxes, each with two drawers. Each drawer of one box contains a gold coin, each drawer of the second box contains a silver coin, and one drawer

of the third box contains a silver coin while the other drawer contains a gold coin. If a box is chosen at random, a drawer opened, and a gold coin found, what is the probability that the coin in the other drawer of the same box is silver?

19. Three jars are black and two jars are white. Each black jar is filled with red chips and blue chips in the ratio 4 : 1. Each white jar is filled with red chips and blue chips in the ratio 1 : 4. A jar is chosen at random and a chip drawn from the jar. The chip is blue. What is the probability that it was drawn from a white jar?

20. A large population of mice contains 70 percent deer mice and 30 percent white-footed mice with the mice nearly indistinguishable. Tail length is an important characteristic of these mice and 30 percent of the deer mice have "long" tails while 50 percent of the white-footed mice have "long" tails. A mouse is chosen at random and found to have a long tail. What is the probability that it is a deer mouse?

21. A test for detecting cancer has been developed. Suppose it was found that 98 percent of the patients having cancer reacted positively to the test, whereas only 4 percent of those not having cancer did so. If 3 percent of the patients in the hospital have cancer, what is the probability that a patient selected at random, who reacts positively to the test, will actually have cancer?

22. A coin is tossed. If it comes up heads, a die is thrown and you are paid the number showing in dollars. If it comes up tails, two dice are thrown and you are paid the sum showing in dollars. What is the probability that you will be paid at most $4?

23. Let A and B be two events, each with positive probability. Show that it is not always true that $P(A \mid B) + P(A \mid B') = 1$.

24. In a T-maze experiment with rats, the animals can go left and get food or go right and receive punishment in the form of a mild electrical shock. On the first trial, the animal will go left or right with equal probability. If the animal receives food on a trial, the probabilities for the next trial are left, 0.6; right, 0.4. If the animal receives the electrical shock on a trial, the probabilities for the next trial are left, 0.7; right, 0.3. What is the probability that the animal goes right on the second trial? On the third trial?

25. A simple genetic model for a characteristic of rats is as follows: each rat has two genes for the characteristic, either (AA), (Aa), or (aa), with (AA) called doubly dominant, (Aa) called heterozygous, and (aa) called doubly recessive. A large collection of male rats is 50 percent doubly dominant (AA) and 50 percent heterozygous (Aa), and one of the collection is bred with a double recessive (aa) female. If the male is double dominant, then the offspring will

always exhibit the dominant characteristic. If the male is hetero-zygous, then the offspring will exhibit the dominant characteristic with a probability of $\frac{1}{2}$. There are three offspring in the litter, and all exhibit the dominant characteristic. What is the probability that the male rat is double dominant (AA)?

26. Show that

$$(B \cap A) \cap (B \cap A') = \varnothing$$

for any set B. (*Hint*: Use set-theory identities.)

27. At the beginning of a year, three economic theories for the American economy are proposed. On the basis of existing evidence, the theories appear equally likely. At the end of the year the actual state of the economy is evaluated in light of the three theories. It is concluded that the probability of the occurrence of the actual state of the economy under the hypothesis of the first theory would be 0.6; under the second, 0.4; and under the third, 0.2. How does this change the original assumption on the probabilities of the three theories?

28. A man has four routes he can travel between home and work. The probabilites that the routes can be traversed in less than 15 minutes are 0.4, 0.3, 0.2, and 0.1. If the route is selected at random and the trip is made in less than 15 minutes, what is the probability that the first route was selected? The second route?

29. Of the 995 students enrolled in college, 300 are freshmen, 250 are sophomores, 235 are juniors, and 210 are seniors. It is estimated that 30 percent of the freshmen and sophomores and 65 percent of the juniors and seniors wear glasses. If a student is selected at random, what is the probability that he wears glasses? If a student selected at random wears glasses, what are the a posteriori probabilities?

30. Shoes are produced by two machines, I and II. Sixty percent of the shoes are produced by machine I with an estimate of 10 percent defective. On machine II 20 percent of the shoes produced are defective. A random sample of three shoes, where two are non-defective and one defective, is taken. What is the probability that all three of these shoes were produced by machine I?

31. A box contains four bad and six good tubes. A tube is drawn at random and tested, with the process being repeated until all four bad tubes are located. What is the probability that the fourth bad tube will be located

(a) on the fifth test?   (b) on the tenth test?

4.3 Independent Events

When E and F are two events, each with positive probability, we have seen that in general $P(E|F)$ and $P(E)$ are not equal. We have also seen that if E and F are mutually exclusive (that is, $E \cap F = \varnothing$, then $P(E|F) = 0$, while if F is a subset of E, then $P(E|F) = 1$.   But these are not all the possibilities. Sometimes we shall find that $P(E) = P(E|F)$, which means that we can write $P(E \cap F) = P(E)P(F)$.   Example 1 shows how this might happen.

Example 1

Toss a pair of fair dice, where one die is red and the other green.   Take the usual sample space of 36 2-tuples for this random experiment.   Let F be the event "the red die is even" and E the event "the green die is 1 or 2."   The probabilities are as follows: $P(F) = \frac{18}{36}$, $P(E) = \frac{12}{36}$, $P(E \cap F) = \frac{6}{36}$, and $P(E|F) = \frac{6}{18}$.   We see that $P(E|F) = \frac{6}{18} = P(E)$ and that $P(E \cap F) = \frac{6}{36} = P(E)P(F)$.   ∎

When $P(E|F) = P(E)$, a reasonable interpretation of the relation is that although F occurs, it does not affect the occurrence of E.   We might say that, E is "*independent*" of F.   Since $P(F|E)P(E) = P(E \cap F)$, the assumption that $P(E|F) = P(E)$ gives us $P(F|E)P(E) = P(E \cap F) = P(E|F)P(F) = P(E)P(F)$.   Therefore, $P(F|E) = P(F)$, and we have shown that event F is "independent" of event E, if event E is "independent" of event F.   Hence, if we can write $P(E \cap F) = P(E)P(F)$, we can say that E and F are "independent" events.   Our approach, however, has been by the definition of conditional probability, which requires the assumption that all the probabilities are positive.   We can give a formal definition of independence that has no special requirements on the probabilities of the events, and that can easily be extended to the continuous probability case.   See [6] for a discussion of this point.

Definition 1 : Independent events

Let E and F be events of the same probability space.   The events are said to be independent if $P(E \cap F) = P(E) \cdot P(F)$.

Two events that are not independent are called *dependent events*.   Sometimes independent events are referred to as "statistically independent" or "stochastically independent."   Mathematically, the statement that two events are independent means only that the product of their probabilities is the probability of their intersection.

Sometimes the use of the term independent events can erroneously be interpreted to mean "unrelated" events. We must be careful to avoid such pitfalls, as Example 2 will show.

Example 2

Toss a fair coin. Let A be the event "head on the first toss" and let B be the event "exactly one tail." If the coin is tossed twice, and we use the usual sample space of four 2-tuples with the uniform probability function, we find that $P(A) = \frac{1}{2}$, $P(B) = \frac{1}{2}$, and $P(A \cap B) = \frac{1}{4}$. Thus $P(A \cap B) = P(A) \cdot P(B)$. If the coin is tossed three times, using the usual sample space of eight 3-tuples and the uniform probability function, we get $P(A) = \frac{1}{2}$, $P(B) = \frac{3}{8}$, and $P(A \cap B) = \frac{1}{4}$. This time $P(A \cap B) \neq P(A) \cdot P(B)$. Consequently, on the probability space for two tosses, the events "head on the first toss" and "exactly one tail" are independent, but on the probability space for three tosses, the same events are not independent. ∎

Whether there are two or three tosses, it might seem that the two events ought to have the same independence. However, the mathematical definition of independence rules otherwise.

Example 3

A city council takes action on two proposals for new projects. Assume that the action on successive proposals is independent and that the probability of favorable action on any proposal is 0.5. Let A be the event "the first proposal is passed," and let B be the event "the action on the two proposals was the same." Are these events independent?

A sample space of four 2-tuples with the uniform probability function is a model for this random experiment. If s denotes favorable action (or success) and f denotes unfavorable action (or failure), the sample space S is the set $\{(s, s), (s, f), (f, s), (f, f)\}$. The event A is the subset $\{(s, s), (s, f)\}$ with $P(A) = \frac{1}{2}$. The event B is the subset $\{(s, s), (f, f)\}$ with $P(B) = \frac{1}{2}$. The event $A \cap B = \{(s, s)\}$, and since

$$P(A \cap B) = P(A)P(B) = \tfrac{1}{4},$$

we can assert that these events are independent. ∎

Example 4

A sample space S has five outcomes: $S = \{e_1, e_2, e_3, e_4, e_5\}$ with the probabilities of the elementary events: $P(E_1) = \frac{2}{6}$, $P(E_2) = \frac{1}{12}$, $P(E_3) = \frac{1}{6}$, $P(E_4) = \frac{2}{6}$, and $P(E_5) = \frac{1}{12}$. Let $A = \{e_3, e_4\}$ and $B = \{e_1, e_4\}$, so $A \cap B = \{e_4\}$. Since $P(A) \cdot P(B) = \frac{1}{2} \cdot \frac{2}{3} = \frac{1}{3} = P(A \cap B)$,

A and B are statistically independent. Now let the probability function be the uniform probability function. Then $P(A) = P(B) = \frac{2}{5}$, so $P(A) \cdot P(B) = \frac{4}{25} \neq P(A \cap B)$, and the events A and B are not statistically independent. ∎

If E and F are independent events in the sense that the occurrence of event E in no way affects the occurrence of event F, then surely the occurrence of E' in no way affects the occurrence of F or F'. We might therefore express events E' and F' to be independent of E and F. Theorem 1 states the facts.

Theorem 1

If A and B are independent events, then

(a) A' and B',
(b) A and B',
(c) A' and B

are independent events.

PROOF: To prove (a) we need to show that $P(A' \cap B') = P(A') \cdot P(B')$. By De Morgan's law for sets, the event $A' \cap B' = (A \cup B)'$, and by Theorem 3 of Section 1.5,

$$P((A \cup B)') = 1 - P(A \cup B).$$

Using Theorem 4 of Section 1.5 on $A \cup B$, we obtain

$$P((A \cup B)') = 1 - [P(A) + P(B) - P(A \cap B)].$$

Hence we may write

$$P(A' \cap B') = 1 - P(A) - P(B) + P(A \cap B),$$

and since A and B are independent,

$$P(A' \cap B') = 1 - P(A) - P(B) + P(A)P(B)$$
$$= [1 - P(A)][1 - P(B)]$$
$$= P(A')P(B')$$

as desired. The proofs of (b) and (c) are left as exercises (see Problem 26). ∎

Unfortunately, Definition 1 for the independence of two events cannot be extended directly to three or more events of the same probability space. Examples 5 and 6 illustrate the problems that arise when there are more than two events.

Example 5

Let $S = \{1, 2, 3, 4\}$ be a sample space with the uniform probability function. If A is the event $\{1, 2\}$, B the event $\{1, 3\}$, and C the event $\{1, 4\}$, then

$$P(A) = \tfrac{1}{2}, \qquad P(B) = \tfrac{1}{2}, \qquad P(C) = \tfrac{1}{2},$$
$$P(A \cap B) = \tfrac{1}{4} = P(A) \cdot P(B),$$
$$P(A \cap C) = \tfrac{1}{4} = P(A) \cdot P(C),$$
$$P(B \cap C) = \tfrac{1}{4} = P(B) \cdot P(C),$$
$$P(A \cap B \cap C) = \tfrac{1}{4} \neq P(A) \cdot P(B) \cdot P(C).$$

Even though the events are pairwise independent, this example shows that the product of the three probabilities is not equal to the probability of the intersection of the three events. ∎

Example 6

Let $S = \{1, 2, 3, 4, 5, 6, 7, 8\}$ be a sample space with the uniform probability function. If A is the event $\{1, 2, 3, 4\}$, B the event $\{2, 3, 4, 5\}$. and C the event $\{2, 6, 7, 8\}$, then

$$P(A) = \tfrac{1}{2}, \qquad P(B) = \tfrac{1}{2}, \qquad P(C) = \tfrac{1}{2},$$
$$P(A \cap B \cap C) = \tfrac{1}{8} = P(A) \cdot P(B) \cdot P(C),$$
$$P(A \cap B) = \tfrac{3}{8} \neq P(A) \cdot P(B).$$

In this case, the product of the three probabilities is the probability of the intersection of the three events, but the events A and B are not independent. ∎

Any definition of three or more independent events must avoid these two anomalies. Observe how this is done in

Definition 2: Three independent events

Let the events A, B, and C be defined on the same probability space. The events are said to be independent if

$$P(A \cap B) = P(A) \cdot P(B),$$
$$P(A \cap C) = P(A) \cdot P(C),$$
$$P(B \cap C) = P(B) \cdot P(C),$$

and

$$P(A \cap B \cap C) = P(A) \cdot P(B) \cdot P(C),$$

The extension of the definition to n events (where n is a positive integer) is now a straightforward generalization of the definition for three events. All subsets of the set of n events must also be independent.

Example 7

Three people vote in an election. Denote the events that the person voted the Republican ticket by $A1$, $A2$, and $A3$. Furthermore, suppose that

$$P(A1) = 0.6, \qquad P(A2) = 0.5, \qquad P(A3) = 0.4.$$

If it is assumed that each person voted independently, what is the probability that exactly one person voted Republican?

The one Republican vote can be cast by any of the three, so the required probability is

$$P(A1 \cap A2' \cap A3') + P(A1' \cap A2 \cap A3') + P(A1' \cap A2' \cap A3).$$

Under the assumption that the events are independent, and using an extension of Theorem 1,

$$P(A1 \cap A2' \cap A3') = P(A1) \cdot P(A2') \cdot P(A3')$$
$$= (0.6)(0.5)(0.6)$$
$$= 0.18.$$

Similarly,

$$P(A1' \cap A2 \cap A3') = (0.4)(0.5)(0.6) = 0.12,$$
$$P(A1' \cap A2' \cap A3) = (0.4)(0.5)(0.4) = 0.08,$$

so the probability of exactly one Republican vote is 0.38.   ∎

Example 8

A state senator has introduced three separate bills to the legislature. He estimates the probability that bill I will pass is 0.6, that bill II will pass is 0.7, and that bill III will pass is 0.5. The bills are unrelated so it is reasonable to assume that passage is independent. What is the probability that at least one bill is passed?

Let A, B, and C, respectively, be the events that bills I, II, and III are passed. The event "at least one bill is passed" is the union $A \cup B \cup C$. However, instead of calculating $P(A \cup B \cup C)$, let us consider the complement $(A \cup B \cup C)'$, which is the intersection $A' \cap B' \cap C'$. Since we have assumed A, B, and C to be independent, A', B', and C' are independent. Thus the probability of the intersection can be easily calculated and the product obtained:

$$P(A' \cap B' \cap C') = P(A')P(B')P(C').$$

Therefore,

$$P(A \cup B \cup C) = 1 - (0.4)(0.3)(0.5)$$
$$= 0.94;$$

that is, the probability is 0.94 that at least one bill is passed.   ∎

Consider a system with two components in series or in sequence, such as a water pipe with two electronically operated valves that control the flow of the water. Suppose that the probability that each component will function properly is known. What is the probability that the system operates? If the operation of each component can be assumed to be independent, then the answer is the product of the probabilities for the components. Otherwise, there is little to say. The assumption of independence is quite common in applications such as this and is usually reasonable. However, if the components of the system for the problem just illustrated are dams on the same river, where "to function properly" is interpreted to mean "to hold the water back," then the assumption of independence may not be justified.

Problems

1. A red die and a green die are tossed. If E is the event "five on the red die" and F is the event "six on the green die," show that E and F are independent events.

2. A fair coin is tossed three times. Show that the events "the same side turns up all three times" and "at most one head turns up" are independent.

3. Three coins are tossed. Are the events "heads on the first coin" and "tails on the last two" independent? Are the events "two coins heads" and "three coins heads" independent?

4. A pair of dice, one red, one green, is tossed twice. What is the probability that on the second toss each die show spots different from those it showed on the first toss? (Assume independence of the outcomes of the two tosses.)

5. Suppose that A and B are independent events with $P(A) = \frac{1}{6}$ and $P(B) = \frac{1}{4}$. Determine

    (a) $P(A \cup B)$ (b) $P(A' \cap B)$ (c) $P(A' \cup B')$

6. Suppose A and B are independent events with $P(A) = \frac{1}{6}$ and $P(B) = \frac{1}{4}$. Determine the probability that

    (a) precisely one of A and B occurs.
    (b) at most one of A and B occurs.
    (c) neither A nor B occurs.

7. There are two parts to a question on an exam, each of which can be answered by "yes" or "no." If a student does not know the answers, what is the probability that he will guess correctly on both parts? Assume all events to be equally likely.

8. One man draws two cards from an ordinary deck of cards. At the same time another man tosses two coins. What is the probability of obtaining one head and two cards from the same suit?

9. A man owns two cars, a station wagon and a sports car. The probability that the station wagon will not start on a cold morning is 0.1 and the probability that the sports car will not start is 0.2. On a cold morning, what is the probability that

(a) neither car will start?   .02
(b) one or the other (but not both) will start?   .26

10. The respective probabilities that the Republican candidate for governor in three states will win the election are 0.6, 0.7, and 0.9. What is the probability that at least of two the governors elected are Republican?

11. If a coin is tossed twice, the sample space S consists of four outcomes. There are 16 events for S. List the events, and find all pairs of independent events.

12. Suppose a die is thrown three times in succession. Which of the following pairs of events are independent:

(a) two on the first toss and three on the second toss?
(b) two on the first toss and at least two tosses of two?

13. An urn contains three red, two white, and five black balls. If three balls are drawn successively, find the probability of a red, white, and black ball in that order if

(a) the ball drawn is replaced after each drawing.
(b) the ball drawn is not replaced after each drawing.

14. Two rats run a T-maze in successive trials with all outcomes equally likely. A is the event "both went left" and B is the event "both went right." Are these events mutually exclusive? If the runs are independent, find $P(A \cup B)$.

15. Two rats run a T-maze in successive trials with all outcomes equally likely. Let A be the event "first rat went left," B the event "second rat went left," and C the event "both rats went the same way. Are these events pairwise independent? Are they completely independent?

16. Let A and B be two events. Suppose that $P(A)=0.4$ and $P(A \cup B)=0.7$. Let $P(B)=p$.

(a) For what choice of p is $P(A \cap B)=0$?
(b) For what choice of p are A and B independent?

17. Consider a lot of 1000 light bulbs and assume that 10 percent are defective. Two light bulbs are selected. If the first light bulb is replaced before the second one is chosen, determine the probability of two nondefective light bulbs. If the first one chosen is not replaced, calculate the probability of two nondefective bulbs. Compare the results with regard to the assumption of independence.

18. Each of two persons tosses three fair coins. What is the probability that they obtain the same number of heads?

19. A die is tossed and a card is drawn at random from a standard deck of 52 cards. What is the probability that

    (a) the die shows an even number and the card is from a red suit?
    (b) the die shows an even number or the card is from a red suit? $3/4$

20. Three people work independently at decoding a message. The respective probabilities that they will succeed are $\frac{1}{5}$, $\frac{1}{4}$, and $\frac{1}{3}$. What is the probability that the message will be decoded?

21. Two people are shooting at a target. If the probability is $\frac{3}{4}$ that the first person hits the target and $\frac{4}{7}$ that the second person hits the target, what is the probability that the target is hit at least once when both shoot simultaneously?

22. A man owns a house in town and a cabin in the mountains. In any one year, the probability of the house being burgalized is 0.01 and the probability of the cabin being burgalized is 0.05. For any one year, what is the probability that one or the other, but not both, will be burgalized?

23. Two machines, I and II, being operated independently, may have a number of breakdowns each day, with the corresponding probability function as listed below:

Machine	\multicolumn{7}{c}{No. of breakdowns}
	0	1	2	3	4	5	6
I	0.1	0.2	0.3	0.2	0.1	0.05	0.05
II	0.2	0.2	0.1	0.1	0.15	0.15	0.1

Compute the following probabilities:

(a) I and II have the same number of breakdowns.
(b) The total number of breakdowns is fewer than 4.
(c) The maximum number of breakdowns of either of the two machines is 2.

24. A pair a dice, one red and the other white, are tossed. Let A be the event "6 on the red die," B the event "6 on the white die," and C the event "the sum of the numbers is odd." Show that A, B, and C are pairwise independent but that they are not three independent events.

25. At the grain commodity market a speculator in the market simultaneously contracted to buy wheat for future delivery and agreed to deliver corn in the future; that is, he bought future wheat and sold short in corn. Both transactions were made at the current price, and the speculator estimates the probabilities that the two grains will rise in price are 0.1 and 0.8, respectively. What is the probability that he will lose in both transactions?

26. Finish the proof of Theorem 1: If A and B are independent events, then A and B', and A' and B, are independent events.

27. If A and B are independent events, show that

$$P(A \cup B) = 1 - P(A')P(B').$$

28. If $A_1, A_2, \ldots, A_n$ are independent events and $P(A_i) = p_i$ for $i = 1, 2, \ldots, n$, show that the probability that none of the events will occur is $(1 - p_1)(1 - p_2) \cdots (1 - p_n)$.

4.4 Repeated Trials

In many of the previous problems and exercises the same experiment was performed several times; an example is tossing a coin three times. Or a sequence of different experiments was undertaken, such as tossing a coin first and then throwing a die. These are examples of repeated random experiments. We now develop the mathematical theory for a countable number of repeated experiments. As we formalize our intuitive ideas about them you will see how extensively we use the assumption of independence.

Let us start with a sample space $S = \{e_1, e_2, \ldots\}$ as a model of a single random experiment and a probability function $P(\cdot)$ defined on the elementary events of S. Perform the random experiment twice under identical conditions so that neither the sample space S nor the probability function $P(\cdot)$ needs to be changed. This process manufactures a new random experiment, which we shall call a *compound experiment*. We shall continue to call each repetition of the original experiment a *trial*, so the compound experiment can be described as two trials under identical conditions. The sample space S^* for the compound experiment is given by all pairs of outcomes (elementary events) of S; that is $S^* = \{(e_i \, e_j) \mid e_i, e_j \in S\}$.

Example 1

If three coins are tossed in a single random experiment, then a sample space S for the experiment is $S = \{\text{HHH, HHT, HTH, THH, HTT, THT, TTH, TTT}\}$. The sample space might also be written as

$\{e_1, e_2, \ldots, e_8\}$, where $e_1 = \text{HHH}$, $e_2 = \text{HHT}, \ldots$. If the toss of three coins is repeated, the sample space for the two trials is

$$S^* = \{(e_1, e_1), (e_1, e_2), \ldots, (e_8, e_8\}$$
$$= \{(e_i, e_j) \mid i \text{ and } j = 1, 2, \ldots, 8\}.$$

Of course, there are $8 \cdot 8 = 64$ elementary events in S^*. ∎

Example 2

One question on a sociological poll was as follows: Which parent decides on the purchase of a car? The allowable answers are (a) mother always, (b) mother more than father, (c) equal, (d) father more than mother, (e) father always, (f) not applicable. A sample space for this question is $S = \{$a, b, c, d, e, f$\}$, which could also be written as $S = \{e_1, e_2, e_3, e_4, e_5, e_6\}$. If the question is asked of 70 students, this can be interpreted as 70 repeated trials, and the new sample space S^* is the collection of 70-tuples, where each component can be any one of the six outcomes in S. There are 6^{70} elementary events in S^*. ∎

Once the new sample space S^* has been defined, we complete the definition of the probability space by assigning a probability to the pairs (e_i, e_j), the outcomes of S^*. Theoretically, we could do this in many ways. However, since the pairs are the result of repeating an experiment under identical conditions, there is a "natural" assignment of probability. Any outcome of the first trial should not influence any outcome of the second trial. This is equivalent, at least intuitively, to the mathematical property of the two trials being independent events.

In S^*, the event $A = \{\text{outcome } e_i \text{ on the first trail}\}$ consists of all pairs of S^* with e_i in the first component of the pair. Let the probability of E_i, which is $\{e_i\}$, with respect to the original sample space S be p_i, so $P(E_i) = p_i$. Then define $P^*(A) = p_i$, where $P^*(\cdot)$ denotes the probability function on S^*. By choosing this probability we are essentially saying that no matter what the result is on the second trial, the probability of outcome e_i should be the same in S and in S^*. Similarly, if $B = \{\text{outcome } e_j \text{ on the second trial}\}$, define $P^*(B) = P(E_j) = p_j$. Now the elementary event $\{(e_i, e_j)\}$ of S^* is the intersection of A and B, so

$$P^*(A \cap B) = P^*(\{(e_i, e_j)\}).$$

On the other hand, in order for the two trials described by A and B to be independent we must have

$$P^*(A \cap B) = P^*(A)P^*(B).$$

Consequently, to capture the independence implied in repeated trials of an experiment we define

$$P^*(\{e_i, e_j\}) = p_i p_j.$$

This assigns a nonnegative number to every elementary event of S^*, but before we can claim that the assignment is actually a probability function, we must show that the sum of these numbers, obtained as products of pairs of numbers, is in fact 1. To do this we first need a little more notation and algebra.

Let $a_1, a_2, \ldots, a_n$ and $b_1, b_2, \ldots, b_n$ be two sets of real numbers. We define

(1)
$$\sum_{i,j=1}^{n} a_i b_j = \sum_{i=1}^{n} \sum_{j=1}^{n} a_i b_j$$

and observe that the right side is identical to

$$a_1 \sum_{j=1}^{n} b_j + a_2 \sum_{j=1}^{n} b_j + \cdots + a_n \sum_{j=1}^{n} b_j,$$

which we can write as

$$(a_1 + a_2 + \cdots + a_n) \sum_{j=1}^{n} b_j.$$

We therefore obtain

(2)
$$\sum_{i,j=1}^{n} a_i b_j = \left(\sum_{i=1}^{n} a_i \right) \left(\sum_{j=1}^{n} b_j \right),$$

which simply generalizes the algebraic identity $a_1 b_1 + a_1 b_2 + a_2 b_1 + a_2 b_2 = (a_1 + a_2)(b_1 + b_2)$. Different subscripts are used to indicate that the sums are formed independently of each other.

Now we shall use this relationship to investigate the sum of the numbers we have assigned to the elementary events in S^*. If we put $P(E_i) = p_i$ and $P(E_j) = p_j$, and assume that S has n elementary events, then $P^*(\{e_i, e_j\}) = p_i p_j$, and we must show that all these number products sum to 1; that is,

(3)
$$\sum_{i,j=1}^{n} p_i p_j = 1.$$

Using the relationship demonstrated in (2) we know that

(4)
$$\sum_{i,j=1}^{n} p_i p_j = \sum_{i=1}^{n} p_i \sum_{j=1}^{n} p_j,$$

and since for all elementary events in S,

$$\sum_{i=1}^{n} P(E_i) = \sum_{i=1}^{n} p_i = 1,$$

each factor on the right side of (4) has the value 1, so (3) obtains. This argument demonstrates that the assignment of probabilities to the elementary events of S^* is acceptable (by virtue of the reasoning in Section 2.2 for finite probability spaces). Hence S^* with $P^*(\cdot)$ forms a probability space. Our justification uses the fact that S is finite. If S is countable and infinite, the sums are infinite sums, which we discuss in Chapter 8. The same results obtain, however.

Example 3

A fair die with three faces marked with a 3, two faces marked with a 2, and one face marked with a 1 is tossed. A sample space is the set $S = \{1, 2, 3\}$, and if the elementary event e_i corresponds to the outcome i for $i = 1, 2, 3$, a probability function for S is $P(E_1) = \frac{1}{6}$, $P(E_2) = \frac{1}{3}$, and $P(E_3) = \frac{1}{2}$. Now toss the die twice, and assume independence for the tosses. The new sample space S^* will be the set $S^* = \{(1, 1), (1, 2), (1, 3), (2, 1), (2, 2), (2, 3), (3, 1), (3, 2), (3, 3)\}$. For the elementary event $\{(2, 3)\}$ of S^*,

$$P(\{(2, 3)\}) = \frac{1}{3} \cdot \frac{1}{2} = \frac{1}{6}. \qquad \blacksquare$$

Example 4

Thirty percent of a political science class claim a Democratic affiliation, 50 percent claim to be Republican, and the remainder are classified as Independent. In each class meeting for five meetings one member of the class is chosen by lot to rebut the instructor. What is the probability that the students chosen alternate between Republican and Democrat?

A sample space for the experiment to be repeated is $S = \{D, R, I\}$ with the probabilities

$$P(\{D\}) = 0.3,$$
$$P(\{R\}) = 0.5,$$
$$P(\{I\}) = 0.2.$$

The new sample space S^* consists of 5-tuples of outcomes of S, which in this case are $3^5 = 729$ in number. There are exactly two outcomes in S^* where D and R alternate, namely, (R, D, R, D, R) and (D, R, D, R, D). As elementary events of S^*,

$$P(\{(R, D, R, D, R)\}) = (0.5)(0.3)(0.5)(0.3)(0.5) = 0.01125,$$
$$P(\{(D, R, D, R, D)\}) = (0.3)(0.5)(0.3)(0.5)(0.3) = 0.00675,$$

so the desired probability is the sum; that is,

$$P(\text{D and R alternate}) = 0.0180. \qquad \blacksquare$$

In general, an event of a compound experiment sample space S^* is a subset of pairs of outcomes of the basic sample space S. In particular, if the occurrence or nonoccurrence of an event A^* of S^* depends only on the outcome of the first trial, then A^* is said to be *an event depending on a trial*. This means that the decision as to whether or not the outcome (e_i, e_j) of S^* belongs to A^* depends only on whether or not the outcome e_i of S belongs to some event A of S.

The probability of an event A^* depending on a trial can be calculated in terms of the event A of S that defines A^*. If

$$A = \{e_{a_1}, e_{a_2}, \ldots, e_{a_k}\},$$

then

$$A^* = \{(e_{a_i}, e_j) \mid e_{a_i} \in A, e_j \in S\}.$$

By the definition of the probability function for S^*,

$$P^*(A^*) = \sum_{j=1}^{n} \sum_{i=1}^{k} p_{a_i} p_j,$$

so using (4) we obtain

$$P^*(A^*) = \left(\sum_{i=1}^{k} p_{a_i}\right)\left(\sum_{j=1}^{n} p_j\right)$$

$$= \sum_{i=1}^{k} p_{a_i}$$

$$= P(A).$$

In an analagous way, if B^* is an event in the compound experiment sample space that depends only on the second trial, then we obtain $P^*(B^*) = P(B)$, where we put $B = \{e_{b_1}, e_{b_2}, \ldots, e_{b_m}\}$. The intersection of A^* and B^* is the subset of all ordered pairs of S^*, where the first component is an outcome in A and the second component is an outcome in B. Clearly, $A^* \cap B^*$ is a subset of S^* and is therefore an event whose probability we can compute.

Using PA2 we can write

$$P^*(A^* \cap B^*) = P(\{\cup \, (e_{a_i}, e_{b_j})\})$$

$$= \sum_{i=1}^{k} \sum_{j=1}^{m} P^*(\{(e_{a_i}, e_{b_j})\}).$$

From the definition of $P^*(\cdot)$, it follows that

$$P^*(A^* \cap B^*) = \sum_{i=1}^{k} \sum_{j=1}^{m} p_{a_i} \cdot p_{b_j}$$

$$= \left(\sum_{i=1}^{k} p_{a_i}\right)\left(\sum_{j=1}^{m} p_{b_j}\right)$$

$$= P^*(A^*)P^*(B^*).$$

This shows that the events A^* and B^* are independent.

Example 5

(Refer to Example 3.)   Let A^* be the event defined by the statement "2 or 3 on the first trial" and B^* the event for the statement "2 on the second trial":

$$A^* = \{(2, 1), (2, 2), (2, 3), (3, 1), (3, 2), (3, 3)\},$$
$$B^* = \{(1, 2), (2, 2), (3, 2)\}.$$

The probabilities of the events can be calculated easily.

$$\begin{aligned}
P^*(A^*) &= \tfrac{1}{3}\tfrac{1}{8} + \tfrac{1}{3}\tfrac{1}{3} + \tfrac{1}{3}\tfrac{1}{2} + \tfrac{1}{2}\tfrac{1}{6} + \tfrac{1}{2}\tfrac{1}{3} + \tfrac{1}{2}\tfrac{1}{2} \\
&= \tfrac{1}{3}(\tfrac{1}{6} + \tfrac{1}{3} + \tfrac{1}{2}) + \tfrac{1}{2}(\tfrac{1}{6} + \tfrac{1}{3} + \tfrac{1}{2}) \\
&= \tfrac{1}{3} + \tfrac{1}{2} \\
&= \tfrac{5}{6}.
\end{aligned}$$

Similarly,

$$P^*(B^*) = \tfrac{1}{3}.$$

The intersection of A^* and B^* is the set $\{(2, 2), (3, 2)\}$, and the probability of $(A^* \cap B^*)$ is $\tfrac{5}{18}$, which is the same as the product $P^*(A^*)P^* \times (B^*) = (\tfrac{5}{6})(\tfrac{1}{3}) = \tfrac{5}{18}$.   ∎

The discussion so far has been confined to just two trials of a random experiment.   We shall now put those results into formal statements, which will enable us to generalize to a larger number of repetitions of the same experiment.   As a matter of convenience we drop the * on the sample space, the events, and the probability function.   Although the examples and the arguments have been restricted to finite sample spaces, the formal statements will be given for countable sample spaces.   Remember that the expression "repeated under identical conditions" implies that the events are independent.

Definition 1: Probability space of two repeated independent trials

For a probability space with a countable sample space S, elementary events $E_1, E_2, \ldots$, and probability function $P(E_i) = p_i$ for $i = 1, 2, \ldots$, the probability space of two repeated independent trials consists of a sample space of 2-tuples (e_i, e_j) to which the probabilities

$$P(\{(e_i, e_j)\}) = p_i \cdot p_j$$

are assigned.

Theorem 1

If A_1 and A_2 are two events of a probability space of two repeated independent trials such that A_1 depends only on the first trial and A_2 only on the second, then the events are independent.

Since the proof of this theorem is similar to the argument given before Example 4, it will not be repeated here.

The generalization of these ideas to an arbitrary number of trials follows in an almost obvious way. Repeat the random experiment in question n times. As before, keep the identical conditions, so that the assumption of independence is sensible. We then form a new sample space of n-tuples of elementary events of the original sample space. Finally, assign a probability function to the sample space by the following device:

$$P(e_{i_1}, e_{i_2}, \ldots, e_{i_n}) = p_{i_1} p_{i_2} \cdots p_{i_n},$$

where $p_{i_j} = P(E_{i_j})$ is the probability of an elementary event E_{i_j} in the original sample space. The double subscript E_{i_j} denotes an arbitrary element in the jth trial. This definition of the probability function is valid because the sum of the probabilities can again be shown to be equal to 1. Thus we have a description of a probability space of n independent repeated trials. If A_1, $A_2, \ldots, A_r$ are r events of a probability space of n independent repeated trials, where A_i depends on the ith trial, then it can be shown in a manner similar to the argument for Theorem 1 that the A_i are mutually independent events.

Example 6

(Refer to Example 4, which is a random experiment of five repeated, independent trials.) The event $A_1 = \{$first student chosen is Independent$\}$ depends only on the first trial, and $P(A_1) = 0.2$. Similarly, the event $A_2 = \{$second student chosen is Independent$\}$ depends only on the second trial. However, the event $B = \{$exactly one Independent is chosen the first or second day$\}$ does not depend on any one trial. ∎

A succession of different experiments can be treated in the same way as repeated trials of one experiment. The essential feature is the assumption of independence. If $S_1 = \{e_1, e_2, \ldots, e_n\}$ and $S_2 = \{f_1, f_2, \ldots, f_m\}$ are two sample spaces, then the probability space of pairs of elementary events that result from executing experiments 1 and 2, and with probabilities assigned to the pairs by

$$P(\{(e_i, f_j)\}) = P(E_i)P(F_j),$$

represents the compound experiment of two successive independent experiments. The argument necessary to justify the assignment of the probabilities is entirely similar to that given for two repetitions of the same experiment. Furthermore, generalizing this notion to n different experiments can be done exactly as we generalized from two repetitions to n repeated experiments.

Example 7

A fair die is tossed and a ball is selected from an urn with seven red, four white, and three green balls.   If we assume that the experiments are independent, then the probability of an even number on the die and a white ball selected is the product $(\frac{1}{2})(\frac{4}{14}) = \frac{1}{7}$.   ∎

The fact that a random phenomenon is a sequence of trials does not necessarily imply that the trials are independent.   The notion of dependent trials is important in many applications; for our purposes, an example will suffice.

Example 8: Polya urn scheme

(Suggested by [5], pp. 109–111.)   Let an urn contain r red balls and b black balls, indistinguishable except for color.   A ball is drawn, the color is noted, and the ball is replaced in the urn.   In addition, one ball of the the color drawn is added to the urn.   A new ball is drawn from the $r + b + 1$ balls in the urn, the color noted, and the ball replaced with another ball of the color drawn added to the urn.   The procedure can be repeated indefinitely; but if it is done three times, for example, the sample space S would be the set of triples

$$\{(B, B, B), (B, B, R), (B, R, B), (R, B, B),$$
$$(B, R, R), (R, B, R), (R, R, B), (R, R, R)\}.$$

The uniform probability function does not apply, nor does the probability of drawing a particular color remain the same from trial to trial. In particular, let B_1 be the event "black on first trial" and B_2 be the event "black on second trial."   Then

$$P(B_1) = \frac{b}{r+b},$$

while $P(B_1 \cap B_2) = P(B_2 | B_1)P(B_1)$ becomes

$$P(B_1 \cap B_2) = \frac{b+1}{r+b+1} \cdot \frac{b}{r+b}.$$

Let A_1 be the event "red on first trial"; then

$$P(A_1) = \frac{r}{r+b}$$

and

$$P(B_2 | A_1) = \frac{r+1}{r+b+1} \cdot \frac{r}{r+b}.$$

Using Theorem 1, we find that

$$P(B_2) = P(B_2 \mid A_1)P(A_1) + P(B_2 \mid B_1)P(B_1)$$

$$= \frac{r+1}{r+b+1} \cdot \frac{r}{r+b} + \frac{b+1}{r+b+1} \cdot \frac{b}{r+b}$$

and clearly $P(B_1 \cap B_2) \neq P(B_1) \cdot P(B_2)$.

Since the probability of picking a colored ball changes after each draw is made, the trials are not independent. The Polya urn scheme has been found to provide a reasonable model for analyzing the spread of contagious diseases. For a full treatment of the problem of dependent trials, see [3] or [5]. ∎

∎ As an application of independent trials and conditional probability to a problem in biology, we shall consider an example in genetics, a special case of random Mendelian mating. Before setting up the model, we shall present necessary facts in genetics (suggested by [7], pp. 123–129).

Hereditary characteristics depend on special carriers called genes, which usually appear in pairs because the chromosomes on which they are found occur in pairs. In the simplest case, two gene forms (alleles) affect a characteristic; we shall call these two alleles A and a. Three pairs are possible: AA, Aa, and aa. We shall assume that every organism in the population under consideration can be classified by the pair that is present. These classifications are called genotypes. Each pair of genes refers to but one hereditary characteristic.

The reproductive cells (gametes) that carry the genes that determine what an offspring will be like are formed by a process whereby one gene is separated from a paternal pair of genes and one gene is separated from a maternal pair. The union of these genes in the fertilized egg then determines the genotype of the new organism. We shall study genotypes and the hereditary characteristic determined by one pair of genes where there are three genotypes. The color of the blossom of the flowering plant known as a four o'clock is an example. If A is the gene for a red blossom and a the gene for a white blossom, then the genotypes and their corresponding *phenotypes* (appearance of the blossom) are AA (red), Aa (pink), and aa (white).

Suppose that three genotypes occur in the male and female organisms of the population in the same proportion. The selection of a male can be considered as the selection of a sample of size 1 from a population where

$$u = \text{probability the organism is genotype AA,}$$
$$2v = \text{probability the organism is genotype Aa,}$$
$$w = \text{probability the organism is genotype aa,}$$

and, of course,

$$u + 2v + w = 1.$$

We select a female organism the same way, with the same probabilities. The genotypes of the offspring depend on a chance process. Let us assume that each parental gene has a probability of $\frac{1}{2}$ of being transmitted and that the mating is random, or, in other words, that the uniform probability function applies to males and females with the selection independent. Thus, given a population with probabilities u, $2v$, and w, what are the genotype probabilities u_1, $2v_1$, and w_1 for the next generation?

Let E_1 be the event "paternal gene is A," and let F_2 be the event "maternal gene is a." E_1 will occur with genotype AA and may occur with genotype Aa (where gene A alternatively could be from the maternal parent). If we let B_1 denote the event "genotype AA," B_2 the event "genotype Aa," and B_3 the event "genotype aa," we can write

$$E_1 = (E_1 \cap B_1) \cup (E_1 \cap B_2).$$

From the definition of conditional probability and the fact that B_1 and B_2 are disjoint, we can write

$$P(E_1) = P(E_1 \mid B_1)P(B_1) + P(E_1 \mid B_2)P(B_2),$$

or

$$P(E_1) = 1 \cdot u + \tfrac{1}{2} \cdot 2v = u + v.$$

In a similar fashion,

$$P(F_2) = P(F_2 \mid B_2)P(B_2) + P(F_2 \mid B_3)P(B_3),$$

$$P(F_2) = \tfrac{1}{2} \cdot 2v + 1 \cdot w = v + w.$$

A male parent is selected at random, and by the process of sperm formation a single gene is selected at random from the two genes the male parent carries. A female parent is selected at random, and by the process of egg formation a single gene is selected from the two genes the female parent carries. The genotype of the new organism for the type of blossom is then determined by the union of the gametes—the sperm and the egg. Since we assumed that the male and female genes could be selected in the same way, if we let E_2 be the event "male gene is a" and F_1 be the event "female gene is A," we have $P(F_2) = v + w$ and $P(F_1) = P(E_1) = u + v$. For the offspring to be of genotype AA, both events E_1 and F_1 must occur, and

$$P(\text{offspring genotype AA}) = P(E_1 \cap F_1).$$

The assumption of independence gives us

$$P(\text{offspring genotype AA}) = P(E_1) \cdot P(F_1) = (u + v)^2.$$

Similarly,

$$\begin{aligned}
P(\text{offspring genotype aa}) &= P(E_2 \cap F_2) \\
&= P(E_2) \cdot P(F_2) \\
&= (v + w)^2,
\end{aligned}$$

$$P(\text{offspring genotype Aa}) = P((E_1 \cap F_2) \cup (E_2 \cap F_1))$$
$$= P(E_1 \cap F_2) + P(E_2 \cap F_1)$$
$$= P(E_1)P(F_2) + P(E_2)\,P(F_1)$$
$$= 2(u + v)(v + w),$$

where the fact that the events are both disjoint and independent is used. Since u_1, $2v_1$, and w_1 are the probabilities for the genotypes in the next generation, we have the result

$$u_1 = P(\text{genotype AA}) = (u + v)^2,$$
$$2v_1 = P(\text{genotype Aa}) = 2(u + v)(v + w),$$
$$w_1 = P(\text{genotype aa}) = (v + w)^2.$$

Observe that

$$u_1 + v_1 = (u + v)^2 + (u + v)(v + w)$$

or

$$u_1 + v_1 = (u + v)(u + v + v + w)$$
$$= u + v,$$

because $u + 2v + w = 1$. Thus the probability of the event "male gene is A" remains constant from one generation to the next under our hypotheses regarding the selection of parents and genes.

Problems

1. Three fair coins are tossed twice. What is the probability that
   (a) exactly one tail appears on each toss?
   (b) a total of two tails appears?

2. A fair die with three faces marked with a 3, two faces marked with a 2, and one face marked with a 1 is tossed three times. What is the probability that the sum of the spots on the uppermost faces will be 5?

3. An urn contains 70 green balls, 20 white balls, and 10 black balls. Four balls are selected one at a time with replacement before the next selection. What is the probability of drawing a white, a black, a green, and a green in that order?

4. Let a sample of size 4 be drawn with replacement from an urn containing six balls, of which four are white. What is the probability that exactly two white balls are drawn?

5. The probability is 0.2 that a person will hit a balloon with a single throw of a dart. Find the probability that in five throws he will have
   (a) no hits.   (b) exactly one hit.

6. A test consists of 10 multiple-choice questions with five possible answers to each question, of which only one is correct. If a student guesses (selects at random) his answer to each question, what is the probability that he will get at least nine correct?

7. If one assumes that the four times a baseball player comes to bat in a game are four independent trials, where the probabilities are estimated to be 0.3 of getting a hit, 0.2 of getting a walk, and 0.5 of being out, what is the probability that the player gets

   (a) two hits and two walks?   .0216
   (b) two hits, one walk, and put out once?

8. Consider three urns: urn I contains one white and two black balls, urn II contains three white and two black balls, and urn III contains two white and three black balls. One ball is drawn from each urn. What is the probability that among the balls drawn there will be

   (a) one white and two black balls?
   (b) more black than white balls?

9. Each of three urns contains four identical balls numbered from 1 to 4. One ball is drawn from each urn, and we shall assume that the drawings are independent. What is the probability that 3 is the greatest number drawn?

10. An urn contains one red and two green balls. A ball is drawn and replaced by a ball of the opposite color. Then another ball is drawn from the urn. Find the (conditional) probability that the first ball drawn was green, given that the second ball drawn was green.

11. Find the proportion of the three genotypes in the first generation of a random Mendelian mating, (that is, find u_1, v_1, w_1) if (1) $u = 0$, $2v = \frac{1}{2}$, $w = \frac{1}{2}$; (2) $u = w = 0$, $2v = 1$.

12. Consider a man who has made two tosses of a die. Let A_1 be the event that the outcome of the first throw is a 1 or 2. Let A_2 be the event that the outcome of the second throw is a 1 or 2. Let B_1 be the event that the sum of the outcomes is 7. Determine whether each of the following statements is true or false.

   Statement 1: A_1 depends on the first throw.
   Statement 2: A_1 and A_2 are mutually exclusive events.
   Statement 3: B_1 depends on the first throw.

13. A fair coin is tossed, a fair die is rolled, and a card is selected from a standard deck of 52 cards. Assume that the three experiments are independent. Find the probability that

   (a) an even number and spade occur.
   (b) a head, even number, and spade occur.

14. At a benefit carnival three men, Art, Bill, and Carl, toss pennies at a saucer floating in a tub of water. If the probabilities of a hit are 0.5, 0.3, and 0.4, respectively, what is the probability that

    (a) all will hit?
    (b) at least one will hit?
    (c) Bill and Carl will hit?

15. An urn contains seven red, five green, and six white balls. A ball is drawn, the color noted, and then the ball is replaced in the urn. If the procedure is repeated five times, determine the probability that

    (a) none of the balls drawn is white,
    (b) all the balls drawn are green,
    (c) exactly two balls are red or green.

16. If five people each toss a coin simultaneously and independently, what is the probability that either exactly one of the coins will fall heads, or that exactly one of the coins will fall tails?

17. What is the probability that of the first five persons encountered on a given day, at least four of them were born on a Saturday?

18. A contractor estimates that if he accepts three jobs simultaneously, he has a probability of 0.8 of completing a job on schedule, whereas if he accepts four jobs simultaneously, the probability is 0.7 of completing a job on schedule. If he accepts five jobs simultaneously, then the probability if 0.6 that a job will be completed on schedule. His goal is to complete three jobs on schedule. How many jobs should he accept for the best chance of achieving his goal?

19. Find the proportions of the three genotypes in the second generation of a random Mendelian mating of

    (1) $u = \frac{1}{2}$, $2v = \frac{1}{2}$, $w = 0$,
    (2) $u = \frac{1}{2}$, $2v = 0$, $w = \frac{1}{2}$.

20. In an election for state senator, candidate I received 45 percent of the vote, candidate II received 40 percent, and candidate III received 15 percent. Five voters are selected at random. What is the probability that two voted for I, two for II, and one for III?

21. A coin is to be thrown four times. What is the most probable number of heads that will occur?

22. A biologist searches caves for a particular species of animal. He estimates the probability of finding a male to be 0.7, the probability of finding a female to be 0.2, and the probability of finding neither to be 0.1. Three caves are searched. If the trials are assumed to be independent, find the probability that an animal is found in two of the three caves.

23. If a rat goes left in a T-maze he receives food, but if he goes right he receives a mild electrical shock. On trial 1 the probabilities of a left turn and a right turn are equal. On any trial after the rat

received food, the probabilities are 0.6 for a left turn and 0.4 for a right turn, whereas if the rat received the electric shock on the previous trial, the probabilities are 0.8 for a left turn and 0.2 for a right turn.   What is the probability that the rat will turn left on trial 2?   What is the probability that the rat will turn left on trial 3?

24. In a mice colony of 250 animals there are 75 carriers of a contagious disease.   Of those mice that come in contact with the carriers 90 percent are infected but do not become carriers.   If in the course of a 6-hour period each individual mouse has contact with one other mouse, what is the probability that a now uninfected mouse will be infected at the end of 6 hours?   At the end of two 6-hour periods?   Suppose that 50 percent of those newly infected become carriers immediately.   How does this change the probabilities?

4.5 Bernoulli Trials and the Binomial Probability Function

In Section 4.4 we saw that the notion of repeated independent trials provides a mathematical model for many common random phenomena.   If the random experiment being repeated has only two outcomes, such as (male, female), or (on, off), or (head, tail), then we have a particularly important case of repeated trials, known as *Bernoulli trials*, named after James Bernoulli (1654–1705), one of the founders of probability theory.   Intuitively, a sequence of Bernoulli trials is no more complicated than 100 tosses of a coin.   We shall obtain a new probability function by finding the probability that r heads appear in a sequence of 100 tosses.

We shall start with any random experiment that has exactly two outcomes and denote the outcomes s for "success" and f for "failure."   The sample space for one experiment is then $\{s, f\}$.   We denote the probability function by $P(\{s\}) = p$ and $P(\{f\}) = q$, where necessarily $p + q = 1$.   The term success is used here in a broad sense, for it may be the head in the toss of a coin, the occurrence of a machine breakdown, or the selection of a defective light bulb.

Suppose that the same random experiment (with two possible outcomes) is repeated n times.   Suppose further that each new trial is not influenced by the results of the other trials (that is, that the trials are independent).   The probabilities for the outcomes at each trial then remain the same, namely, $P(\{s\}) = p$ and $P(\{f\}) = q$.   These conditions comprise a compound experiment of repeated independent trials whose outcomes are the set of n-tuples (for n trials), where each entry in an n-tuple is either s or f.   The probability of any elementary event is the product of the probabilities for the components of the n-tuple and is thus a power of p multiplied by a power of q, with the sum of the powers equal to n.   In general, if the n-tuple has r entries that are s, the probability of the n-tuple as an elementary event is $p^r q^{n-r}$.   Any random phenomenon for which this mathematical model is appropriate is called a sequence of Bernoulli trials.

Definition 1 : Bernoulli trials

Repeated trials of a random experiment are called Bernoulli trials if
(a) there are only two outcomes possible for each trial,
(b) the probabilities of the outcomes are the same for each trial,
(c) the trials are independent.

A sequence of repeated trials with more than two outcomes is often converted to the case of Bernoulli trials by considering the occurrence of event A as success and the occurrence of A' as failure, with $p = P(A)$ and $q = P(A')$. For example, in measuring bolts produced by a machine, a bolt is considered defective if the length does not fall between certain limits.   Thus, if we are counting the number of defective bolts, a bolt is considered a "success" if it is defective.   The bolts are thereby separated into two classes, $A = \{$defective bolts$\}$ and $A' = \{$good bolts$\}$.

Example 1

Throw a single die 100 times.   In the sample space for a single throw, let A be the event "1 or 2."   Thus success is the occurrence of a 1 or 2, and failure occurs when the die shows 3, 4, 5, or 6.   Clearly, $p = P(A) = \frac{1}{3}$ and $q = \frac{2}{3}$.   Since each throw is independent of the others, we have a random experiment of 100 Bernoulli trials with a sample space of 2^{100} 100-tuples as outcomes.   ▎

What is the probability of obtaining exactly k successes in n Bernoulli trials?   The answer is simple; find all those n-tuples each of which has an s in exactly k of its entries, and add their individual probabilities.   To do this we observe that the probability of any one n-tuple having s in exactly k entries is $p^k q^{n-k}$.   The number of different n-tuples having s in exactly k components can be determined by selecting a k-combination (without replacement) from the n-set of components.   By Section 3.3, there are $\binom{n}{k}$ distinct ways in which this can be done.   Every n-tuple of the 2^n possible n-tuples that has s

occurring in exactly k places and f in $n - k$ places has been included. Therefore, the probability of exactly k successes in n Bernoulli trials with a probability of p for success is

$$\binom{n}{k} p^k q^{n-k}.$$

Since the number of success in n Bernoulli trials can be any integer from 0 to n, we can think of n Bernoulli trials as a new random experiment whose sample space is the set $\{0, 1, 2, \ldots, n\}$. The probability for the elementary event $\{k\}$ will be†

$$P(k) = \binom{n}{k} p^k q^{n-k},$$

as we have already determined, and to prove that we have a probability space we must show that

$$\sum_{k=0}^{n} P(k) = 1.$$

We need only our determination of $P(k)$ and the binomial theorem for the proof:

$$\sum_{k=0}^{n} P(k) = \sum_{k=0}^{n} \binom{n}{k} p^k q^{n-k}$$

$$= (q + p)^n = 1.$$

Hence we have proved

Theorem 1

If a random experiment consists of n Bernoulli trials with p the probability of success on each trial, the probability function for the sample space of the number of successes in n trials is

$$P(k) = \binom{n}{k} p^k q^{n-k}.$$

The function $P(\cdot)$ is called the *binomial probability function*.

This theorem completes a cycle in the development of probability models for random experiments. From a single Bernoulli trial that has a sample space of just two outcomes, we generated a new sample space of 2^n elementary

---

† Strictly speaking, the symbol for the probability for the elementary event $\{k\}$ should be written $P(\{k\})$. However, the distinction between the outcome k and the elementary event $\{k\}$ will be dropped at this point.

events (n-tuples) by taking n repetitions of the experiment. Still another sample space has now been obtained by a contraction of the preceding sample space, using the number of successes in n Bernoulli trials as outcomes. In each interpretation the probability function was defined for new elementary events as a consequence of the assumption that the original trials were independent.

Example 2 shows how the probability space of n Bernoulli trials differs from the probability space of exactly k successes.

Example 2

A fair coin is tossed eight times and a head is considered a success. The sample space for the compound experiment is the set of 8-tuples of s's and f's, and there are $2^8 = 256$ n-tuples. A typical 8-tuple with five heads and three tails is (s, f, s, s, f, s, f, s). The probability of this or any such 8-tuple having exactly five heads and three tails is $(\frac{1}{2})^5(\frac{1}{2})^3$. The probability of exactly five successes in eight trials is $\binom{8}{5}(\frac{1}{2})^5(\frac{1}{2})^3 = \frac{7}{32}$.

On the other hand, the probability of a head on each of the first five tosses and a tail on each of the next three is precisely $(\frac{1}{2})^5(\frac{1}{2})^3$, since this event is exactly one of the 2^8 elementary events in the eight Bernoulli trials space. ∎

Example 3

The relationship between the clutch size and the survival of the eggs of the American eider has been studied and it is estimated that for a clutch of two eggs the probability that the nest is unaffected is 0.25. Six nests are observed on a field trip. What is the probability that four or more of the nests have all eggs intact?

The model for this random phenomenon is that of six Bernoulli trials with a probability of $p = \frac{1}{4}$ of success on each trial, success being the equivalent of unaffected nests. The event "four or more intact" is the subset $\{4, 5, 6\}$, so the required probability is

$$P(4) + P(5) + P(6).$$

Since $n = 6$ and $p = \frac{1}{4}$, this sum can be written as

$$\sum_{k=4}^{6} \binom{6}{k}\left(\frac{1}{4}\right)^k \left(\frac{3}{4}\right)^{6-k}.$$

The calculation is

$$\binom{6}{4}\frac{3^2}{4^6} + \binom{6}{5}\frac{3}{4^6} + \binom{6}{6}\frac{1}{4^6} = 0.0376. \quad ∎$$

The binomial probability function must have two quantities specified before any calculations are possible. These are n, the number of trials, and

p, the probability of success on one trial. These numbers are known as *parameters* and are characterized by the fact that they are constant throughout one discussion or problem, but they will usually change from problem to problem. Tables of values of $P(k)$ for a large set of values of n and p are available. A small table is listed in Appendix E and an extensive table is available in *Tables of the Binomial Probability Distribution*, National Bureau of Standards.

Carrying out n repetitions of a random experiment is equivalent to selecting an n-permutation with replacement from a population where a fraction p are of one kind and a fraction q are of a second kind. In this way Bernoulli trials can be used in certain sampling problems, as we show in Examples 4 and 5. This interpretation of Bernoulli trials is similar to the approach we used in sampling in Examples 2 and 3 of Section 3.4.

Example 4

One fourth of the voters living in a county are registered as Democrats. In a sample of 20 voters chosen at random, what is the probability that exactly five are Democrats? What is the probability that at least five are Democrats?

The selection of a sample can be considered as a sequence of 20 Bernoulli trials with the selection of a Democrat classified as a success. Since the total voter population is large, it is reasonable and safe to view the selection as sampling with replacement. The probability p for success is $\frac{1}{4}$, so the number required is $P(5)$, which is

$$P(5) = \binom{20}{5} \left(\frac{1}{4}\right)^5 \left(\frac{3}{4}\right)^{15}.$$

The probability of selecting at least five Democrats is the probability of the event $\{5, 6, \ldots, 20\}$. Let A be this event; then $A' = \{0, 1, 2, 3, 4\}$ and $P(A) = 1 - P(A')$. The probability of the complement is given by

$$P(0) + P(1) + P(2) + P(3) + P(4),$$

which is equal to

$$\sum_{k=0}^{4} \binom{20}{k} \left(\frac{1}{4}\right)^k \left(\frac{3}{4}\right)^{20-k} = 0.4148.$$

Thus $P(A) = 0.5852$. ∎

Example 5

An urn contains 18 balls, of which 10 are red and 8 are green. A sample of 5 balls is drawn with replacement. What is the probability that exactly 3 red balls are drawn?

This random experiment can be considered as a sequence of five Bernoulli trials, with the selection of a red ball counted as a success. The probability p for success is $\frac{5}{9}$, so the number required is $P(3)$, which is

$$P(3) = \binom{5}{3} \left(\frac{5}{9}\right)^3 \left(\frac{4}{9}\right)^2.$$

We can also ask for the probability that at most 3 balls are red. The probability of the event "at most 3 red balls" is the probability of picking 0, 1, 2, or 3 red balls and is given by

$$P(0) + P(1) + P(2) + P(3) = \sum_{k=0}^{3} \binom{10}{k} \left(\frac{2}{5}\right)^k \left(\frac{3}{5}\right)^{10-n}$$
$$= 0.5138. \qquad \blacksquare$$

Example 6

A certain newsboy loses money if he sells fewer than 50 papers a day. He estimates that the probability of selling fewer than 50 papers is $\frac{1}{10}$. In a 30-day month, what is the probability that he will lose money on at most 3 days? This problem can be interpreted as 30 Bernoulli trials with a probability of success on a trial of $\frac{1}{10}$, where success is the sale of fewer than 50 papers a day. The event is $\{0, 1, 2, 3\}$, so the desired probability is

$$\sum_{k=0}^{3} P(k) = \sum_{k=0}^{3} \binom{30}{k} \left(\frac{1}{10}\right)^k \left(\frac{9}{10}\right)^{30-k}$$
$$= 0.647. \qquad \blacksquare$$

Although Bernoulli trials have been discussed primarily in connection with the toss of a coin, many physical phenomena can be idealized as a Bernoulli scheme of trials, as, for example, in testing the effectiveness of a new drug. ([5], p. 139.)
■ If there should be more than two choices available each of which can be assigned a probability, then the binomial probability function generalizes to what is called the *multinomial* probability function. Examples 7 and 8 contain the essential facts.

Example 7

Assume that a person who votes Democratic in one election will vote Democratic in the next election with a probability of $p_1 = 0.8$, vote Republican with a probability of $p_2 = 0.15$, and vote Independent with a probability of $p_3 = 0.05$. This is a model of a random phenomenon with a sample space $S = \{D, R, I\}$ and the probability function as assigned above.

We observe the voting preference of seven individuals (all of whom voted Democratic in the last election) and assume that the voting behavior of the individuals is independent of one another. This is another case of independent trials, where on each trial there are three possible outcomes with the probability of each outcome constant from trial to trial. The sample space for the experiment with seven people consists of 3^7 outcomes, each of which is a 7-tuple such as (R, D, D, D, I, R, D). The probability associated with each elementary event is the product of the probabilities for each of the seven components, because the trials are independent. Thus, for the 7-tuple above,

$$P(\{(R, D, D, D, I, R, D)\}) = (0.8)^4(0.15)^2(0.05)^1.$$

What is the probability that in the next election, of the seven people, four vote Democratic, two vote Republican, and one votes Independent? To find the probability of the event, it is necessary to count all the 7-tuples where there are four entries D, two entries R, and one entry I mixed up in the 7-tuple, and then to multiply that number by the probability of the elementary event that was found above, $(0.8)^4(0.15)^2(0.05)^1$. Counting the 7-tuples is done by observing that of the seven components, four must be D, and there are $\binom{7}{4}$ ways to select these spots. There are then $\binom{3}{2}$ ways to select the spots for the two R's which leaves $\binom{1}{1}$ spots for the I. The total number of ways in which these three tasks can be accomplished is the product $\binom{7}{4}\binom{3}{2}\binom{1}{1}$, and the desired probability is

$$\binom{7}{4}\binom{3}{2}\binom{1}{1}(0.8)^4(0.15)^2(0.05)^1 = 0.48.$$

This example is a special case of the multinomial probability function. ∎

Example 8

Suppose that a sack contains 20 chips that are identical except for color, with 16 red, 3 white, and 1 blue. One chip is drawn at random and its color noted. A sample space for this random experiment is $S = \{R, W, B\}$ with $P(\{R\}) = 0.8$, $P(\{W\}) = 0.15$, and $P(\{B\}) = 0.05$.

If the chip is returned to the sack before a subsequent draw, then the trials may be considered independent, and the overall random experiment is that of repeated trials. Suppose that seven successive draws are made. What is the probability that four red chips, two white chips, and one blue chip were drawn?

The answer to the question is exactly

$$\binom{7}{4}\binom{3}{2}\binom{1}{1}(0.8)^4(0.15)^2(0.05)^1 = 0.48,$$

the same answer as in the previous example. The probability space and the event are the same for both models, so the answer is the same and there is no need to calculate any further. ∎

In the examples so far the probability of some outcome was known or assumed known, and from this premise the probability of another event was calculated. This is not always the case. In fact, a major statistical problem is estimating probabilities of an event so that the probability of a related event can be computed. As an example of the procedure consider Example 9.

Example 9

(Suggested by [19], pp. 170–174.) A man named Art claims that he can predict the outcome of baseball games, by which he means that he can predict outcomes correctly more than half the time. Bill doubts his claim and they agree to the following test of his prognostic powers. If Art correctly predicts the outcome of 8 or more games out of a set of 10, then Bill will grant his claim. Otherwise, Bill maintains his opinion that Art has no special talent.

Suppose that it is possible to assign a probability p, which is the probability that Art will correctly predict the outcome of a game. Art claims that $p > \frac{1}{2}$, whereas Bill contends $p = \frac{1}{2}$ at best.

If the true value of p is $\frac{1}{2}$, what is the probability that Bill is correct; that is, what is the probability that Art does not pick 8 or more winners? This probability is seen to be

$$1 - \sum_{k=8}^{10} \binom{10}{k} \left(\frac{1}{2}\right)^k \left(\frac{1}{2}\right)^{10-k},$$

and from the table of binomial probabilities the value is 0.945.

Suppose that Art is right and that the true value of p is significantly greater than $\frac{1}{2}$. What is the probability that he does correctly predict 8 or more winners? In this case the probability is the value of

$$\sum_{k=8}^{10} \binom{10}{k} p^k (1-p)^{10-k}.$$

If $p = 0.7$, the value is 0.383, whereas if $p = 0.8$, the value is 0.678. For $p = 0.9$, the probability is 0.930, and so it is not until p is quite close to 1 that the test of 8 or more correct seems to be fair to both.

If Art must name only 7 or more correctly, then the probability that he cannot do so when $p = \frac{1}{2}$ changes to 0.828, and the probability that he can name 7 or more becomes 0.879 when $p = 0.8$. If one thinks that 0.8 is a more realistic estimate of Art's ability, then the experiment with 7 or more correct seems to be a fairer arrangement to Art than the original test. ∎

Problems

1. A coin is tossed six times.   What is the probability of exactly four heads?   Of at most four heads?

2. An urn contains 4 red balls, 4 green balls, and 4 white balls.   A sample of 10 balls is drawn with replacement.   What is the probability that at most 2 white balls are drawn?

3. Seven dice are rolled.   Calling a 5 or 6 a success, find the probability of getting exactly four successes; at most, four successes.

4. If a baseball player with a 0.400 batting average comes to bat five times in a game, what is the probability that he will get exactly two hits?   Less than two hits?

5. In shooting at a target, the probability that a man will hit the target is $\frac{9}{10}$.   Find the probability that he

   (a) hits the target exactly 4 times in 10 shots.
   (b) hits the target the first 4 times and misses all the rest.

6. The probability that a student can solve any problem is estimated to be $\frac{1}{2}$.   In order to pass he is required to solve 7 of 10 problems on an examination.   What is the probability that he will pass?

7. The probability that a subject will complete an experiment is $\frac{3}{4}$. If a group of eight undergoes the experiment, what is the probability that six or more will complete the experiment, assuming the independence of the trials?

8. The probability that an archer will hit a small target on a single shot is 0.20.   If he shoots five times, what is the probability that he will score

   (a) exactly one hit?   (b) at least two hits?

9. If one assumes that the probability of a newborn child being a boy is 0.52, what is the probability that a family with four children will have

   (a) exactly one boy?   (b) exactly one girl?
   (c) at least one boy?

10. The probability that a certain machine will produce a defective item has been estimated to be 0.15.   If a sample of 10 items is taken at random from the output of the machine, what is the probability that there will be

    (a) no defective items?
    (b) at least one defective item?

11. The probability that a rat will die from a disease is $\frac{1}{4}$. If 10 rats are infected with the disease, write an exact expression for the probability that

    (a) exactly 9 will succumb.   (b) at least 5 will succumb.

12. In Podunk City it rains one day out of three. Three dates are selected at random. What is the probability that it will rain on at least one of the three dates?

13. In a city of 5000 voters, a sample of 20 selected at random were questioned on a proposed municipal project. Twelve favored the project and 8 were opposed. If the voters were evenly divided on the question, what would be the probability of obtaining a majority of 12 or more favoring the proposal in a sample of 20?

14. It is estimated that 60 percent of a student body of 4000 favor a proposed change in rules. The student newspaper selects a random sample of 100 students. Write an exact expression for the probability that

    (a) at least 60 of those selected favor the proposal.
    (b) at least 60, but not more than 75, of those selected favor the proposal.

15. Five voters are selected at random and asked their opinion on the construction of a new school. Suppose that only 40 percent of the eligible voters favor the project. What is the probability that a majority of the five polled will support it?

16. Twenty people are given either hand-knit sweaters or factory-knit sweaters in random order. Each person is asked to state whether his sweater is hand-knit or not. If one assumes that each person cannot distinguish the knitting and makes his decision by guessing what is the exact probability that there are at least 15 correct choices?

17. Find the probability of obtaining the most probable number of 6's in five rolls of a die.

18. A coin is tossed repeatedly. What is the probability that the fourth head appears on the eleventh toss?

19. The probability that an archer will hit a small target on a single shot is 0.20. What is the conditional probability of two hits in five shots, given that the archer has scored an even number of hits in five shots?

20. A bead on a horizontal wire is moved under the following set of rules. A fair coin is tossed. If the coin comes up heads, the bead is moved one unit to the right. If the coin comes up tails, the bead is moved one unit to the left. What is the probability that after 10 tosses of the coin the bead will be

    (a) back at the starting point?
    (b) No more than two units from the starting point?

21. Two people, A and B, play a game in which the probability that A wins is $\frac{3}{8}$ and the probability that B wins is $\frac{5}{8}$. A series of games is to be played in which the winner of the series is the first to win four games. If each has won twice, find the probability that A wins the series.

22. If each of two people tosses a coin n times, find the probability that they both get the same numbers of heads. (Assume any independence you need.)

23. A coin is weighted so that a head occurs twice as often as a tail. If the coin is tossed four times, find the probability of k heads for each value of k.

24. In learning a repetitive process, such as the operation of a keypunch machine, the probability for error usually decreases with each trial. Assume that the probability for error on the ith trial is $1/(1+i)$, for $i = 1, 2, \ldots, n$. If four trials are made, find the probability that three were without error? (*Note*: Because the probability of "success" is not constant from trial to trial, the binomial probability function does not apply.)

25. Refer to Example 7. What is the probability that at least one voter of the seven changes in the next election?

26. Refer to Example 7. What is the probability that the number of voters who change is less than the number who do not?

27. Two fair dice are rolled with three outcomes distinguished: Outcome A is when the sum is 5 or less; outcome B when the sum is 6, 7, or 8; and outcome C when the sum is 9 or more. If the dice are rolled four times, what is the probability of one A, two B's, and one C?

chapter five

Random Variables

5.1 Random Variables Defined

5.2 Point and Distribution Functions

5.1 Random Variables Defined

Earlier descriptions of the outcomes of a random experiment usually have been qualitative, as the color of a ball, or heads or tails on a coin. The generalization we used with Bernoulli trials, s or f (for success or failure), began to modify the qualitative descriptions of outcomes, especially when we assigned s for success to a commodity failure or defect, if what we were looking for was defective products. We could just as easily have used the numbers 0 and 1 to represent these particular cases of a two-outcome sample space.

In this chapter we shall represent the outcomes of a random phenomenon by assigning a real number to each outcome. Thus qualitative aspects will be subdued, and we shall find it easier to examine the essential structure of the random phenomenon.

Many random experiments have a natural numerical description of the outcomes, such as the number of letters in a word or the number of spots on the uppermost face of a die. If the random phenomenon does not have a natural numerical description, we shall invent an assignment of a set of real numbers to represent the outcomes. Thus to each elementary event in a sample space we somehow assign a real number, thereby defining a function (Section 1.7).

Definition 1 : Random variable

A function whose domain is a finite or countably infinite sample space and whose codomain is the set of real numbers is a random variable.

A random variable, therefore, is a mathematical function that assigns to each elementary event of the sample space a real number. The usual notation is to let a capital letter, such as X or Y, represent the functional correspondence and let $X(e_i)$ denote the value of the random variable for the elementary event $\{e_i\}$. In this context, the term random variable may seem to be an unfortunate name, as the object it represents is apparently neither random nor a variable. The term is well established through long usage, however, and to change the name here would invite unnecessary confusion. Later we shall interpret X more as the arbitrary representative of a set of real numbers and less as a function. Nevertheless, we reiterate that the random variable X represents a function.

We shall denote by S_X the actual range set of the function X in the co-domain of real numbers. The domain will be denoted by S, the sample space as a set of outcomes. Since S is at most countably infinite, so is S_X. Examples 1 and 2 show how random variables can be defined on a sample space and how different random variables may be defined on the same sample space.

Example 1

Flip a coin four times in succession. This experiment is a sequence of four Bernoulli trials and the basic sample space S is the 16 4-tuples of H and T.

$$
\begin{aligned}
e_1 &= (H, H, H, H), & e_9 &= (T, H, T, H), \\
e_2 &= (H, H, H, T), & e_{10} &= (T, T, H, H), \\
e_3 &= (H, H, T, H), & e_{11} &= (T, H, H, T), \\
e_4 &= (H, T, H, H), & e_{12} &= (H, T, T, T), \\
e_5 &= (T, H, H, H), & e_{13} &= (T, H, T, T), \\
e_6 &= (H, H, T, T), & e_{14} &= (T, T, H, T), \\
e_7 &= (H, T, H, T), & e_{15} &= (T, T, T, H), \\
e_8 &= (H, T, T, H), & e_{16} &= (T, T, T, T).
\end{aligned}
$$

(a) Let us define the random variable X on S to be the number of heads in any outcome. Then $S_X = \{0, 1, 2, 3, 4\}$, and the correspondence for the original sample space is given by

$$
\begin{aligned}
X(e_1) &= 4, & X(e_9) &= 2, \\
X(e_2) &= 3, & X(e_{10}) &= 2, \\
X(e_3) &= 3, & X(e_{11}) &= 2, \\
X(e_4) &= 3, & X(e_{12}) &= 1, \\
X(e_5) &= 3, & X(e_{13}) &= 1, \\
X(e_6) &= 2, & X(e_{14}) &= 1, \\
X(e_7) &= 2, & X(e_{15}) &= 1, \\
X(e_8) &= 2, & X(e_{16}) &= 0.
\end{aligned}
$$

(b) Let us define the random variable Y on S to be $+1$, if there are more H's than T's, -1 if there are fewer H's than T's, and 0 if there are the same number of H's and T's. In this case $S_Y = \{-1, 0, 1\}$, with the functional correspondence given by

$$Y(e_1) = Y(e_2) = Y(e_3) = Y(e_4) = Y(e_5) = +1,$$
$$Y(e_6) = Y(e_7) = Y(e_8) = Y(e_9) = Y(e_{10}) = Y(e_{11}) = 0,$$
$$Y(e_{12}) = Y(e_{13}) = Y(e_{14}) = Y(e_{15}) = Y(e_{16}) = -1. \quad \blacksquare$$

Example 2

A jar is filled with chips that are identical except for a single digit written on each. A chip is drawn and the number noted. The sample space S is $\{0, 1, 2, 3, 4, 5, 6, 7, 8, 9\}$. Even though the outcomes are described numerically, we can still define random variables on S. If X is defined to be the identity function on S so that $X(i) = i$, then S_X coincides with S. Let us define the random variable Y, so that $Y(i) = +10$ for odd $i \in S$, and $Y(i) = -10$ for even $i \in S$. Hence $S_Y = \{-10, +10\}$ and $Y(3) = +10$, while $Y(4) = -10$. As another example, let us define the random variable Z so that if $i \in S$, then $Z(i) = i^2$. Hence $Z(5) = 25$, for instance. Thus $S_Z = \{0, 1, 4, 9, 16, 25, 36, 49, 64, 81\}$. Frequently, the kind of information we seek will influence the choice of the random variable. $\quad \blacksquare$

Example 3

Four films are ranked in order of artistic achievement. If there is no reason to believe that there are discernable differences, the $4! = 24$ permutations of 4 symbols, say 1, 2, 3, and 4, constitute a sample space, with the uniform probability function.

Let us define the random variable X to be 1 whenever film 2 precedes film 4 in the ranking, and 0 otherwise. Then if

$e_1 = 1234,$	$e_7 = 2134,$	$e_{13} = 3124,$	$e_{19} = 4123,$
$e_2 = 1243,$	$e_8 = 2143,$	$e_{14} = 3142,$	$e_{20} = 4132,$
$e_3 = 1324,$	$e_9 = 2314,$	$e_{15} = 3214,$	$e_{21} = 4213,$
$e_4 = 1342,$	$e_{10} = 2341,$	$e_{16} = 3241,$	$e_{22} = 4231,$
$e_5 = 1423,$	$e_{11} = 2413,$	$e_{17} = 3412,$	$e_{23} = 4312,$
$e_6 = 1432,$	$e_{12} = 2431,$	$e_{18} = 3421,$	$e_{24} = 4321,$

$X(e_1) = X(e_2) = X(e_3) = X(e_7) = X(e_8) = X(e_9) = X(e_{10}) = X(e_{11}) = X(e_{12}) = X(e_{13}) = X(e_{15}) = X(e_{16}) = 1$, and $X = 0$ for all other elementary events.

As another example, let us define Z to be the random variable whose values are the number of positions 2 is to the left or the right of 1, with positions to the right indicated by positive numbers and positions to the left by negative numbers.   Then

$$X(e_1) = X(e_2) = X(e_{13}) = X(e_{17}) = X(e_{19}) = X(e_{23}) = 1,$$
$$X(e_3) = X(e_5) = X(e_{14}) = X(e_{20}) = 2,$$
$$X(e_4) = X(e_6) = 3,$$
$$X(e_7) = X(e_8) = X(e_{15}) = X(e_{18}) = X(e_{21}) = X(e_{24}) = -1,$$
$$X(e_9) = X(e_{11}) = X(e_{16}) = X(e_{22}) = -2,$$
$$X(e_{10}) = X(e_{12}) = -3. \quad \blacksquare$$

Since S_X is a set of numerical descriptions of a random phenomenon, S_X can be considered a sample space, and we now try to define a probability function on it.   An elementary event of S_X is a single real number and an event is a subset of the numbers in S_X. (However, each event in S_X arises from a corresponding event in S.)

Let A be an arbitrary event in S_X (that is, $A \subset S_X$), and define B as the event in S such that an outcome e_i of S is in B if and only if $X(e_i)$ is in A. In symbols,

$$(1) \qquad\qquad B = \{e_i \mid e_i \in S, X(e_i) \in A\}.$$

Once again, note that A is an event of the sample space S_X and B is an event of the original sample space S.

The outcomes of S_X are real numbers, which we denote by the symbol x_i. That is, we write

$$S_X = \{x_1, x_2, \ldots, x_n\}$$

for a finite sample space, and

$$S_X = \{x_1, x_2, \ldots\}$$

for a countably infinite sample space.   The single elementary event $\{x_j\}$ also corresponds to an event of the sample space S, which we shall designate by B_j:

$$B_j = \{e_i \mid e_i \in S, X(e_i) = x_j\}.$$

For each x_j in S_X there is a B_j, and any B_j may contain more than one outcome of S.   If S is finite, $\bigcup_{j=1}^{n} B_j = S$.   If S is a countably infinite space, then $\bigcup_{j=1}^{\infty} B_j = S$.   Since the $B_j, j = 1, 2, \ldots, n$, are events of S, $P(B_j)$ is known.   We can use this fact to determine a probability function for S_X.

Theorem 1

If X is a random variable whose domain is the sample space S with probability function $P(\cdot)$, and if S_X is the range set of real numbers for

X with $S_X = \{x_1, x_2, \ldots, x_n\}$, then the function $P_X(\cdot)$ defined for any elementary event $\{x_j\}$ by†

$$P_X(x_j) = P(\{e_i \mid e_i \in S, X(e_i) = x_j\})$$

is a probability function for S_X.

PROOF: We only need to show that $\sum_{j=1}^{n} P_X(x_j) = 1$, for, clearly, the numbers $P_X(x_j)$ satisfy the inequality $0 \le P_X(x_j) \le 1$. By definition of $P(\cdot)$, however,

$$\sum_{j=1}^{n} P_X(x_j) = \sum_{i=1}^{n} P(\{e_i \mid e_i \in S, X(e_i) = x_j\}).$$

Every e_i in S is included in the sum on the right because X is defined on all of S, and every e_i has an image $X(e_i) = x_j$. Moreover, no e_i will appear more than once, for if it did there would be two values of x_j, say x_{j_1} and x_{j_2}, for which $X(e_i) = x_{j_1}$ and $X(e_i) = x_{j_2}$. Unless $x_{j_1} = x_{j_2}$, X cannot be a mathematical function. Consequently, the sum on the right can be interpreted as a partition of S and is equal to

$$\sum_{i=1}^{m} P(E_i),$$

where $S = \{e_1, e_2, \ldots, e_m\}$. Since S is the sample space, the value is 1. ∎

Theorem 1 can be extended to countably infinite S_X by the same device used for finite S_X; however, we cannot justify the procedure until we are in a position to attach some meaning to the notion of infinite sums. We would have to demonstrate, for example, that a countably infinite set of nonnegative numbers has a "sum" of 1. This is a problem in infinite series; see [18, Chap. 8].

If A is an arbitrary event of S_X, then by the reasoning of Section 2.4 it follows that

$$P_X(A) = \sum_{x_i \in A} P_X(x_i).$$

Since an event A of S_X is a subset of real numbers, we shall often describe A as a set of numbers in some interval, such as

$$A = \{x_i \mid x_i \in S_X, a \le x_i \le b\},$$

where a and b are arbitrary real numbers. The notation $a \le X \le b$ represents the event A with the understanding that

$$P_X(a \le X \le b) = P_X(A).$$

---

† Sometimes it is convenient to emphasize that a random variable takes on a value x_j and write $P_X(x_j) = P(X = x_j)$.

Using the notation suggested earlier we can write

$$P_X(a \leq X \leq b) = \sum_{a \leq x_i \leq b} P_X(X = x_i).$$

This notation is used in Examples 4–6.

Example 4

(Suggested by [14], pp. 163–165.) The number of turning points in a permutation of different measurements is an example of a random variable. This topic is often used in the study of economic time series, such as the daily stock-market averages. A time series is a set of observations arranged in the order in which they are made. Usually the numbers will fluctuate, but if there is a continual increase or decrease, a cyclical change, or some other recognizable pattern, it is possible to make some sort of prediction. If among three successive numerical measurements the middle one is the least or the greatest of the three, it is called a turning point of the sequence. Let us take the set of 4-permutations of a 4-set for a sample space S and define X to be the number of turning points in the permutation. Using the functional notation for the permutation 4213, we see that $X(4213) = 1$ since the triple 421 has no turning point, while 1 is a turning point for the triple 213. The complete assignment for S is given in Table 1. Thus

Table 1

	X		X		X		X
1234	0	2134	1	3124	1	4123	1
1243	1	2143	2	3142	2	4132	2
1324	2	2314	2	3214	1	4213	1
1342	1	2341	1	3241	2	4231	2
1423	2	2413	2	3412	2	4312	1
1432	1	2431	1	3421	1	4321	0

$S_X = \{0, 1, 2\}$, with $x_1 = 0$, $x_2 = 1$, and $x_3 = 2$. From the table we deduce that the probability function has the values

$$P_X(0) = \tfrac{2}{24},$$
$$P_X(1) = \tfrac{12}{24},$$
$$P_X(2) = \tfrac{10}{24}.$$

Example 5

Suppose that X is a random variable with a range set

$$S_X = \{-1, 0, \tfrac{1}{4}, \tfrac{1}{2}, 2, \tfrac{5}{2}, 4\}$$

and a probability function defined by

x_j	-1	0	$\tfrac{1}{4}$	$\tfrac{1}{2}$	2	$\tfrac{5}{2}$	4
$P_X(x_j)$	0.1	0.05	0.2	0.15	0.2	0.2	0.1

Then

(a) $P_X(0 < X < 1) = P_X(\tfrac{1}{4}) + P_X(\tfrac{1}{2}) = 0.35,$

(b) $P_X(X \geq 2) = P_X(2) + P_X(\tfrac{5}{2}) + P_X(4) = 0.5,$

(c) $P_X(0 \leq X \leq 2) = P_X(0) + P_X(\tfrac{1}{4}) + P_X(\tfrac{1}{2}) + P_X(2) = 0.6.$ ∎

It should be clear now that we can take S_X as a sample space with real numbers as outcomes and $P_X(\cdot)$ as a probability function on S_X. The new probability space has been induced on the real numbers by the definition of the random variable X. This procedure allows us to cast the probability space of any random phenomenon in terms of a probability space whose outcomes are real numbers.

If the sample space S_X is a finite set, the random variable will be called a *finite random variable*. When S_X is a countably infinite set, the variable will be called a *discrete random variable*. If there are just two outcomes in S_X, the variable will be called a *Bernoulli random variable*. If X is the number of successes in n Bernoulli trials, the random variable X will be called a *binomial random variable*. Other descriptive names are attached to particular random variables, which help to identify the random phenomenon under study.

Example 6

At the conclusion of a coeducational living experiment, a survey of the attitudes of the participants toward the project was conducted. Each question had three possible answers, "opposed," "indifferent," and "favor," which were coded with 0, 1, and 2, respectively. The questions "What is the overall rating of your experience with the project?" and "Would you participate in the project again?" were examined for interrelationship. The results given in Table 2 record the preferences of the 100 students in the survey.

If the survey is interpreted as a random experiment, a sample space with the nine 2-tuples $\{(0, 0), (0, 1), (0, 2), (1, 0), (1, 1), (1, 2), (2, 0), (2, 1), (2, 2)\}$ would be appropriate with the probability function the relative frequency of occurrence. Thus $P(\{(1, 1)\}) = \tfrac{17}{100}$, because 100 students were in the survey.

The random variable X defined as the sum of the codes for the responses will generate the sample space $S_X = \{0, 1, 2, 3, 4\}$. To find $P_X(\cdot)$, construct the events B_j, $j = 1, 2, \ldots, 5$, in S corresponding to $x_j \in S_X$:

$$x_1 = 0, \quad B_1 = \{(0, 0)\}, \qquad\qquad P_X(0) = \tfrac{14}{100},$$

$$x_2 = 1, \quad B_2 = \{(0, 1), (1, 0)\}, \qquad P_X(1) = \tfrac{10}{100},$$

$$x_3 = 2, \quad B_3 = \{(0, 2), (1, 1), (2, 0)\}, \quad P_X(2) = \tfrac{26}{100},$$

$$x_4 = 3, \quad B_4 = \{(1, 2), (2, 1)\}, \qquad P_X(3) = \tfrac{17}{100},$$

$$x_5 = 4, \quad B_5 = \{(2, 2)\}, \qquad\qquad P_X(4) = \tfrac{33}{100}.$$

The sum of the codes provides a numerical measure of the preferences of the participants. For example, if $X = 0$ or 1, then the student was either opposed to both questions or opposed to one and indifferent to the other. Ordered pairs are harder to work with in some ways, as we shall see in Chapter 6.

Table 2

	Question 1		
	Oppose	Indifferent	Favor
Question 2	0	1	2
Oppose 0	14	8	5
Indifferent 1	2	17	10
Favor 2	4	7	33

A similar numerical measure can be the random variable Y, defined as the absolute value of the difference between the codes. The random variable Y will generate the sample space $S_Y = \{0, 1, 2\}$. If the events B_j, $j = 1, 2, 3$, are set up, $P_Y(\cdot)$ can be determined as above:

$$x_1 = 0, \quad B_1 = \{(0, 0), (1, 1), (2, 2)\}, \qquad\qquad P_Y(0) = \tfrac{64}{100},$$

$$x_2 = 1, \quad B_2 = \{(0, 1), (1, 0), (1, 2), (2, 1)\}, \quad P_Y(1) = \tfrac{27}{100},$$

$$x_3 = 2, \quad B_3 = \{(0, 2), (2, 0)\}, \qquad\qquad\quad P_Y(2) = \tfrac{9}{100}. \quad ∎$$

Problems

1. Suppose that four coins are tossed. If X is the number of heads minus the number of tails, list the numbers in S_X and find $P_X(\cdot)$.

2. Two ordinary dice are thrown. Let X denote the total number of spots on the uppermost faces. Find S_X and $P_X(\cdot)$.

3. A sack contains eight apples of which two are rotten. A sample of four apples is drawn without replacement from the sack. Let X denote the number of rotten apples in the sample. Find S_X and $P_X(\cdot)$.

4. A bag contains 50 envelopes of which 10 contain $5 each, 10 contain $1 each, and the others are empty. A single envelope is drawn and X denotes the amount of money obtained. Describe S_X and find $P_X(\cdot)$.

5. Three balls are drawn in succession without replacement from an urn containing eight red and eight black balls. Every black ball drawn is worth one point, and every red ball drawn is worth two points. Let X denote the total number of points for the three balls. Find S_X and $P_X(\cdot)$.

6. Let X denote the number of 7's that occur in five throws of a pair of dice. Find S_X and $P_X(\cdot)$.

7. A radio tube is tested, and let us assume that the probability that it tests positive is $\frac{3}{4}$. Suppose that a large supply of such tubes is being tested. Let X denote the number of tests required to find the first tube that tests positive. Find S_X and $P_X(\cdot)$.

8. Let X be the number of 2's that occur in six independent tosses of a fair die. Determine an exact expression for the probability that X satisfies the inequality $2 \leq X \leq 4$; that is, $P_X(2 \leq X \leq 4) = ?$

9. There are two indistinguishable boxes with currency bills in them. In box I there are nine $1 bills and one $10 bill, while box II contains five $1 bills, seven $5 bills, and three $10 bills. You select a box at random and then a bill from the box. Let X be the amount of money drawn. Find $P_X(X \geq 5)$.

10. Select a number at random from the interers 1 through 20. Let X be the number of its positive divisors (or factors), where an integer is a divisor of another integer if the second is a whole-number multiple of the first, including the special case that an integer is a divisor of itself. What is $P_X(X > 3)$?

11. Toss a coin three times in succession. Let X be the random variable that denotes the gain of a player who wins $2 if the first head occurs on the first toss, wins $1 if the first head occurs on the second toss, loses $1 if the first head occurs on the third toss, and loses $2 if all three tosses are tails. Find S_X and $P_X(\cdot)$.

12. Let $S_X = \{0, 1, 2, 3, 4, 5, 6, 7\}$ and suppose that

X	0	1	2	3	4	5	6	7
$P_X(\cdot)$	0	$2c$	c	$2c$	$3c$	c^2	$3c^2$	$6c^2 + c$

(a) Find c.   (b) Find $P_X(3 \leq X \leq 5)$.

13. If the probability of success on a single Bernoulli trial is p, let X be the number of successes in n Bernoulli trials. Find S_X and $P_X(\cdot)$.

14. A sack contains three black and two red marbles. Find the sample space and the probability function for each of the following random variables:

(a) the number of red marbles in a sample of size 3 that is drawn with replacement.
(b) the number of marbles that are drawn with replacement in order to get a red marble.

15. Let X denote the maximum of the numbers on two balls drawn without replacement from an urn containing six balls numbered 1 to 6. Find $P_X(X \leq 3)$.

16. Let X denote the weight in pounds, measured to the nearest pound, of an individual selected from a certain group of 100 individuals. Describe S_X and explain the meaning of $P_X(100 \leq X < 150)$.

Age of husband	Age of wife							
	15–19	20–24	25–29	30–34	35–39	40–44	45–49	50+
15–19	42	10	3					
20–24	153	504	51	10	1			
25–29	52	271	184	22	7	2		
30–34	5	52	87	69	13	5		
35–39	1	12	27	29	21	2	3	
40–44		1	9	18	17	8	2	1
45–49	1		3	6	16	16	7	1
50+			1	4	11	15	21	43

SOURCE: A. B. Hollingshead, Cultural Factors in the Selection of Marriage Mates, *American Sociological Review*, **15** (1950), p. 622.

17. The ages at marriage of husband and wife for 1848 marriages in New Haven, Connecticut, are listed in the table. Define X to be $+1$ for the husband in an older age group than the wife, 0 for the same group, and -1 in a younger age group. Find $P_X(\cdot)$.

18. The students in a two-semester calculus course receive a grade (A, B, C, D, or F) for each semester. Let grade points 4, 3, 2, 1, and 0 be assigned to the grades, and assume that all pairs of grades are equally likely. If X is the random variable denoting the total grade points earned, find S_X and $P_X(\cdot)$. If Y is the random variable denoting the difference between the grade points for the first semester and the second semester, find S_Y and $P_Y(\cdot)$.

5.2   Point and Distribution Functions

In Section 5.1 we defined a probability function $P_X(\cdot)$ in such a way that for any event $A \subset S_X$, $P_X(A)$ is obtained by summing the appropriate values of $P_X(x_j)$, where $x_j \in A$. The function $P_X(A)$ is therefore defined for all events of S_X and hence is a set function. We also showed in Section 5.1 that $P_X(\cdot)$ is a valid probability function by showing that $\sum P_X(x_j) = 1$, where the sum is taken over all elementary events of S_X. Let us define a function on S_X that coincides with $P_X(\cdot)$ only at the points x_j and write it as $p(\cdot)$; that is, $p(x_j) = P_X(x_j)$. Since $p(x_j)$ is defined only at the points x_j of S_X and not on the events (that is, subsets) of S_X, we call $p(\cdot)$ a *point function*. That is, $p(\cdot)$ is defined only for real numbers x and not for sets of real numbers. This new function is so important that we shall define it formally.

Definition 1: Probability point function

The function $p(\cdot)$ defined for all real numbers of $S_X = \{x_1, x_2, \ldots\}$ by $p(x_j) = P_X(x_j)$ is called the probability point function for the random variable X.†

This shift from a function whose domain is a set of subsets (that is, set function) to a function whose domain is a discrete set of points (that is, point function) brings us to the threshold of the analytical approach to the study of probability. (However, the use of the analytical approach when the sample space is not discrete requires the calculus. We postpone this case to Chapter 9.)

† On occasion the symbol $p_X(\cdot)$ will be used to denote the probability point function for the random variable X, particularly when it is necessary to distinguish among probability point functions for different random variables.

If S_X is a finite sample space with n elements and $p(\cdot)$ is a probability point function defined on S_X, then $0 \leq p(x_j) \leq 1$ for all x_j in S_X, and the sum of the values of $p(\cdot)$ must be 1.   Conversely, any function f defined on the real numbers $y_1, y_2, \ldots, y_m$ and with the two properties

(a) $0 \leq f(y_k) \leq 1, \qquad k = 1, 2, \ldots, m$

(b) $\sum_{k=1}^{m} f(y_k) = 1$

can be considered a probability point function for the random variable Y, which takes on the values $y_1, y_2, \ldots, y_m$.

The function $p(\cdot)$ is sometimes called the *frequency distribution of X*, or the *probability function of the discrete random variable X*.   There are several ways of representing $p(\cdot)$.   One way is to form a table of x and $p(x)$ and thereby obtain a set of ordered pairs $(x_i, p(x_i))$.   A graphical representation is obtained easily by marking the points $(x_i, p(x_i))$ on a pair of coordinate axes. These n points are the graph of the probability point function.   Example 1 illustrates the new terminology.

Example 1

Let X be a random variable with $S_X = \{0, 1, 2, 3\}$.   The probability point function is specified in tabular form as

x_i	0	1	2	3
$p(x_i)$	$\frac{1}{8}$	$\frac{3}{8}$	$\frac{3}{8}$	$\frac{1}{8}$

and is shown graphically in Figure 1.   ▮

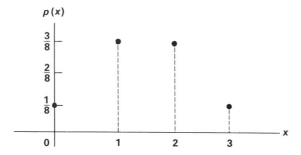

Figure 1

Sometimes it is convenient to extend the definition of $p(\cdot)$ to all real numbers by defining $p(x) = 0$ for those x not in S_X.   Thus if

$$S_X = \{x_1, x_2, \ldots, x_n\},$$

$$p(x) = \begin{cases} x_i, & x = x_i, \ i = 1, 2, \ldots, n, \\ 0, & \text{otherwise.} \end{cases}$$

If A is any event of S_X, the value of $P_X(A)$ can be written as

$$P_X(A) = \sum_{x_i \in A} p(x),$$

where the sum is taken over all points in A such that $p(x) > 0$. The probabilities of events such as $a \leq X \leq b$, where a and b are arbitrary real numbers, can be written in terms of $p(x)$ as

$$P_X(a \leq X \leq b) = \sum_{a \leq x \leq b} p(x),$$

where the sum is taken over all x in the interval such that $p(x) > 0$.

For a sample space S_X generated by a random variable X, a new function can be defined that will also serve as a probability function.

Definition 2: Probability distribution function

The function $F(\cdot)$ whose value for each real number x is given by

$$F(x) = P_X(X \leq x)$$

is called the probability distribution function for the random variable X.†

Each real number x defines an event, namely, the set of elementary events of S_X for which the values of the random variable are less than or equal to x. The event depends on x and the probability of the event depends on x, so we have a function that is defined for all values of x. This function $F(\cdot)$ has as its domain the set of all real numbers and as its range a set of real numbers between 0 and 1 inclusive. The probability distribution function can be specified by the probability point function as follows:

$$F(x) = \sum_{x_j \leq x} p(x_j),$$

where the sum is taken over all points $x_j \leq x$ such that $p(x_j) > 0$. The probability distribution function is often called the *cumulative distribution function*, because $F(x)$ is a sum that includes all probabilities that occur to the left of and including the point x.

Distribution functions have three important properties:

1. $F(x)$ is a nondecreasing function of x. That is, if $u < v$, where u and v are real numbers, then $F(u) \leq F(v)$. If the sample space is $S_X = \{x_1, x_2, \ldots, x_n\}$, where $x_1 < x_2 < \cdots < x_n$, then

† On occasion the symbol $F_X(\cdot)$ will be used to denote the probability distribution function for the random variable X, particularly when it is necessary to distinguish among probability distribution functions for different random variables.

2. $F(x) = 0$ for $x < x_1$, and
3. $F(x) = 1$ for $x \geq x_n$.

These properties are proved easily from the definition of $F(x)$:
1. Let

$$F(u) = \sum_{x_j \leq u} p(x_j),$$

$$F(v) = \sum_{x_j \leq v} p(x_j).$$

If $u < v$, then there are either no points of S_X between u and v, in which case $F(u) = F(v)$, or there is at least one point of S_X between u and v, in which case we add a nonnegative quantity $p(x_k)$ to $F(u)$, so $F(u) \leq F(v)$.

2. If $x < x_1$, then $x < x_j$ for all j; hence $F(x) = 0$.
3. If $x \geq x_n$, then $F(x) = \sum_{j=1}^{n} p(x_j)$, which is the sum over the sample space S_x; hence $F(x) = 1$.

Examples 2 and 3 illustrate these properties.

Example 2

(Refer to Example 1.)   Let the probability point function $p(\cdot)$ be defined for $S_X = \{0, 1, 2, 3\}$ as

$$p(x) = \begin{cases} \binom{3}{x}\left(\frac{1}{2}\right)^x \left(\frac{1}{2}\right)^{3-x}, & x = 0, 1, 2, 3, \\ 0, & \text{otherwise,} \end{cases}$$

That is, for $x = 0$, $p(0) = \frac{1}{8}$; for $x = 1$, $p(1) = 3 \cdot \frac{1}{2} \cdot (\frac{1}{2})^2 = \frac{3}{8}$; and so forth for the other values of $p(x)$ in Table 1 of Example 1.

For the probability distribution function $F(x)$ we have

$$F(x) = \begin{cases} 0, & x < 0, \\ \frac{1}{8}, & 0 \leq x < 1, \\ \frac{4}{8}, & 1 \leq x < 2, \\ \frac{7}{8}, & 2 \leq x < 3 \\ 1, & 3 \leq x. \end{cases}$$

The graph of $F(x)$, given in Figure 2, suggests how $F(\cdot)$ has been defined for all real numbers.   ▌

For a discrete random variable, $F(\cdot)$ will always have a graph that has the appearance of the one in Figure 2.   Such a function is called a *step function*. The steps (or jumps) occur only at those points x_j where $p(\cdot) > 0$, and the magnitude of the jump is precisely the value of $p(\cdot)$.   Thus we can reconstruct the probability point function from the probability distribution function.

The probability distribution function immediately provides the probability of such events as $\{X \leq a\}$, while the probability point function provides the probability of such events as $\{X = b\}$, where a and b are arbitrary real numbers. Consequently, for any events that can be written as set combinations of these particular sets, we can easily find expressions for the probability.

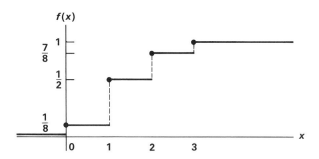

Figure 2

Theorem 1

(a) $P_X(a < X \le b) = F(b) - F(a)$.

(b) $P_X(a \le X \le b) = F(b) - F(a) + p(a)$.

PROOF: The event

$$\{a < X \le b\} = \{X \le b\} - \{X \le a\}.$$

so that

$$\{X \le b\} = \{X \le a\} \cup \{a < X \le b\}.$$

It follows from axiom PA3 (Section 2.2) that

$$P_X(X \le b) = P_X(X \le a) + P_X(a < X \le b).$$

By the definition of the distribution function,

$$F(b) = F(a) + P_X(a < X \le b),$$

and part (a) is proved.

Since $\{a \le X \le b\} = \{X = a\} \cup \{a < X \le b\}$, part (b) follows immediately. ∎

Example 3

In a large population of insects, exactly 40 percent are infected by a given virus W. Samples of five insects are selected at random and each insect is examined for the presence of the virus. The population is so large that the sampling may be considered to be with replacement.

If the random variable X denotes the number of infected insects in the sample, X is a binomial random variable with the parameters $n = 5$ and $p = 0.4$, where a success is the occurrence of an infected insect. Therefore, $S_X = \{0, 1, 2, 3, 4, 5\}$, and

$$P_X(k) = \binom{5}{k}(0.4)^k(0.6)^{5-k}, \qquad k = 0, 1, 2, 3, 4, 5,$$

as was seen in Section 4.5. The probability point function $p(\cdot)$ is

$$p(x) = \begin{cases} \binom{5}{x}(0.4)^x(0.6)^{5-x}, & x = 0, 1, 2, 3, 4, 5, \\ 0, & \text{otherwise.} \end{cases}$$

Explicitly,

$$p(0) = \binom{5}{0}(0.4)^0(0.6)^5 = 0.0778,$$

$$p(1) = \binom{5}{1}(0.4)^1(0.6)^4 = 0.2592,$$

$$p(2) = \binom{5}{2}(0.4)^2(0.6)^3 = 0.3456,$$

$$p(3) = \binom{5}{3}(0.4)^3(0.6)^2 = 0.2304,$$

$$p(4) = \binom{5}{4}(0.4)^4(0.6)^1 = 0.0768,$$

$$p(5) = \binom{5}{5}(0.4)^5(0.6)^0 = 0.0102.$$

These numbers describe the theoretical probability to be attached to the outcomes in the process of selecting five insects from a population with 40 percent infected and ascertaining the number in the sample that are infected. A graphical representation of $p(\cdot)$ is given in Figure 3.

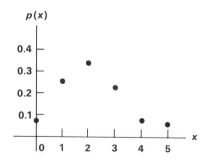

Figure 3

For the probability distribution function $F(x)$,

$$F(x) = \begin{cases} 0, & x < 0, \\ 0.0778, & 0 \le x < 1, \\ 0.3370, & 1 \le x < 2, \\ 0.6826, & 2 \le x < 3, \\ 0.9130, & 3 \le x < 4, \\ 0.9898, & 4 \le x < 5, \\ 1.0000, & 5 \le x. \end{cases}$$

The graph of $F(x)$ is given in Figure 4. ∎

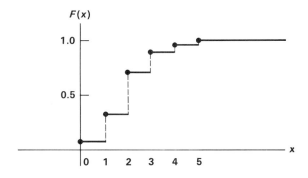

Figure 4

Example 4

If X is the number of successes in n Bernoulli trials with p the probability for success on one trial, then $S_X = \{0, 1, 2, \ldots, n\}$. From Section 4.5 we have

$$(1) \qquad P_X(k) = \binom{n}{k} p^k q^{n-k}, \qquad k = 0, 1, 2, \ldots, n.$$

The probability point function can be written

$$(2) \qquad p(x) = \begin{cases} \binom{n}{x} p^x q^{n-x}, & x = 0, 1, 2, \ldots, n, \\ 0, & \text{otherwise.} \end{cases}$$

The probability distribution function is given by

$$(3) \qquad F(x) = \sum_{k=0}^{[x]} \binom{n}{k} p^k q^{n-k},$$

where $[x]$ denotes the greatest integer less than or equal to x. For $k > n \ \binom{n}{k}$ is defined to be 0. This summarizes the different ways in which the binomial function can be specified.   ∎

Problems

In Problems 1–5 find (a) the probability point function $p(\cdot)$, and (b) the probability distribution function for the random variable X, and (c) sketch the graph of $F(\cdot)$.

1. Let X be the random variable whose values are the number of heads obtained in four tosses of a fair coin.

2. Let X be the random variable whose values are the number of spots on the uppermost face in the toss of a fair die.

3. Let X be the total number of spots on the uppermost faces in the toss of a pair of dice.

4. Let X be the number of 3's or 4's that occur in four independent tosses of a fair die.

5. Let X be the number of defectives found in a sample of size 3 selected without replacement from a lot of eight items, three of which are defective.

6. A random variable X has a probability point function of the following form:

$$p(x) = \begin{cases} 2k, & x = 0, \\ k, & x = 1, \\ 3k, & x = 2, \\ 0, & \text{otherwise.} \end{cases}$$

(a) Determine the value of k.
(b) Find the probability distribution function $F(\cdot)$.
(c) Find $P_X(X \leq 2)$.
(d) Find $P_X(0 < X < 2)$.

7. The probability of at least one accident on a Boy Scout campout is 0.2 for each troop. Six troops are out, and let X denote the number of troops for which at least one accident occurs. Assume that the accidents are independent and write an expression for the probability point function of X.

8. If the probability of a marriage ending in divorce within 20 years is $\frac{1}{3}$, let X denote the number of divorces in a group of five couples selected at random. Assume that the population of married couples is large enough so that an assumption of sampling with replacement is reasonable. Write an expression for the probability point function of X.

9. A random variable X has a probability point function of the following form:

$$p(x) = \begin{cases} 0.1, & x = 0, \\ kx, & x = 1 \text{ or } 2, \\ k(5 - x), & x = 3 \text{ or } 4, \\ 0, & \text{otherwise.} \end{cases}$$

(a) Determine the value of k.
(b) Find the probability distribution function $F(\cdot)$.
(c) Find $P_X(X \geq 2)$.
(d) Find $P_X(X \leq 2)$.

10. Suppose that there are two white and four red balls in urn I, eight white and four red balls in urn II, and one white and three red balls in urn III, and one ball is drawn from each urn. Let X be the number of white balls in the sample. Find $p(\cdot)$ and $F(\cdot)$ for X.

11. Let X be the number of divisors of an integer selected at random from the integers 1 through 20. Find $p(\cdot)$ and $F(\cdot)$ for X.

12. A city council consists of nine members, four of party A, three of party B, and two of party C. Each week one member is selected by lot to preside over the council meetings. Let X denote the number of times in two consecutive weeks that the council member chosen is from party B. Find the probability point function and the probability distribution function for X.

13. Let X be a random variable with a probability point function defined by

x	-3	-1	0	1	2	3	6	7	8
$p(x)$	0.05	0.15	0.2	0.1	0.05	0.05	0.1	0.15	0.15

(a) Find $P_X(X \leq 0)$. (b) Find $P_X(1 \leq X \leq 8)$.
(c) Find the probability that X is even.
(d) Find $P_X(X > 2)$.

14. Let X be a random variable with a probability distribution function defined by

$$F(x) = \begin{cases} 0, & x < -1, \\ 0.1, & -1 \leq x < 0, \\ 0.3, & 0 \leq x < 0.5, \\ 0.4, & 0.5 \leq x < 1.5, \\ 0.6, & 1.5 \leq x < 2, \\ 0.65, & 2 \leq x < 3, \\ 1.0, & 3 \leq x. \end{cases}$$

(a) Find $P_X(\frac{1}{4} \leq X \leq 2.5)$.   (b) Find $P_X(X \geq 1.3)$.

(c) Find $P_X(X \leq \frac{1}{10})$.   (d) Find $p(\cdot)$.

15. Four cards are selected without replacement from an ordinary deck of playing cards. Let X be the number of aces drawn and Y the number of spades drawn. Find the probability point function for each random variable.

In Problems 16–18 show that each of the functions qualifies as a probability point function and sketch its graph.

16.
$$p(x) = \begin{cases} \frac{2}{3}(\frac{1}{3})^{x-1}, & x = 1, 2, 3, 4, \\ (\frac{1}{3})^4, & x = 0, \\ 0, & \text{otherwise.} \end{cases}$$

17.
$$p(x) = \begin{cases} \frac{1}{100}, & x = 1, 2, \ldots, 100, \\ 0, & \text{otherwise.} \end{cases}$$

18.
$$p(x) = \begin{cases} \dfrac{\binom{8}{x}\binom{4}{6-x}\binom{12}{6}}{}, & x = 2, 3, \ldots, 6, \\ \\ 0, & \text{otherwise.} \end{cases}$$

19. The number of books checked out of a public library in an afternoon can be represented by a random variable X with the probability point function
$$p(x) = \begin{cases} kx, & x = 1, 2, \ldots, 50, \\ k(100 - x), & x = 51, \ldots, 100, \\ 0, & \text{otherwise.} \end{cases}$$

(a) Find the value of k.   (b) Find $P_X(X > 50)$.

(c) Find $P_X(40 \leq X \leq 60)$.

(d) Denote the events in (b) and (c) by A and B, respectively. Find $P_X(B \mid A)$.

20. The number of days in succession that the probability professor gets to class on time can be represented by a random variable X with the probability point function
$$p(x) = \begin{cases} (\frac{1}{3})^5, & x = 0, \\ k(\frac{1}{3})^{x-1}, & x = 1, 2, \ldots, 5, \\ 0, & \text{otherwise.} \end{cases}$$

(a) Find the value of k.   (b) Find $P_X(X \geq 2)$.

(c) Find $P_X(1 \leq X \leq 3)$.

(d) Denote the events in (b) and (c) by A and B, respectively. Find $P_X(B \mid A)$.

chapter six

Mathematical Expectation and Variance

6.1 Concepts of "Average"

The word *average* implies the existence of a number that in some sense represents a set of quantities. Sometimes both the average and the set of numbers are only hypothetical, assigned in an imprecise way to "measure" qualities in another set. Women's magazines, for example, attempt to characterize husbands by listing a set of qualities graded on a numerical scale. If the reader ticks off a total score that lies in a certain range, the husband is called "average." The stock broker who refers to the Dow–Jones stock averages as a measure of the economy for that day, and the weatherman who convinces a farmer that a particular month was not lacking in moisture because the average daily rainfall was the same as in previous years, are using the word average according to Definition 1.

In this section we shall discuss some of the ways in which average can be defined. We start with the most familiar one, the arithmetic average of a set of n numbers, $\{x_1, x_2, \ldots, x_n\}$. Since it may happen easily that two or more of the numbers in the set of numbers are the same, we shall think of the n numbers as the results of n distinct observations or trials. In this way, finding repeated numbers in a set will not contradict our convention that the members of a set are unique. From here on, therefore, the term "set of numbers" is to be taken in this way. Thus we shall think of the numbers $\{77, 81, 77\}$ as the values for the set of observations $\{x_1, x_2, x_3\}$.

Definition 1: Arithmetic average

Let $x_1, x_2, \ldots, x_n$ be a set of numbers. The arithmetic average, denoted by $\bar{x}$, is the number

(1)
$$\bar{x} = \left(\sum_{i=1}^{n} x_i \right) / n.$$

The arithmetic average is also called the *arithmetic mean*.

Example 1

On five rounds of golf at a particular course (or on five successive math examinations—it does not matter which), a student had scores of 77, 80, 79, 89, and 85. The arithmetic average of the set of numbers is $(77 + 80 + 79 + 89 + 85)/5 = 82$, an excellent golf score but only a good math score. ∎

Whenever some of the numbers in a set of numbers occur repeatedly, the arithmetic average can be calculated by first determining the frequency with which each distinct number occurs. Then instead of n numbers $x_1, x_2, \ldots, x_n$, we obtain k pairs of numbers (x'_i, n_i), where n_i denotes the number of times x'_i appears. By rearranging the expression for $\bar{x}$, we can now write (1) as

(2)
$$\bar{x} = \left(\sum_{i=1}^{k} x'_i n_i \right) / n = \frac{x'_i n_i + \cdots + x'_k n_k}{n}.$$

The numbers $n_1/n, n_2/n, \ldots, n_k/n$ resemble the numbers that appear in the relative frequency interpretation of probability. The integer n_i is the actual number of times that the number x_i occurs and, n_i/n represents the frequency of occurrence relative to the observed set of numbers. By defining the function

$$f(x'_i) = n_i/n, \qquad i = 1, 2, \ldots, k,$$

the expression for $\bar{x}$ can be written as

(3)
$$\bar{x} = \sum_{i=1}^{k} x'_i f(x'_i).$$

Example 2

The scores on an examination for a class of 20 students are

$$100, \quad 100, \quad 95, \quad 90, \quad 90, \quad 90, \quad 90, \quad 85, \quad 85, \quad 85,$$
$$80, \quad 80, \quad 80, \quad 80, \quad 75, \quad 75, \quad 75, \quad 75, \quad 70, \quad 65.$$

The average for these scores is the sum of scores divided by 20. Thus $\bar{x} = 1600/20 = 80$.

Using the notation and procedure described in the paragraph preceding this example, we can summarize the same scores as follows:

x_i'	100	95	90	85	80	75	70	65
n_i	2	1	4	3	4	4	1	1
$f(x_i')$	$\frac{2}{20}$	$\frac{1}{20}$	$\frac{4}{20}$	$\frac{3}{20}$	$\frac{4}{20}$	$\frac{4}{20}$	$\frac{1}{20}$	$\frac{1}{20}$

Thus

$$\bar{x} = \sum_{i=1}^{8} x_i' f(x_i')$$
$$= 100(\tfrac{2}{20}) + 95(\tfrac{1}{20}) + 90(\tfrac{4}{20}) + 85(\tfrac{3}{20}) + 80(\tfrac{4}{20})$$
$$+ 75(\tfrac{4}{20}) + 70(\tfrac{1}{20}) + 65(\tfrac{1}{20}) = 80. \quad \blacksquare$$

Statisticians say that the arithmetic mean is a measure of *central tendency*; that is, the observed values of an experiment that is repeated many times tend to cluster near the mean. In another sense the *mean* is that number which identifies the *center* of the observed values.

Another measure of the center for a set of numbers is the *median*, which can be roughly defined as that number which has an equal number of numbers above it and below it. This definition is not appropriate for the countably infinite or continuous case, so we shall define it again later. The median denotes a position in a sequence of values, and in this sense it is a positional average.

An additional measure of central tendency for a set of numbers is the *mode*, which is that value, if any, that occurs most frequently. In terms of the relative frequency of occurrence, the mode is that value for which $f(x_i')$ is the largest or, what is the same thing, the most probable observed value. In this sense the mode is a probability average.

Example 3

   (*a*) The median of the set of golf scores 80, 81, 81, 82, and 86 is 81 and the mode is also 81. The mean is 82.

   (*b*) The median of the set of scores 79, 80, 80, 81, 81, and 81 is 80.5 and the mode is 81, while $\bar{x} = 80\frac{1}{3}$.

   (*c*) If the golf score of 96 is included in the scores of part (*a*), the mode is still 81, the median becomes 81.5, and the mean is now 84. $\blacksquare$

Although the median and the mode give some idea about the location of the midpoint of the observed values, and are somewhat similar to the arithmetic mean, in practice the median and mode are not used as much. One reason is that there is no simple rule for calculating either the median or the mode as has been given for the arithmetic mean.

Unfortunately, the mean, median, and mode do not completely summarize a set of observed values. The golf scores 75, 81, 81, 86, and 87 have a *mean* of 82, a *mode* of 81, and a *median* of 81. These latter values coincide exactly with those for the set of numbers in part (*a*) of Example 3. Therefore, additional characteristics are needed to summarize data of this sort.

Although the mean, median, and mode are useful measures of location, the way in which the set of numbers is spread, dispersed, or distributed about one of them characterizes the set more fully. The arithmetic mean is particularly important and is used in defining two such averages, the mean-absolute dispersion and the mean-square dispersion, defined as follows:

Definition 2

For a set of numbers

$$x_1, x_2, \ldots, x_n,$$

(4) $$mean\text{-}absolute\ dispersion = \frac{1}{n} \sum_{i=1}^{n} |x_i - \bar{x}|;$$

(5) $$mean\text{-}square\ dispersion = \frac{1}{n} \sum_{i=1}^{n} (x_i - \bar{x})^2.$$

In terms of the relative frequency notation, the expressions in (4) and (5) can be written as

(6) $$mean\text{-}absolute\ dispersion = \sum_{i=1}^{k} |x_i' - \bar{x}| f(x_i');$$

(7) $$mean\text{-}square\ dispersion = \sum_{i=1}^{k} (x_i' - \bar{x})^2 f(x_i').$$

Example 4

Twenty-one measurements were taken in a biological laboratory with the results as shown in Table 1. Thus $\bar{x} = 62.4/21 = 2.97$. Table 2 can be computed easily. If the original data are represented graphically the meaning of these two numbers is more evident. One useful type of graph is a set of contiguous columns whose heights are proportional to the frequencies of the observations. It is called a *histogram*, and the histogram for these data is shown in Figure 1. If the intervals of length 0.144 are marked off from $\bar{x} = 2.97$, the significance of the number 0.144 as a measure of dispersion can be seen. ∎

Table 1

x_i'	n_i'	$f(x_i')$	$x_i' f(x_i')$
2.6	1	1/21	2.6/21
2.7	1	1/21	2.7/21
2.8	3	3/21	8.4/21
2.9	5	5/21	14.5/21
3.0	5	5/21	15.0/21
3.1	3	3/21	9.3/21
3.2	1	1/21	3.2/21
3.3	1	1/21	3.3/21
3.4	1	1/21	3.4/21

Table 2

x_i'	$|x_i' - \bar{x}|$	$(x_i' - \bar{x})^2$	$|x_i' - \bar{x}| f(x_i')$	$(x_i' - \bar{x})^2 f(x_i')$
2.6	0.37	0.1369	0.37/21	0.1369/21
2.7	0.27	0.0729	0.27/21	0.0729/21
2.8	0.17	0.0289	0.51/21	0.0867/21
2.9	0.07	0.0049	0.35/21	0.0245/21
3.0	0.03	0.0009	0.15/21	0.0045/21
3.1	0.13	0.0169	0.39/21	0.0507/21
3.2	0.23	0.0529	0.23/21	0.0529/21
3.3	0.33	0.1089	0.33/21	0.1089/21
3.4	0.43	0.1849	0.43/21	0.1849/21

mean-absolute dispersion $= 3.03/21 = 0.144$,
mean-square dispersion $= 0.7229/21 = 0.034$

Since both measures are the sums of positive numbers, namely, the distance and the square of the distance of an observed value from the mean, the higher the value of the measure, the more widely the observed values are dispersed about the mean.

Evidently a set of numbers can be characterized by many different kinds of averages. Our choice of average will usually be influenced by what we intend to do with the average.

The mean, mean-absolute dispersion, and mean-square dispersion (but not the median and mode) are obtained by computing certain expressions. We now extend the idea of the averages of a set of numbers to obtain the average of a function with respect to a set of numbers.

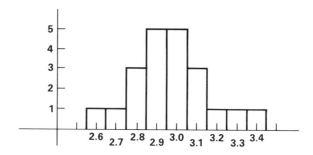

Figure 1

By taking particular functions for $g(x)$, equations (1), (4), and (5) become special cases of Definition 3. For example, if $g(x) = x$, we obtain the arithmetic mean. If $g(x) = (x - \bar{x})^2$, we obtain the mean-square dispersion, while $g(x) = |x - \bar{x}|$ will give the mean-absolute dispersion.

As we have seen, an average not only provides a characterization of a set of numbers, but also summarizes in some ways the data that are represented by the numbers.

Problems

In problems 1–4 find the average with respect to the data given for the functions (a) $g(x) = x$; (b) $g(x) = (x - \bar{x})^2$, where $\bar{x}$ is the answer obtained for part (a); (c) $g(x) = x - \bar{x}$; and (d) $g(x) = |x - \bar{x}|$.

1. Three fair coins were tossed 10 times and the numbers of heads recorded for each toss were

$$2, \quad 2, \quad 2, \quad 0, \quad 2, \quad 2, \quad 1, \quad 3, \quad 2, \quad 2.$$

2. The golf scores for 10 rounds of 18 holes were

$$84, \ 84, \ 86, \ 87, \ 92, \ 87, \ 95, \ 84, \ 83, \ 85.$$

3. The numbers of mice in 20 litters in a breeding experiment were

$$7, \ 3, \ 4, \ 8, \ 6, \ 5, \ 2, \ 6, \ 4, \ 6,$$
$$3, \ 9, \ 8, \ 1, \ 3, \ 6, \ 5, \ 4, \ 4, \ 5.$$

4. The results of 30 throws of a pair of dice were

$$8, \ 3, \ 12, \ 10, \ 10, \ 6, \ 4, \ 8, \ 9, \ 7, \ 7, \ 11, \ 8, \ 7, \ 9,$$
$$8, \ 9, \ 6, \ 8, \ 6, \ 6, \ 12, \ 6, \ 8, \ 7, \ 5, \ 5, \ 4, \ 6, \ 9.$$

5. On an examination in a mathematics course the following grades were made:

$$96, \ 68, \ 86, \ 89, \ 83, \ 79, \ 75, \ 78, \ 86, \ 72,$$
$$86, \ 85, \ 94, \ 50, \ 72, \ 98, \ 90, \ 78, \ 86, \ 98.$$

Find the mean, median, and mode of these grades.

6. An urn contains 10 balls numbered $1, 2, \ldots, 10$. The results of 20 draws with replacement are as follows:

$$8, \ 6, \ 3, \ 8, \ 5, \ 5, \ 9, \ 9, \ 3, \ 1,$$
$$5, \ 1, \ 10, \ 3, \ 8, \ 8, \ 2, \ 8, \ 3, \ 4.$$

Find the mean, median, and mode of these numbers.

7. The *harmonic mean* is the reciprocal of the arithmetic mean of the reciprocals of the numbers in the set. If an airplane flies the sides of a square whose side is 100 miles long, taking the first side at 100 mph, the second at 200 mph, the third at 300 mph, and the fourth at 400 mph, find the harmonic mean of the speeds, and compare the results with the arithmetic mean of the speeds.

8. The *geometric mean* of the set $x_1, x_2, \ldots, x_n$ is defined to be $\sqrt[n]{x_1 \cdot x_2 \cdot \ldots \cdot x_n}$. Find the arithmetic and geometric mean of

(a) 2, 4, 8   (b) 3, 4, 5, 6, 7, 8
(c) 2, 8   (d) 5, 5, 5, 10, 10, 10

9. Prove that the logarithm of the geometric mean of a set of numbers is the arithmetic mean of the logarithms of the numbers.

10. The family size of 100 families was as follows:

Size	1	2	3	4	5	6	7	8	9	10
Number	20	24	17	15	9	5	4	3	1	2

Find the mean, median, mode, and mean-square dispersion of these numbers.

11. The sum of the deviations from the mean is formula (4) without the absolute values. Show that the sum of the deviations from the mean is zero.

12. Use a computer to compute the mean, mean-absolute dispersion, and mean-square dispersion of the following data, the murder and nonnegligent manslaughter rate per 100,000 inhabitants in selected standard metropolitan stastistical areas.

Abilene, Tex.	5.3	Lincoln, Neb.	1.2
Albany, Ga.	1.7	Manchester, N.H.	1.5
Altoona, Pa.	1.4	Miami, Fla.	11.3
Ann Arbor, Mich.	1.9	Nashville, Tenn.	12.2
Atlantic City, N.J.	6.6	Ogden, Utah	4.7
Bay City, Mich.	0.9	Pensacola, Fla.	3.8
Boise, Idaho	1.0	Peoria, Ill.	2.8
Brockton, Mass.	3.0	Pueblo, Col.	3.2
Cedar Rapids, Ia.	0.0	Racine, Wis.	1.2
Champaign, Ill.	6.7	Reading, Pa.	3.4
Charlotte, N.C.	17.0	Rockford, Ill.	5.9
Chicago, Ill.	9.5	Salem, Ore.	2.2
Columbus, Ohio	9.3	San Antonio, Tex.	10.1
Decatur, Ill.	4.7	San Jose, Cal.	2.5
Eugene, Ore.	1.5	Savannah, Ga.	10.5
Evansville, Ind.	4.4	Shreveport, La.	7.6
Fargo, N.D.	0.0	South Bend, Ind.	2.8
Fort Smith, Ark.	3.6	Springfield, Ill.	5.7
Grand Rapids, Mich.	3.9	Tacoma, Wash.	3.8
Harrisburg, Pa.	3.8	Terre Haute, Ind.	2.4
Honolulu, Ha.	2.8	Topeka, Kan.	2.6
Jackson, Mich.	4.2	Utica, N.Y.	1.1
Johnstown, Pa.	2.2	Waterloo, Ia.	0.8
Lansing, Mich.	2.6	Wichita, Kan.	5.1
Lawton, Okla.	8.2	York, Pa.	1.3

SOURCE: *Crime in the U.S.—Uniform Crime Reports—*1967, Department of Justice, Federal Bureau of Investigation, Washington, D.C.

6.2 Mathematical Expectation of a Random Variable

Two players agree to play the following game. Player A pays player B $1 and then rolls a fair die. If the uppermost face of the die is a 6, player B pays player A $5, otherwise player A receives nothing. The net exchange for player A on any play of the game is either $+\$4$ or $-\$1$. Suppose that the game is played 20 times with the following outcomes:

$$
\begin{array}{ccccc}
-\$1, & -\$1, & -\$1, & -\$1, & -\$1, \\
+\$4, & -\$1, & -\$1, & -\$1, & -\$1, \\
-\$1, & -\$1, & -\$1, & -\$1, & -\$1, \\
-\$1, & -\$1, & -\$1, & +\$4, & +\$4.
\end{array}
$$

The average net exchange is $[-17 + 3(4)]/20 = -\$0.25$.

On any one toss player A can expect to exchange $+\$4$ with probability $\frac{1}{6}$, and to exchange $-\$1$ with probability $\frac{5}{6}$, so he can expect an exchange of

$$
\tfrac{1}{6}\$4 - \tfrac{5}{6}\$1 = -\$0.17.
$$

The amount $-\$0.25$ is an actual average exchange for 20 plays of the game, while the amount $-\$0.17$ is a theoretical expected exchange. The notion of expected values for games with monetary outcomes was one of the earliest developments in the history of probability theory. It has led to the definition of a number that can be associated with a random variable and that gives information about the probability function of the random variable. This number helps to characterize the probability function much as the slope of a line helps to characterize the line, or the center of a circle helps to characterize the entire circle. (Neither the slope of a line nor the center of a circle tells everything about the line or the circle.)

Although the definition will be made only for random variables with a finite set of values, it carries over to a random variable with a countably infinite set of values. However, here again we need to know about infinite sums (see [18, Chap. 8]). The definition for the countably infinite case is stated in Chapter 8.

Definition 1: Mathematical expectation

Let X be a random variable with possible values $S_X = \{x_1, x_2, \ldots, x_n\}$ and a probability point function $p(\cdot)$. The mathematical expectation of X, denoted by $E(X)$, is defined to be

(1)
$$
E(X) = \sum_{i=1}^{n} x_i p(x_i).
$$

Example 1

Let X be a random variable with a set of values S_X and a probability point function $p(\cdot)$ as given in

x_i	0	1	2	3
$p(x_i)$	0.2	0.1	0.4	0.3

The mathematical expectation according to (1) is

$$E(X) = \sum_{i=1}^{4} x_i p(x_i)$$

$$= 0(0.2) + 1(0.1) + 2(0.4) + 3(0.3)$$

$$= 1.8.$$

The practical interpretation of the number 1.8 will be apparent after the next example. ∎

Example 2

If Y denotes the number of heads uppermost on three tosses of a fair coin, then from the sample space of tossing three coins we find that (see Example 1 of Section 5.2)

y_i	0	1	2	3
$p(y_i)$	$\frac{1}{8}$	$\frac{3}{8}$	$\frac{3}{8}$	$\frac{1}{8}$

The mathematical expectation is

$$E(Y) = \sum_{i=1}^{4} y_i p(y_i)$$

$$= 0(\tfrac{1}{8}) + 1(\tfrac{3}{8}) + 3(\tfrac{3}{8}) + 3(\tfrac{1}{8})$$

$$= \tfrac{3}{2} \text{ or } 1.5. \quad ∎$$

Many names are used for the number $E(X)$; *mean value of X*, the *mean of X*, or *expected value* of X are among the more common ones. Sometimes the adjective "mathematical" is omitted and $E(X)$ is referred to as the expectation of X. We shall use either the phrase "expected value" or "expectation."

We shall use the symbol μ when we refer to the expected value of a random variable with a given probability distribution, and the symbol $E(X)$ when it is necessary to refer to the expected value of a random variable. Often the symbols μ_x, $E(x)$, and μ are used interchangeably.

Although there is some justification for using the phrase "the expected value of X," the value of $E(X)$ need not be an outcome of X, and usually will not be. Expected value is a particularly apt term in the context of games of chance, for the expected value is an indication of how much should be set as an entrance fee in order to participate in a game. If $E(X) = 0$, the game is called "fair," and if $E(X)$ is positive, then that amount might be asked of a player as an entrance fee in order to make the game fair.

If expected value is used in the sense of average, then $E(X)$ does represent the average value we should expect if the experiment were conducted over a long sequence of trials. The arithmetic average of the sequence of outcomes and $E(X)$ are close together, which in some sense explains the connection between mathematical expectation and the notion of the average of a set of numbers. Example 3 illustrates the relationship.

Example 3

If Y denotes the number of heads that occur on three tosses of a fair coin, then $E(Y) = 1.5$, as we calculated in Example 2. Suppose that we throw three coins 20 times and get the following results for the number of heads:

$$2, \quad 0, \quad 2, \quad 1, \quad 3, \quad 0, \quad 3, \quad 1, \quad 0, \quad 2,$$
$$1, \quad 1, \quad 2, \quad 2, \quad 1, \quad 2, \quad 1, \quad 2, \quad 2, \quad 1.$$

Here is a set of 20 numbers for which the arithmetic average can be calculated from Table 1, using the technique of Section 6.1.

Table 1

x_i'	n_i'	$f(x_i')$
0	3	$\frac{3}{20}$
1	7	$\frac{7}{20}$
2	8	$\frac{8}{20}$
3	2	$\frac{2}{20}$

$$\bar{x} = \sum_{i=1}^{4} x_i' f(x_i')$$
$$= 0(\tfrac{3}{20}) + 1(\tfrac{7}{20}) + 2(\tfrac{8}{20}) + 3(\tfrac{2}{20}) = 1.45$$

Let us repeat the entire procedure again; that is, we throw the three coins 20 more times. Suppose that this time we get the outcomes

$$2, \quad 1, \quad 3, \quad 1, \quad 1, \quad 3, \quad 2, \quad 1, \quad 2, \quad 2,$$
$$0, \quad 1, \quad 2, \quad 2, \quad 2, \quad 0, \quad 1, \quad 2, \quad 1, \quad 1.$$

We obtain the results listed in Table 2.

Table 2

x_i'	n_i'	$f(x_i')$
0	2	$\frac{3}{20}$
1	8	$\frac{7}{20}$
2	8	$\frac{8}{20}$
3	2	$\frac{2}{20}$

$$\bar{x} = \sum_{i=1}^{4} x_i' f(x_i')$$
$$= 0(\tfrac{3}{20}) + 1(\tfrac{7}{20}) + 2(\tfrac{8}{20}) + 3(\tfrac{2}{20}) = 1.5 \quad \blacksquare$$

In the first case of Example 1 the arithmetic average is not the same as the expected value of X, whereas it is in the second case. In general, the value of $\bar{x}$ will *not* coincide with $E(X)$, but as more and more trials are used, the probability that $\bar{x}$ will be close to $E(X)$ increases. This idea parallels the notion that the relative frequency of the occurrence of an event A in a sequence of trials closely approximates the true (or theoretical) value of $P(A)$. In Chapter 7 we shall show that $E(X)$ and $\bar{x}$ are closely connected.

The mathematical expectation can be calculated explicitly for random variables with standard probability functions.

Theorem 1

If X is a random variable $S_X = \{x_i, \ldots, x_n\}$ and $p(x_i) = 1/n$ is the uniform probability function, then

$$E(X) = \left(\sum_{i=1}^{n} x_i \right) \bigg/ n.$$

PROOF: Since $S_X = \{x_1, x_2, \ldots, x_n\}$ and $p(x_i) = 1/n$ for $i = 1, 2, \ldots, n$, then

$$E(X) = \sum_{i=1}^{n} x_i p(x_i)$$

$$= \sum_{i=1}^{n} x_i \frac{1}{n} = \frac{1}{n} \sum_{i=1}^{n} x_i. \quad \blacksquare$$

This shows that for the random variable with a uniform probability function, the mathematical expectation is exactly the arithmetic mean of the values of the random variable X.

Theorem 2

If X is a Bernoulli random variable $S_X = \{0, 1\}$ and $p(1) = p$, then $E(X) = p$.

PROOF: From the definition of $E(X)$ we have

$$E(X) = \sum_{i=0}^{1} x_i p(x_i)$$

$$= 0 \cdot p(0) + 1 \cdot p(1) = p. \quad \blacksquare$$

Theorem 3

If X is a binomial random variable for n trials with $S_X = \{0, 1, 2, \ldots, n\}$, then $E(X) = np$, where p denotes the probability of success on a trial.

PROOF: Since $S_X = \{0, 1, \ldots, n\}$ and $p(k) = \binom{n}{k} p^k q^{n-k}$ for $k = 0, 1, \ldots, n$, a direct application of $E(X)$ gives us

$$E(X) = \sum_{k=0}^{n} k p(k) = \sum_{k=0}^{n} k \binom{n}{k} p^k q^{n-k}$$

$$= \sum_{k=0}^{n} k \frac{n!}{k!(n-k)!} p^k q^{n-k}.$$

Cancelling the k and omitting the term for $k = 0$, we get

$$(2) \qquad E(X) = \sum_{k=1}^{n} \frac{n!}{(k-1)!(n-k)!} p^k q^{n-k}.$$

Every term in (2) contains p ad n as a factor; therefore, we can write

$$E(X) = np \sum_{k=1}^{n} \frac{(n-1)!}{(k-1)!(n-k)!} p^{k-1} q^{n-k}.$$

We now show that the last sum has the value 1. Observe that $n - k = (n-1) - (k-1)$ and that

$$\frac{(n-1)!}{(k-1)!(n-k)!} = \binom{n-1}{k-1};$$

hence

$$(3) \qquad E(X) = np \sum_{k=1}^{n} \binom{n-1}{k-1} p^{(k-1)} q^{(n-1)-(k-1)},$$

$$= np \sum_{j=0}^{n-1} \binom{n-1}{j} p^j q^{(n-1)-j},$$

where we have changed the index of summation from k to j by putting $k - 1 = j$. Since k runs from 1 to n, j runs from 0 to $n - 1$. The last sum is precisely the sum of the probabilities for j successes, $j = 0, 1, \ldots, n - 1$, in $n - 1$ trials with the probability of success on a trial being p. This sum necessarily has the value 1. Consequently,

$$E(X) = np. \quad \blacksquare$$

We point out that the sum in (3) can also be evaluated easily using the binomial theorem. The expansion of $(p + q)^{n-1}$ is indeed the sum in (3) and, since $p + q = 1$, the result is immediate.

Example 4

Let X be the number of 5's or 6's in 10 tosses of a die. Thus X is a binomial random variable with the values of the parameters $n = 10$ and $p = \frac{1}{3}$. By the result of Theorem 3, $E(X) = 10 \cdot \frac{1}{3} = \frac{10}{3}$. Therefore, the expected number of 5's or 6's in 10 tosses is $\frac{10}{3}$. ∎

At the beginning of this section we implied that the expectation of a random variable could be used to characterize the probability function. Let us see how we might use the expectation as a way of comparing two probability functions under the following circumstances. Suppose that the set of values for two random variables X and Y are identical, that is, $S_X = S_Y$, but that the probabilty point functions are different. Example 5 will indicate a way to proceed.

Example 5

Suppose that two marksmen fire at a target where three different outcomes are recorded; three points are awarded for a center hit, two points for a hit in the outer ring, and one point for a miss. We can let X be the random variable equal to the score of the first marksman, and Y the corresponding random variable for the second. Further, let us suppose that we can estimate probabilities for each random variable and that the estimates give the probability point functions listed in Table 3. Which of the two marksmen is the better?

Table 3

x_i	$p_X(x)$	$p_Y(y_i)$
3	0.5	0.3
2	0.1	0.6
1	0.4	0.1

The first marksman is more likely to hit the center than the second; but the first is also more likely to miss altogether, whereas the second marksman will usually hit the outer ring.

Let us calculate the expected value for each random variable and compare the numbers.

$$E(X) = 3(0.5) + 2(0.1) + 1(0.4) = 2.1,$$

$$E(Y) = 3(0.3) + 2(0.6) + 1(0.1) = 2.2.$$

Because $E(Y)$ is slightly larger than $E(X)$, we can assert with only some confidence that the second marksman will usually have a better score than the first. The expected values of X and Y are ideal numbers in nature, but they can provide a comparison between two sets of numbers. ∎

Example 6

T. J. Scheff ([41], pp. 97–107) argues that an alternative to the medical practitioner's apparent strategy of the continuation of treatment when uncertainty enters the diagnosis would be to use expected value analysis.

Suppose that the random variable X denotes the value of treatment and the random variable Y denotes the value of no treatment. Both X and Y are defined for $\{0, 1\}$, where 0 denotes a patient with the disease and 1 denotes a healthy patient. If the probability that a patient actually has the disease is $1 - p$, then the expected values are

$$E(X) = (1 - p)X(0) + pX(1),$$
$$E(Y) = (1 - p)Y(0) + pY(1).$$

$X(0)$ is the value of the treatment for a healthy person, $X(1)$ the value of the treatment for a sick person. $Y(0)$ and $Y(1)$ are the corresponding values for nontreatment. For example, if $X(0) = -2$, $X(1) = 10$, $Y(0) = 5$, $Y(1) = -15$, and $p = \frac{1}{5}$, then

$$E(X) = \tfrac{4}{5}(-2) + \tfrac{1}{5}(10) = 0.4,$$
$$E(Y) = \tfrac{4}{5}(5) + \tfrac{1}{5}(-15) = 1.0.$$

In this case the recommended decision would be no treatment, because the expected value to the patient for nontreatment is higher than for treatment.

We point out that the values for $X(0)$ and $X(1)$ are not numbers that occur in a medical textbook but are part of a scale of values that a physician might assign to the disease. The specific figures above reveal that giving the treatment to a healthy patient may make him ill $[X(0) = -2]$ and that the treatment for a sick patient is the most beneficial $[X(1) = 10]$. The value of p is a physician's estimate of the treatment. In other words, this loose arrangement of numbers, when subjected to the expected value analysis, does give some indication of what to do. ∎

Example 7

A farmer is considering, as an expansion of his operation, feeding either beef cattle or hogs for the next year. The possible states of the livestock market are (1) increase in prices, (2) decrease in prices, and (3) no change in prices. His estimates of the net returns per dollar invested under the different states are given in Table 4. From government

Table 4

	Increase	Decrease	No change
Cattle	10	-3	2
Hogs	5	-1	3

agriculture sources the farmer assumes that the probability is 0.4 that market prices will rise, 0.3 that market prices will fall, and 0.3 that no change will occur. What should he decide to do?

One possibility is to compare the expected value of the alternative decisions. Let X be a random variable whose values are the estimated returns for cattle, and Y the random variable whose values are the returns for the hog investment. Applying the definition of expected value,

$$E(X) = 10(0.4) + (-3)(0.3) + 2(0.3) = 3.5,$$
$$E(Y) = 5(0.4) + (-1)(0.3) + 3(0.3) = 2.6.$$

If the farmer wishes to maximize his net expected return, he should decide to invest in cattle. ▮

Another interpretation of $E(X)$ is obtained if a probability point function is interpreted as the distribution of a physical unit of mass on a thin rod. Each element x_i of $S_X = \{x_1, x_2, \ldots, x_n\}$ is considered as a spot on the rod, x_i units of length from some specific point called the *zero point*; and $p(x_i)$ is thought of as the quantity of mass placed at the spot x_i. The product $x_i p(x_i)$, for each i, is a number that in physics is called the *moment* about the zero point for the mass $p(x_i)$ at x_i. The set of points $(x_i, p(x_i))$ is called a *mass system*, and the sum of the moments is called the *first moment* of the mass system. Hence $E(X)$ is the first moment of the mass system $\{(x_i, p(x_i))\}$. If a unit mass is located at a position $E(X)$ units from the zero point, then the first moment will be exactly that of the original system. The number $E(X)$ is often called the *center of mass* of the mass system, and the mass system will be in equilibrium if it is supported at the center of mass (Figure 1).

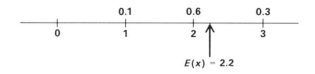

Figure 1

Problems

1. If X denotes the number of heads that occurs on four tosses of a fair coin, find $E(X)$.

2. Let X be the number of 1's that occurs in six independent tosses of a fair die. Find $E(X)$.

3. Ten balls numbered with the ten numbers 1, 2, 2, 3, 3, 4, 5, 5, 6, and 7 are placed in an urn and one ball is selected at random. Let X be the random variable whose value, for each ball selected, is the number of the ball. Find $E(X)$.

4. The green light at an intersection is on for a 45-second interval, the yellow for a 5-second interval, and the red for a 25-second interval. Suppose that "making the green light" is a chance event. Find the expected value of the number of successes in 25 independent trials.

5. There are two defectives in a lot of eight articles. A sample of four articles is drawn at random without replacement from the lot. Let X denote the number of defectives in the sample. Find $E(X)$.

6. What is the expected number of 7's in 100 throws of a pair of dice?

7. Let X denote the sum of the numbers on two fair dice. Find $E(X)$.

8. Let X be the larger of the two scores in two independent throws of a fair die. Find $E(X)$.

9. A bridge hand of 13 cards is dealt from a standard deck of 52 cards. Let X denote the number of spades in the hand. Find an expression for $E(X)$.

10. Toss a coin three times in succession. Let X be the random variable that denotes the gain of a player who wins \$2 if the first head occurs in the first toss, wins \$1 if the first head occurs on the second toss, loses \$1 if the first head occurs on the third toss, and losses \$2 if all three tosses are tails. Find $E(X)$.

11. An urn contains three green and two red balls. Let X be the number of red balls in a random sample of three balls drawn without replacement, and let Y be the number of red balls in a random sample of three balls drawn with replacement. Find $E(X)$ and $E(Y)$.

12. Arnold agrees to play chess until he wins one game or loses three. Assuming that the probability of Arnold's winning is $\frac{1}{2}$, find the expected number of games played.

13. Chances for a lottery are $1 each, with a first prize of $400, three prizes of $100 each, and six prizes of $25 each. One thousand chances are sold. Let X be the random variable that denotes the net gain for the purchase of one chance. Find $E(X)$.

14. A dealer has four television sets of which two are defective. The sets are tested one at a time until the defectives are discovered. Let X be the random variable that denotes the number of sets tested when the second defective is found. Find $E(X)$.

15. A used car salesman reaches his car in a dark parking lot with five ignition keys, entirely similar. He tries the keys one at a time, and he is able to eliminate unsuccessful keys from subsequent selections. If X denotes the number of keys he tries in order to find the proper key, find $E(X)$.

16. (a) Let X denote the number selected at random from the set $\{1, 2, 3, 4, 5\}$. Find $E(X)$.
    (b) Let X denote the maximum of two numbers selected at random with replacement from the set $\{1, 2, 3, 4, 5\}$. Find $E(X)$.

17. A game played by two persons is said to be "fair" if the expectation of return is zero for each player. A and B toss a balanced die, and A agrees to pay B $6 if the score recorded is less than or equal to 2. How much should B pay A when the score is greater than 2 if the game is to be fair?

18. Let X be a random variable with a sample space S_X such that for each $x_j \in S_X$, $x_j \geq c$. Show that $E(X) \geq c$.

19. An ice cream vendor has discovered that he can sell 100 ice cream bars on a warm day. The probability of a warm day occurring is $\frac{2}{3}$. However, on a chilly day he will sell only 35 ice cream bars, and the probability is $\frac{1}{3}$ for this occurrence. Further, suppose he makes a profit of $0.05 on every ice cream bar sold and he loses $0.05 on each bar not sold. What is his expected profit on an arbitrary day if he orders 35 bars? 50 bars? 100 bars?

20. Suppose that the odds are asserted to be $r : s$ in favor of a baseball team winning the World Series. That is, the probability that the team will win is $r/(r + s)$, and the probability that the team will lose is $s/(r + s)$. Further, suppose that one receives s dollars if the team wins and pays r dollars if the team loses. If the random variable X denotes his net gain, find $E(X)$.

21. A mining company owns two mines but can operate only one mine at a time. The mines produce three different grades of ore, but in any one week of operation the first mine produces 18 tons of high-grade ore, 27 tons of medium-grade ore and 10 tons of low-grade ore, while the second mine produces 12 tons of high-grade ore, 35 tons of medium-grade and 23 tons of low-grade ore. Each ton of high-

grade ore sold will yield a profit of $25, each ton of medium-grade ore a profit of $15, and each ton of low-grade ore a profit of $7. The manager must decide each week which mine to operate depending on the market for the ores produced. If he estimates the probability to be 0.3 that all the high-grade ore produced will be sold, 0.5 that all the medium grade ore will be sold, and 0.2 that all the low-grade ore will be sold, which of the two mines should be operated?

22. Consider the following simple game for two players. One player is given two cards, a red 5 and a black 5; the second player is given a black 3, a red 3, a black 1, and a red 1. At a given signal each player exposes one of his cards. If the cards are the same color, the first player receives from the second player the difference between the numbers on the cards. If the cards are not the same color, the second player receives from the first player the difference between the numbers. Suppose that the second players follows the strategy of rolling a die and playing the black 3 if a 1 comes up, the black 1 if a 2 occurs, the red 1 if a 5 or 6 is uppermost, and the red 3 the rest of the time. Under these conditions, which card should the first player expose?

6.3 Expectation of a Function of a Random Variable

It is often the case that the values x_k of a random variable X are not as important as some other numbers that are computed from the x_k. For example, the random variable X may represent the number of refrigerators sold in one week, but the more important number is $25X$, the profit from the sale of X refrigerators, where $25 is the profit from the sale of a single unit. In this context, consider Example 1.

Example 1

A game is devised for two players wherein player A pays player B $3.50 and then rolls a fair die. Player B then pays player A the same number of dollars as there are spots on the uppermost face.

If X is the random variable whose values are the outcomes of the roll of a fair die, $S_X = \{1, 2, 3, 4, 5, 6\}$ and $p_X(k) = \frac{1}{6}$ for all values of x_k in S_X. However, the amount of money that player A gains or loses can be represented as a random variable on the same sample space. Instead of returning to the sample space for the development, let us represent the net amount of money exchanged by player A in terms of X. Let Y denote this random variable; then

$$Y = X - 3.5,$$

because player A pays \$3.50 and receives X dollars in return.   The sample space for Y is

$$S_Y = \{-2.5, -1.5, -0.5, 0.5, 1.5, 2.5\},$$

and $p_Y(y_k) = \frac{1}{6}$ for all y_k in S_Y, because there is a one-to-one correspondence between the values for X and the values for Y.   If $f(x) = x - 3.5$, then it is consistent with the function definition to write $Y = f(X)$.   ∎

In this section we shall generate new random variables from a given random variable using functions ordinarily defined on a set of real numbers.   The probability function defined for the new random variable will be given in terms of the probability function of the original random variable and hence will enable us to extend probability functions to more complicated random variables.

Let S be a sample space with outcomes e_i, so that $S = \{e_1, e_2, \ldots, e_n\}$, and let X be a random variable whose associated sample space is $S_X = \{x_1, x_2, \ldots, x_n\}$.   As usual, the random variable X is a function whose domain is S and whose range is S_X; that is, $X(e_i) = x_i$.

Now suppose that we let f be any function whose domain includes S_X and whose codomain is a set of real numbers.   For example, we might choose f so that $f(x) = x^2$ or $f(x) = x + 5$.   Let y_i denote the value of f when $x = x_i$ in S_X.   Hence we may write $y_i = f(x_i)$, where, of course, the y_i are real numbers.   However, since $x_i = X(e_i)$, we can also write $y_i = f(X(e_i))$. Alternatively, by letting $Y(e_i)$ represent the functional values y_i when e_i determines x_i, we can write

$$Y(e_i) = f(X(e_i)).$$

The function values y_i are a set of real numbers, which we denote by

$$S_Y = \{Y(e_1), Y(e_2), \ldots, Y(e_n)\} = \{y_1, y_2, \ldots, y_n\}.$$

Since S_Y is a set of real numbers each of whose values corresponds to an outcome $e_i \in S$, Y is a new random variable defined on S.

The new feature of the random variable Y is that its values originate through the function values of the random variable X.   After the values of Y are thus determined, they can be related to the original space S, and Y can be considered as a function whose domain is S.   Therefore, we shall write

$$Y = f(X).$$

A symbolic representation of this functional relationship looks like

$$S = \{e_1, \ldots, e_n\} \xrightarrow{X} S_X = \{x_1, \ldots, x_n\} \xrightarrow{Y} S_Y = \{y_1, \ldots, y_n\},$$

where $Y = f(X)$.

It may happen that some of the y_i are the same.   Then we can write $S_Y = \{y_1, y_2, \ldots, y_m\}$, where all the y_i are distinct numbers and $m < n$.

Actually, the random variable Y is a *function of a function* and is called a *composite* function.   For a discussion of composite functions, see [18, Chap. 8].

Example 2

A cattleman is interested in the probability that he will make a profit. He buys 100 head of Hereford calves in the fall.   The calves that survive the winter can be sold in the spring for a profit of $40 a head.   Each calf that dies, however, represents a loss of $500.   From his experience with cattle, he knows that the probability that a calf will survive the winter is 0.95.   The profit on the purchase of 100 calves is then a random variable.

Let X denote the number of calves that survive the winter.   In fact, X is a binomial random variable with $S_X = \{0, 1, \ldots, 100\}$ and $p = 0.95$. For each value of X, the profit Y can be computed as follows:

$$Y = 50X - 500(100 - X)$$

$$= 550X - 50{,}000.$$

The probability that the cattleman makes a profit is the same as the probability of the event $\{Y \geq 0\}$.   The probability that this is so is the probability of the event $\{X \geq 91\}$; that is,

$$P(Y \geq 0) = P(X \geq 91) = \sum_{k=91}^{100} \binom{100}{k}(0.95)^k(0.95)^{100-k}. \qquad \blacksquare$$

In order to have a probability space defined on this sample space we must produce a probability function defined either by $p(\cdot)$ or $F(\cdot)$.   If we know the original sample space S for Y, we could determine the probability point function directly from it and completely ignore the probability functions for X.   In Example 1 we had to use $P_X(\cdot)$.   In Example 3 we show how to determine the probability functions for Y directly from the probability functions for X.

Example 3

Suppose that the random variable X has the values and probability point function $p_X(\cdot)$ defined as follows:

x_i	-1	0	1	2
$p_X(x_i)$	0.2	0.3	0.4	0.1

Let us define a new random variable by $Y = X^2 + 1$. Then $S_Y = \{2, 1, 5\}$ and we can calculate the probability point function $p_Y(\cdot)$ for Y as follows:

$$p_Y(1) = P_Y(Y = 1) = P_X(X = 0) = p_X(0) = 0.3,$$
$$p_Y(2) = P_Y(Y = 2) = P_X(X = -1 \text{ or } X = 1) = p_X(-1) + p_X(-1) = 0.6,$$
$$p_Y(5) = P_Y(Y = 5) = P_X(X = 2) = p_X(2) = 0.1.$$

The results are summarized in the table

y_i	1	2	5
$p_Y(y_i)$	0.3	0.6	0.1

Now that we know what the probability point function for the random variable $Y = f(X)$ means, it is easy to calculate the expected value of Y directly from the definition of $E(Y)$. But X and Y are related through the function f, and, since $E(X)$ is known, it is not unreasonable to expect that $E(Y)$ can be obtained from $E(X)$. Theorem 1 shows how.

Theorem 1

If X is a random variable with $S_X = \{x_1, x_2, \ldots, x_n\}$, $E(X) = \sum_{i=1}^{n} x_i p_X(x_i)$ and $Y = f(X)$ is a random variable, then

$$E(Y) = E(f(X)) = \sum_{i=1}^{n} f(x_i) p_X(x_i).$$

PROOF: Observe that $E(Y)$ is obtained from $E(X)$ by replacing x_i by $f(x_i)$. Let $S_Y = \{y_1, y_2, \ldots, y_m\}$, where y_j, $j = 1, 2, \ldots, m$, are the distinct values of $f(x_i)$, $i = 1, 2, \ldots, n$, with $m \leq n$. The probability point function $p_Y(\cdot)$ can be explicitly written as

$$p_Y(y_j) = \sum_{i} p_X(x_i),$$

where the sum is taken over all the x_i for which $f(x_i) = y_j$. The expected value of Y is

(1) $$E(Y) = \sum_{j=1}^{m} y_j p_Y(y_j)$$
$$= \sum_{j=1}^{m} y_j \left(\sum_{i} p_X(x_i) \right),$$

where the second sum on the right side of (1) is also taken over all the x_i for which $f(x_i) = y_j$. To evaluate the double sum we observe: (a) every x_i, for $i = 1, 2, \ldots, n$, is associated with exactly one y_j and so will appear once in the entire sum, and (b) if x_i is associated with y_j, then

$y_j = f(x_i)$. If the sum is expanded, we shall have each of the y_j's, for $j = 1, 2, \ldots, m$, multiplied by the sum of the numbers $p_X(\cdot)$ for those x_i such that $f(x_i) = y_j$. A typical term in the expanded sum looks like

$$y_j(p_X(x_{i_1}) + p_X(x_{i_2}) + \cdots + p_X(x_{i_k})),$$

which we can write as

$$f(x_{i_1})p_X(x_{i_1}) + f(x_{i_2})p_X(x_{i_2}) + \cdots + f(x_{i_k})p_X(x_{i_k}),$$

because

$$f(x_{i_1}) = f(x_{i_2}) = \cdots = f(x_{i_k}) = y_j.$$

There will be exactly n such terms appearing in the sum (1), one for each x_i. Hence we can write

$$E(Y) = \sum_{i=1}^{n} f(x_i)p_X(x_i) = E(f(X)). \qquad \blacksquare$$

Example 4

A contractor is building a house with the provision that, if the house is not ready for occupancy on the agreed date, the contractor will pay a penalty that depends on the number of weeks required to complete the house. Let X be a random variable whose values are $\{0, 1, 2, 3, 4\}$, the number of weeks beyond the agreed date to complete the house. The contractor estimates the probabilities to be such that

x_i	0	1	2	3	4
$p_X(x_i)$	0.75	0.10	0.08	0.05	0.02

The penalty agreed upon can be expressed by the function $f(x) = 8x - x^2$, and if Y denotes the penalty measured in hundreds of dollars, then

$$Y = 8X - X^2, \qquad \text{when } X = 0, 1, 2, 3, 4.$$

Now,

$$E(X) = \sum_{i=1}^{5} x_i p_X(x_i)$$
$$= 0.49,$$

which means that the expected value of the delay in construction is about $\frac{1}{2}$ week. The expected value of the penalty can be calculated from Theorem 1:

$$E(Y) = (f(X)) = \sum_{i=1}^{5} f(x_i)p_X(x_i)$$
$$= f(0)p_X(0) + f(1)p_X(1) + f(2)p_X(2) + f(3)p_X(3) + f(4)p_X(4)$$
$$= 7(0.10) + 12(0.08) + 15(0.05) + 16(0.02)$$
$$= 2.73.$$

Thus the expected penalty is \$273. $\qquad \blacksquare$

A special case of the function f occurs when f is the sum of two other functions g_1 and g_2. For example, $Y = X^2 + 6X$ can be written as $Y = (X^2) + (6X)$; so if $g_1(X) = X^2$ and $g_2(X) = 6X, f(X) = g_1(X) + g_2(X)$.

Theorem 2

Let X be a random variable and let g_1 and g_2 be two functions whose domains include S_X. If $Y = g_1(X) + g_2(X)$, then
$$E(Y) = E(g_1(X)) + E(g_2(X)).$$
PROOF: Let $g_1(X) + g_2(X)$ be the function in Theorem 1, so that
$$E(Y) = \sum_{i=1}^{m} (g_1(x_i) + g_2(x_i)) p_X(x_i),$$
which we can write as
$$E(Y) = \sum_{i=1}^{m} (g_1(x_i) p_X(x_i) + g_2(x_i) p_X(x_i)).$$
Hence
$$E(Y) = \sum_{i=1}^{m} g_1(x_i) p_X(x_i) + \sum_{i=1}^{m} g_2(x_i) p_X(x_i).$$
Applying Theorem 1 to each of the sums on the right gives
$$E(Y) = E(g_1(X)) + E(g_2(X)). \quad \blacksquare$$

Some other useful relations for the expected value of a function of a random variable are proved in Theorems 3 and 4.

Theorem 3

If X is a finite random variable and a and b are arbitrary real numbers, then
$$E(aX + b) = aE(X) + b.$$
PROOF: Let X have a sample space $S_X = \{x_1, x_2, \ldots, x_n\}$ and let $Y = aX + b$. The function g defining Y can be written as $g(x) = ax + b$, where x denotes an arbitrary real number. Then, using previous theorems,
$$E(Y) = \sum_{i=1}^{n} g(x_i) p_X(x_i)$$
$$= \sum_{i=1}^{n} (ax_i + b) p_X(x_i)$$
$$= \sum (ax_i p_X(x_i) + b p_X(x_i))$$
$$= a \sum_{i=1}^{n} x_i p_X(x_i) + b \sum_{i=1}^{n} p_X(x_i)$$
$$= aE(X) + b. \quad \blacksquare$$

Theorem 4

If X is a finite random variable and a and b are arbitrary real numbers, then
(a) $E(aX) = aE(X)$.
(b) $E(X + b) = E(X) + b$.
(c) $E(b) = b$.
(d) $E(X - E(X)) = 0$.

PROOF: We shall only prove part (a). Putting $b = 0$ in Theorem 3, (a) is then obvious. The other parts of the theorem can be proved as easily, using Theorem 3. ∎

For completeness, we point out that Theorems 2, 3, and 4 are valid when the random variable takes on a countable infinite set of values or a continuum of values, as we shall show in Chapter 9.

Problems

For Problems 1–6 use the following data:

x_i	-2	-1	0	1	2
$p_X(x_i)$	0.2	0.1	0.2	0.4	0.1

1. Compute $E(X)$.

2. Let $Y = 3X - 1$. Find $p_Y(\cdot)$ and compute $E(Y)$. Check your answer using Theorem 2.

3. Let $Y = X^2$. Find $p_Y(\cdot)$ and compute $E(Y)$. Check your answer using Theorem 1.

4. Let $Y = X^2 + X$. Find $p_Y(\cdot)$ and compute $E(Y)$. Check your answer using Theorem 1.

5. Let $Y = (2X - 5)^2$. Compute $E(Y)$ using Theorem 2 and then Theorem 1.

6. Let $Y = (X - E(X))^2$. Compute $E(Y)$ using Theorem 1. Show that your answer is equal to $E(X^2) - (E(X))^2$.

7. A cattleman buys 100 head of Hereford calves in the fall. The calves that survive the winter can be sold in the spring for a profit of \$50 per head. Each calf that dies, however, represents a loss of \$500. What is the expected profit if the probability that a calf survives the winter is 0.95?

8. Ten balls numbered with the ten numbers 1, 2, 3, 3, 4, 5, 6, 6, 6, and 7 are placed in a box, and one ball is drawn at random. Let X be the random variable whose value, for each ball selected, is the number on the ball. Find $E(X)$, $E(X^2)$, and $E(|X - E(X)|)$.

9. Let X denote the sum of the numbers obtained when two fair dice are rolled. Find $E(X^2)$. Is $E(X^2) = (E(X))^2$?

10. A grocer regularly orders a perishable product—fresh Dover sole— which must be discarded if not sold. The fish costs him $1.25 each and he sells it for $2.00 each. Let X be the random variable denoting the number of items his customers demand in one day. The probability point function has been estimated to be as follows:

x_i	0	1	2	3
$p_X(x_i)$	0.1	0.3	0.4	0.2

What is the expected net profit if the grocer orders two fish?

11. How many fresh Dover sole should the grocer order to have the maximum expected value of his net profit?

12. Art and Bob play the following game. Art pays Bob $1 and then three coins are tossed. Bob pays Art as follows: $1 if one tail occurs, $2 if two tails occur, $4 if three tails occur, and nothing if no tail occurs. Is the game fair? If not, what entrance fee should Art pay Bob for the game to be fair? Recall that a game between two persons is said to be fair if the expected value to both persons is zero.

13. An emergency ward of a hospital is such that it can handle only 100 patients per day at a cost (to the hospital) of $20 per patient. Also, if more than 100 such patients arrive, the hospital is required to ship this excess to another hospital at a cost (to the hospital) equal to $25 per patient. It is reasonable to assume that the number of patients that arrive at the hospital per day can be described by a binomial random variable with parameters $n = 125$ and $p = 0.02$. On this basis determine an expression for the expected value of the total cost to the hospital.

14. A man has one quarter and five nickels in his pocket and he plans to buy a 10-cent newspaper. The newsboy offers him a paper in exchange for one coin selected at random from the six coins. Is this a fair proposition? If not, to whom is it favorable?

15. A tennis ball manufacturer has a new process under consideration wherein the probability that a defective ball will be produced is 0.05. In the present process this probability is 0.10. The cost to produce a tennis ball by the new process will be 75 cents, while the cost under the present process is 50 cents. Defective tennis balls cannot be sold, but good tennis balls are sold for $1. Which process should be used to maximize the expected profit?

16. Show that if $Y = aX + b$, then $p_Y(\cdot) = p_X(\cdot)$; that is, if $y_i = ax_i + b$, then $p_Y(y_i) = p_X(x_i)$ for $i = 1, 2, \ldots, n$.

17. The demand X, per week, of a certain product is a random variable with a probability point function

x_i	0	1	2	3	4	5	6
$p(x_i)$	0.1	0.1	0.15	0.1	0.4	0.05	0.1

Suppose that the cost to the supplier is k_1 dollars per item, and he sells the item for k_2 dollars. Any item not sold at the end of the week must be stored at a cost of k_3 dollars per item. If the supplier decides to produce n items at the beginning of the week, what is his expected profit per week?

18. If X is a random variable with mean $E(X)$, find the mean of $Y = af(X)$, where a is an arbitrary real number.

6.4 Variance of a Random Variable

So far we have seen that the expected value of a random variable often can be computed easily, and we can take advantage of some of its special features to gain some insight into a sample space. However, the expected value can be almost useless without additional information. It does not, for example, give any clue as to how the individual values of the random variable are spread out, just as the arithmetic mean (Section 6.1) does not indicate the dispersion of the values that determine it, as in the case of two sets of golf scores, {80, 81, 81, 82, 86} and {75, 81, 81, 86, 87}. Both have an arithmetic mean of 82. For random variables, consider Example 1.

Example 1

Let X and Y be random variables with values and probability point functions specified by the following tables:

x_i	0	1	2	3	4
$p_X(x_i)$	0.1	0.2	0.3	0.1	0.3

y_i	-3	1	8
$p_Y(y_i)$	0.2	0.5	0.3

The expected value of X is 2.3 and $E(Y)$ is also 2.3. However, it is clear that the random variables differ in several ways. There are five values for X and only three for Y. On the other hand, the difference between the largest and smallest value for X is 5 units, while for Y it is 11 units, so that in one sense of the word Y is more spread out than X. The difference between the largest value of X and $E(X)$ is 1.7, while the same calculation for Y yields 5.7. Similar computations can be made for the smallest values of X and Y. In short, X and Y disperse differently around the expected value.   ∎

Hence we are interested in more information than $E(X)$ alone provides about the values of X. In Section 6.1 the mean-absolute dispersion and the mean-square dispersion were defined as "averages" that describe the property of dispersion for a set of numbers. This notion can be applied to the values of a random variable. Although we shall not develop an abstract definition of the dispersion of a set of numbers, we shall present a scheme whereby a real number can be assigned to the set of values of a random variable so that a small positive number will indicate less scatter or variance than a large positive number.

A very rough measure of the dispersion is the numerical difference between the largest and the smallest value of a random variable X. This number, although meaningful and of some use in statistical applications, is at best a crude estimate and not amenable to much manipulation.

A more useful description of dispersion arises when we study the deviation of the values of X from a fixed value, which is usually taken to be $E(X)$. Let us examine several possible ways of taking differences between values of X and $E(X)$. Suppose that X has values $\{x_1, x_2, \ldots, x_n\}$ and a probability point function $p(\cdot)$.

(a) Let $Y = X - E(X)$, so that Y represents the deviation of X from its expected value $E(X)$. Clearly, Y is a random variable with a different set of values than X. We compute $E(Y)$:

$$E(Y) = E(X - E(X)) = E(X) - E(X) = 0.$$

Regardless of what particular random variable X may be, $E(Y) = 0$, indicating an ineffective method for obtaining information about Y.

(b) But let $Z = |X - E(X)|$, so that Z denotes the absolute deviation of X from its expected value $E(X)$. We can write $Z = |Y|$ and calculate $E(Z)$:

(1) $$E(Z) = E(|X - E(X)|) = \sum_{i=1}^{n} |x_i - E(X)| \, p(x_i).$$

Every term in the sum is nonnegative, so if the sum is small, all the terms must be small. Thus, if there is a value of X for which $|x_i - E(X)|$ is large, say x_k, then the corresponding probability value $p(x_k)$ must be quite small. Consequently, if the expected value of the absolute deviations is small, it is unlikely that the random variable X differs greatly from $E(X)$.

The expected value of the absolute deviations from $E(X)$ has all the features of a successful numerical characteristic, except that it is algebraically unmanageable; that is, there is no way to calculate the right side of (1) except by brute force.

(c) We can maintain a positive quantity and introduce algebraic tractability if we square $x_i - E(X)$. Let $W = (X - E(X))^2$, or $W = Y^2$, where $Y = X - E(X)$. Once again we have a bona fide random variable. Using the results of Section 6.2, we obtain

$$(2) \qquad E(W) = E((X - E(X))^2) = \sum_{i=1}^{n} (x_i - E(X))^2 p(x_i).$$

Each term of the sum is nonnegative. So we may conclude as in (b) that widely scattered values of X are improbable when $E(W)$ is small. The numerical characteristic $E(W)$ may seem somewhat artificial, but is is easy to compute and it effectively indicates the dispersion of the values of a random variable. Its importance in theoretical discussions will soon be evident.

Example 2

Refer to Example 5 in Section 6.2, where X and Y are the random variables for the success of two marksmen firing at similar targets. The values of the random variables and the corresponding probabilities are repeated in the following table:

x_i	1	2	3
$p_X(x_i)$	0.4	0.1	0.5
y_i	1	2	3
$p_Y(y_i)$	0.1	0.6	0.3

In Example 5 we found that the expected value for X was $E(X) = 2.1$, and the expected value for Y was $E(Y) = 2.2$. Using equation (1), we find the expected value of the absolute deviations for X to be

$$E(|X - E(X)|) = \sum_{i=1}^{3} |x_i - 2.1| p_X(x_i)$$

$$= (1.1)(0.4) + (0.1)(0.1) + (0.9)(0.5) = 0.90.$$

The corresponding computation for Y yields

$$E(|Y - E(Y)|) = \sum_{i=1}^{3} |y_i - 2.2| p_Y(y_i)$$

$$= (1.2)(0.1) + (0.2)(0.6) + (0.8)(0.3) = 0.48.$$

The expected values of the squared deviations for X and Y can be computed using equation (2).

$$E((X - E(X))^2) = \sum_{i=1}^{3} (x_i - 2.1)^2 p_X(x_i)$$

$$= (1.1)^2(0.4) + (0.1)^2(0.1) + (0.9)^2(0.5) = 0.890,$$

$$E((Y - E(Y))^2) = \sum_{i=1}^{3} (y_i - 2.2)^2 p_Y(y_i)$$

$$= (1.2)^2(0.1) + (0.2)^2(0.6) + (0.8)^2(0.3) = 0.360.$$

The fact that the values of the expected absolute deviation and of the expected squared deviation are considerably smaller, for the second marksman may be interpreted to mean that the second marksman will be more consistent in his shooting than the first. That is, the first marksman's shots will most likely be more scattered than the second marksman's during any sequence of trials. ∎

The expected value of the squared deviation is formally denoted as the variance of the random variable.

Definition 1: Variance

Let X be a random variable with possible values $S_X = \{x_1, x_2, \ldots, x_n\}$, a probability point function $p(\cdot)$, and an expected value $E(X)$. The variance of the random variable X, denoted by $V(X)$, is defined to be

$$V(X) = E((X - E(X))^2) = \sum_{i=1}^{n} (x_i - E(X))^2 p(x_i).$$

If the random variable is given as X units, then $(X \text{ units} - E(X) \text{ units})^2 = (X - E(X))^2$ units^2. The square root of the variance thereby becomes $\sqrt{\sum (x_i - E(X))^2 p(x_i)}$ units; that is, X and $\sqrt{V(X)}$ are in the same units. Because of its importance and practicality we define $\sqrt{V(X)}$ formally.

Definition 2: Standard deviation

The square root of the variance of a random variable X is called the standard deviation of X and denoted by σ_X:

$$\sigma_X = \sqrt{V(X)}.$$

Example 3

If X denotes the uppermost face when a fair die is tossed, then the expected value of X and the variance of X are computed as follows:

$$E(X) = \sum_{i=1}^{6} x_i p(x_i)$$

$$= 1 \cdot \tfrac{1}{6} + 2 \cdot \tfrac{1}{6} + 3 \cdot \tfrac{1}{6} + 4 \cdot \tfrac{1}{6} + 5 \cdot \tfrac{1}{6} + 6 \cdot \tfrac{1}{6}$$

$$= \tfrac{21}{6}$$

$$= 3.5,$$

$$V(X) = \sum_{i=1}^{6} (x_i - 3.5)^2 p(x_i)$$

$$= [(2.5)^2 + (1.5)^2 + (0.5)^2 + (0.5)^2 + (1.5)^2 + (2.5)^2]\tfrac{1}{6}$$

$$= \tfrac{35}{12}.$$

The standard deviation is $\sigma_X = \sqrt{\tfrac{35}{12}}$. ∎

Although we can always calculate $V(X)$ from the definition, Theorem 1 gives us an easier way to compute it.

Theorem 1

If X is a finite random variable with a variance $V(X)$, then

$$V(X) = E(X^2) - (E(X))^2.$$

PROOF: If we start with the definition of $V(X)$ and use appropriate theorems from Section 6.3 on the expectation of a random variable, then we can write the following sequence of statements:

$$V(X) = E((X - E(X))^2)$$
$$= E(X^2 - 2XE(X) + (E(X))^2)$$
$$= E(X^2) + E(-2XE(X)) + E((E(X))^2)$$
$$= E(X^2) - 2E(X)E(X) + (E(X))^2$$
$$= E(X^2) - (E(X))^2. \quad ∎$$

In terms of σ_X we have

$$\sigma_X^2 = E(X)^2 - (E(X))^2,$$

which may be verbalized as "The variance equals the expected value of the square of the variable less the square of the expected value."

Example 4

Sixty percent of the student body favors a proposal for a new meal arrangement. Five students are to be selected at random for an interview on the question. If the random variable X denotes the number of those selected in favor of the proposal, find the expected value and variance of X.

Since X is a binomial random variable with $n = 5$ and $p = 0.6$, the probability point function can be obtained from a table of binomial probabilities:

x_i	0	1	2	3	4	5
$p_X(x_i)$	0.010	0.077	0.230	0.346	0.259	0.078

In Section 6.2 we found that the expected value of a binomial random variable is the product np (Theorem 3, Section 6.2), so $E(X) = 3$. To calculate $V(X)$ by the formula of Theorem 1, we find the value of $E(X^2)$ using the formula of Theorem 1 of Section 6.2 as follows:

$$E(X^2) = \sum_{i=0}^{5} x_i^2 \, p_X(x_i)$$

$$= 10.2.$$

Thus

$$V(X) = E(X^2) - (E(X))^2$$

$$= 10.2 - 9.0$$

$$= 1.2,$$

and

$$\sigma_X = \sqrt{1.2} = 1.09. \quad \blacksquare$$

Several properties of the variance which can be deduced directly from the definition are stated in the following theorems.

Theorem 2

If X is a finite random variable and a and b are arbitrary real numbers, then

$$V(aX + b) = a^2 V(X).$$

PROOF: The definition of the variance applied to the random variable $aX + b$ gives

$$V(aX + b) = E(((aX + b) - E(aX + b))^2).$$

By Theorem 3 of Section 6.3, we know that

$$E(aX + b) = aE(X) + b.$$

Hence

$$V(aX + b) = E((aX - aE(X))^2) = E(a^2(X - E(X))^2).$$

Using Theorem 4 of Section 6.3, we obtain

$$E(a^2(X - E(X))^2) = a^2 E((X - E(X))^2).$$

Therefore,

$$V(aX + b) = a^2 V(X). \quad \blacksquare$$

Theorem 3

If X is a finite random variable and a and b are arbitrary real numbers,

(a) $V(X + b) = V(X)$.
(b) $V(aX) = a^2 V(X)$.

These results are a direct consequence of Theorem 2, and the proof is left to the reader. The definition of the variance together with the properties stated in Theorems 1, 2, and 3 are valid for countable sample spaces and continuous sample spaces. The extension to countable spaces occurs in Chapter 8 and that for continuous spaces in Chapter 14.

The variance for some standard random variables can be calculated easily. We state the results as theorems.

Theorem 4

If X is a random variable with a uniform probability function for n values, then

$$V(X) = \frac{1}{n} \sum_{i=1}^{n} (x_i - E(X))^2 = \frac{1}{n} \sum_{i=1}^{n} x_i^2 - (E(X))^2.$$

PROOF: Let $S_X = \{x_1, x_2, \ldots, x_n\}$ and $p(x_j) = 1/n$ for $j = 1, 2, \ldots, n$. Then, by Definition 1,

$$V(X) = \sum_{i=1}^{n} (x_i - E(X))^2 \frac{1}{n},$$

which by Theorem 1 becomes

$$V(X) = \sum_{i=1}^{n} x_i^2 \frac{1}{n} - (E(X))^2. \quad \blacksquare$$

Example 5

A digit is chosen at random from the digits $0, 1, \ldots, 9$. Let X be the digit chosen. Find $V(X)$.

The random variable is a uniform random variable for $n = 10$ values, so that $p_X(k) = \frac{1}{10}$ for $k = 0, 1, 2, \ldots, 9$. For the expected value we find that

$$E(X) = \sum_{i=1}^{10} x_i p(x_i)$$

$$= \sum_{k=0}^{9} k \frac{1}{10}$$

$$= \frac{45}{10} = 4.5.$$

The variance is

$$V(X) = \frac{1}{n} \sum_{i=1}^{10} x_i^2 - (E(X))^2$$

$$= \frac{1}{10} \sum_{k=0}^{9} k^2 - (4.5)^2$$

$$= 28.5 - 20.25$$

$$= 8.25. \quad \blacksquare$$

Theorem 5

If X is a Bernoulli random variable with values $\{0, 1\}$, then $V(X) = pq$, where $p = p(1)$ and $q = 1 - p$.

PROOF: The random variable X takes on the two values 0 and 1 such that $p(0) = 1 - p = q$ and $p(1) = p$. By Theorem 2 of Section 6.2, $E(X) = p$. Therefore,

$$V(X) = E(X^2) - p^2$$

$$= \left(\sum_{i=1}^{2} x_i^2 p_X(x_i) \right) - p^2$$

$$= 0 \cdot (1 - p) + 1 \cdot p - p^2$$

$$= p - p^2$$

$$= p(1 - p)$$

$$= pq. \quad \blacksquare$$

Theorem 6

If X is a binomial random variable for n trials, where p is the probability of success on a trial, then

$$V(X) = np(1 - p) = npq,$$

where $p + q = 1$.

PROOF: The set of values for X is $\{0, 1, \ldots, n\}$ and the probability point function is

$$p(k) = \binom{n}{k} p^k q^{n-k}, \qquad k = 0, 1, \ldots, n.$$

According to Theorem 1,

$$V(X) = E(X)^2 - (E(X))^2,$$

and, by Theorem 3 of Section 6.2, $E(X) = np$. We shall have proved the theorem when we evaluate $E(X^2)$,

$$E(X^2) = \sum_{k=0}^{n} k^2 p(k) = \sum_{k=0}^{n} k^2 \binom{n}{k} p^k q^{n-k}.$$

The technique that was used in Theorem 3, Section 6.2, can be used to evaluate this sum. However, an easier way to find $E(X^2)$ is to compute

$$E(X(X-1)) = E(X^2 - X) = E(X^2) - E(X).$$

Now

$$E(X(X-1)) = \sum_{k=0}^{n} k(k-1) p(k)$$

$$= \sum_{k=0}^{n} k(k-1) \binom{n}{k} p^k q^{n-k}$$

$$= \sum_{k=0}^{n} \frac{k(k-1)n!}{(n-k)!k!} p^k q^{n-k}$$

$$= \sum_{k=2}^{n} \frac{n!}{(n-k)!(k-2)!} p^k q^{n-k}$$

$$= p^2 n(n-1) \sum_{k=2}^{n} \frac{(n-2)!}{(n-k)!(k-2)!} p^{k-2} q^{n-k}.$$

Since the summation is to run from 2 to n, put $t = k - 2$ and the summation will run from 0 to $n - 2$:

$$E(X^2 - X) = p^2 n(n-1) \sum_{t=0}^{n-2} \binom{n-2}{t} p^t q^{n-2-t}$$

$$= p^2 n(n-1) \cdot 1.$$

But now we can write

$$E(X^2) - E(X) = n(n-1)p^2,$$
$$E(X^2) = n(n-1)p^2 + E(X)$$
$$= n(n-1)p^2 + np$$
$$= np[(n-1)p + 1].$$

Hence

$$V(X) = np[(n-1)p + 1] - (np)^2$$
$$= (np)^2 - np^2 + np - (np)^2$$
$$= np(1 - p)$$
$$= npq. \quad \blacksquare$$

The proof of Theorem 6 was based on the definition of variance and required a tricky manipulation of sums. In Chapter 7 we shall obtain a much simpler proof of this theorem.

Example 6

In producing electrical fuses the manufacturing process over long periods results in 15 percent defective items. If X is the number of defective fuses in a sample of 10 fuses selected at random, find $E(X)$ and $V(X)$.

Let us consider the sample as 10 independent repeated Bernoulli trials, with the probability of success (a defective fuse) 0.15 at each trial. Thus X is a binomial random variable with the values of the parameters $n = 10$ and $p = 0.15$; so $E(X) = 10(0.15) = 1.5$ and $V(X) = 10(0.15)(0.85) = 1.275.$ $\quad \blacksquare$

For every random variable X with expectation $E(X)$ and standard deviation σ_X, a random variable X^* can be defined in such a way that X^* has the same kind of probability function as X; but $E(X^*) = 0$ and $V(X^*) = 1$, as we now show.

Definition 3: Standardized random variable

If X is any random variable with mean $E(X)$ and standard deviation $\sigma_X > 0$, the random variable

$$X^* = \frac{X - E(X)}{\sigma_X}$$

is called the standardized random variable corresponding to X.

We can show easily that $E(X^*) = 0$ and $V(X^*) = 1$. Clearly,

$$E(X^*) = E\left(\frac{1}{\sigma_X} X - \frac{1}{\sigma_X} E(X)\right)$$

$$= \frac{1}{\sigma_X} E(X) - \frac{1}{\sigma_X} E(X) = 0,$$

while

$$V(X^*) = V\left(\frac{1}{\sigma_X} X - \frac{1}{\sigma_X} E(X)\right)$$

$$= \left(\frac{1}{\sigma_X}\right)^2 V(X)$$

$$= 1.$$

In particular, if X is a binomial random variable with parameters n and p, then the standardized random variable is

$$X^* = \frac{X - np}{\sqrt{npq}}.$$

It may not seem necessary to require $\sigma_X > 0$ in Definition 3, for in general this is the case. However, if X has only one value x_1, so that $E(X) = x_1$, then $\sigma_X = 0$. Hence the requirement eliminates this possibility.

Problems

For Problems 1–4 use the following data:

x_i	-2	-1	0	1	2
$p(x_i)$	0.2	0.1	0.1	0.3	0.3

1. Compute $V(X)$.

2. Let $Y = 3X - 1$. Find $p_Y(\cdot)$ and compute $V(Y)$. Demonstrate that $V(Y) \neq 3V(X) - 1$.

3. Let $Y = X^2$. Find $p_Y(\cdot)$ and compute $V(Y)$. Show that $V(X)^2 \neq (V(X))^2$.

4. Let $Y = X^2 + X$. Compute $V(Y)$.

5. In a precinct, 70 percent of the voters are registered Republican and 30 percent are registered Democrat. A voter is selected at random and we let $X = 0$ if he is a Republican and $X = 1$ if he is a Democrat. Find $E(X)$ and $V(X)$.

6. A number is chosen at random from each of the following sets of numbers. Find the variance of the number chosen.

   (a) $\{-1, 0, 1, 2, 3\}$ (b) $\{-5, 0, 5, 10, 15\}$

7. Twenty-five percent of the student body has attended a short course in computer programming. Twenty are chosen at random and interviewed. Let X denote the number who have attended the computer course. Find $E(X)$ and $V(X)$.

8. In a student body of 1500, a referendum on a change in dining room procedures passed 500 to 250. After the change was implemented, 25 students were randomly selected to determine their reaction. If X denotes the number of those selected who originally opposed the measure, find $V(X)$.

9. In a study of the length of stay measured in days of patients at a hospital, the following data were obtained:

Stay	1	2	3	4	5	6
Number	18	24	22	16	12	8

Let X denote the length of stay of a person randomly selected from this study. Find $V(X)$.

10. The results of a survey of the monthly allowance given to each of 30 tenth-grade male students are listed below:

Amount	0	$4	$10	$12	$15	$20	$25
Number	2	5	4	6	8	2	3

Let X denote the amount of the allowance of a person randomly selected from the group. Find $V(X)$.

11. A fair coin is tossed 10 times. Let X denote the number of heads that appear. Find $E(X)$ and $V(X)$.

12. Let X be the total score in two independent throws of a die. Find $V(X)$.

13. Twenty independent tosses of a single die are made. Let X be the number of 1's or 6's. Find $V(X)$.

14. A balanced die is thrown. If the number that appears is less than 4, it is tossed again. Let X be the sum of the numbers that appear if the die is tossed twice, or the number that appears on the first toss if it is thrown only once. Find $E(X)$ and $V(X)$.

15. Refer to Problem 7 of Section 6.3. Find the variance of the random variable that denotes the profit.

16. Refer to Problem 8 of Section 6.3. Find the variance of the random variable that is described in the problem.

For each of the random variables in Problems 17–19, calculate the expected value, the variance, the standard deviation, and the expected absolute deviation.

17.

x_i	-1	0	1
$p_X(x_i)$	0.4	0.2	0.4

18.

x_i	1	2	3	4
$p_X(x_i)$	0.1	0.3	0.4	0.2

19.

x_i	-2	-4	-6	4	2
$p_X(x_i)$	0.1	0.2	0.4	0.2	0.1

20. Prove that the value of c which makes $E((X - c)^2)$ a minimum is $c = E(X)$.

21. You toss a coin three times and are paid n dollars if heads comes up for the first time on the nth throw, $n = 1, 2, 3$. Find the expected value of your gain and its variance.

22. A random variable X has mean 50 and variance 75. If X^* is the standardized random variable corresponding to X,

(a) what value of X^* corresponds to each of the following values of X: 65, 110, and 73?

(b) what value of X corresponds to each of the following values of X^*: $+5$, -3, and -2.3?

chapter seven

Sums of Random Variables

7.1 Joint Random Variables

More than one random variable can be defined on the same sample space as we did in Example 1 of Section 5.1, where the sample space was the 16 outcomes of the four tosses of a fair coin. The random variable X was the number of heads and the random variable Y took the value 1 if the number of heads exceeded the number of tails, the value -1 if the converse was true, and the value 0 if the numbers were the same. That example was presented as an illustration of the notion of a random variable as a function whose domain was the sample space. Now we shift our attention and consider the joint behavior of the variables X and Y.

The idea of two variables being related so that one of them varies in a way that is connected in a functional manner to the other was formally introduced in Section 6.3 for random variables. We now examine two related variables whose correspondence may not be explicitly formulated. For instance, a biologist compares the overall length of an animal and the length of its tail. An operation analyst compares the duration of a patient's stay in a hospital and his age, or a psychologist compares the rating given to a teacher by a randomly selected student and the grade the same student received in the course.

The relationship between two characteristics (or measurements) of a single outcome of a random experiment is fundamental to analysis of empirical studies and often occurs in the form of cross-classification or *joint frequency tables*, another important idea in statistical analysis. The theoretical basis for the joint frequency table is found in the notion of the joint (or bivariate, as it is frequently called) random variable. As an introduction to the idea, consider Examples 1 and 2.

Example 1

Eight chips are put into a jar. The chips are indistinguishable except for color and a number on the chip. Two chips are red and have the numbers 1 and 2, respectively; two chips are white with the numbers 2 and 3, respectively, and four blue chips have the numbers 1, 2, 3, and 4, respectively. A chip is drawn from the jar. A sample space S for the random experiment can be written as the eight pairs

$$S = \{(R, 1), (R, 2), (W, 2), (W, 3), (B, 1), (B, 2), (B, 3), (B, 4)\}.$$

Let X be the random variable whose value is the number on the chip so that $S_X = \{1, 2, 3, 4\}$. The events $\{X = i\}$, $i = 1, 2, 3, 4$, are identified below.

$$\{X = 1\} = \{(R, 1), (B, 1)\},$$
$$\{X = 2\} = \{(R, 2), (B, 2), (W, 2)\},$$
$$\{X = 3\} = \{(B, 3), (W, 3)\},$$
$$\{X = 4\} = \{(B, 4)\}.$$

The probability point function $p_X(\cdot)$ is

x_i	1	2	3	4
$p_X(x_i)$	$\frac{2}{8}$	$\frac{3}{8}$	$\frac{2}{8}$	$\frac{1}{8}$

Let Y be the random variable whose value is 0 if the chip is white and $+1$ if the chip is not white. Then

$$\{Y = 0\} = \{(W, 2), (W, 3)\},$$
$$\{Y = 1\} = \{(R, 1), (R, 2), (B, 1), (B, 2), (B, 3), (B, 4)\},$$

and

y_j	0	1
$p_X(y_j)$	$\frac{2}{8}$	$\frac{6}{8}$

Since both X and Y take on values that relate to the single draw of a chip, we can ask the question: What is the probability that on a single draw

X will be 2 and Y will be 0? Because there is only one white chip with the number 2, the probability is $\frac{1}{8}$, which can be written as

$$P(X = 2,\ Y = 0) = \tfrac{1}{8}.$$

The comma inside the parentheses indicates the intersection of the two events $\{X = 2\}$ and $\{Y = 0\}$, and $P(\cdot)$ is understood to be the probability function for the sample space of chips. Similarly, $P(X = 2, Y = 1) = \frac{1}{4}$, because there are two elementary events in the intersection of $\{X = 2\}$ and $\{Y = 1\}$. The probabilities for all possible pairs of values of X and Y are conveniently arranged in Table 1.

Table 1

	y_j	
x_i	0	1
1	0	$\frac{1}{4}$
2	$\frac{1}{8}$	$\frac{1}{4}$
3	$\frac{1}{8}$	$\frac{1}{8}$
4	0	$\frac{1}{8}$

The event defined by $\{Y = 1\}$, without reference to X, has six elementary events, the six nonwhite chips. The event can also be represented as the union of the events

$$\{X = 1,\ Y = 1\},\ \{X = 2,\ Y = 1\},\ \{X = 3,\ Y = 1\},\ \{X = 4,\ Y = 1\},$$

which are disjoint (mutually exclusive), so that

$$P(Y = 1) = \tfrac{1}{4} + \tfrac{1}{4} + \tfrac{1}{8} + \tfrac{1}{8} = \tfrac{3}{4}.$$

By a similar process we would find that

$$P(Y = 0) = 0 + \tfrac{1}{8} + \tfrac{1}{8} + 0 = \tfrac{1}{4}.$$

A glance at Table 1 reveals that these probabilities are the sums of the two columns. The probabilities for $X = i$ are obtained by taking the sums of the elements in the rows. ∎

Example 2
In a study of the relationship of the length of stay in a hospital to the age of the patient, data on a group of 30 patients were collected. The

patients were classified by age into the categories 55–60, 60–65, 65–70, and 70–75 with the results

Age group	55–60	60–65	65–70	70–75
Number	5	10	11	4

Let X denote the age group in which a randomly selected person falls. Then $p_X(\cdot)$ is defined as follows:

x_i	1	2	3	4
$p_X(x_i)$	$\frac{5}{30}$	$\frac{10}{30}$	$\frac{11}{30}$	$\frac{4}{30}$

On the other hand, the patients were classified by the length (in days) of stay in the hospital into the categories 1, 2, 3, 4, 5, and over 5, with the results listed in Table 2.

Table 2

Age	Length of stay					
	1	2	3	4	5	Over 5
55–60	2	2	1	0	0	0
60–65	1	3	3	2	1	0
65–70	0	2	4	2	2	1
70–75	0	1	2	0	0	1

Now let us consider the cross-classification of the 30 patients by age and length of stay. By the cross-classification we mean a two-way table that separates the data according to the length of stay and age group. In Table 3 the entry in the second row and second column indicates that 3 people in the 60–65 group stayed in the hospital 2 days.

If we replace the characteristics age and length of stay by the random variables X and Y, respectively, and the frequencies by the relative frequencies, the probability function for the pair X, Y has the shape given in Table 3. We now use the table to find the probability of events. The probability that a person selected at random from the 30 patients will have $X = 3$ and $Y = 4$ is given by the value in the third row and fourth column:

$$P(X = 3,\ Y = 4) = \tfrac{2}{30}.$$

Table 3

x_i	1	2	3	4	5	6
			y_j			
1	$\frac{2}{30}$	$\frac{2}{30}$	$\frac{1}{30}$	0	0	0
2	$\frac{1}{30}$	$\frac{3}{30}$	$\frac{3}{30}$	$\frac{2}{30}$	$\frac{1}{30}$	0
3	0	$\frac{2}{30}$	$\frac{4}{30}$	$\frac{2}{30}$	$\frac{2}{30}$	$\frac{1}{30}$
4	0	$\frac{1}{30}$	$\frac{2}{30}$	0	0	$\frac{1}{30}$

This can be interpreted in terms of events to mean that there are exactly two patients in the intersection of the event $\{X=3\}$ and the event $\{Y=4\}$.   ∎

Holding these examples as references, we state formally the general case of two random variables defined on the same sample space.

Definition 1: Joint probability point function

Let X and Y be two random variables defined on the same sample space S, where the probability function is $P(\cdot)$ and where

$$S_X = \{x_1, x_2, \ldots, x_m\}, \qquad S_Y = \{y_1, y_2, \ldots, y_n\}.$$

The function $p(\cdot, \cdot)$ is defined on the set $\{(x_i, y_j) \mid x_i \in S_X, y_j \in S_Y\}$ by

$$p(x_i, y_j) = P(X = x_i, Y = y_j)$$

and is called the joint probability point function for X and Y.

A double-entry table such as Table 4 is a very convenient way to represent a joint probability point function.   To define a joint probability function, the values for X, the values for Y, and the mn values for $p(\cdot, \cdot)$ must be specified.

Since $p(x_i, y_j)$ is the probability of an event $\{X = x_i, Y = y_j\}$ of the sample space S, $p(x_i, y_j) \geq 0$ for $i = 1, 2, \ldots, m$ and $j = 1, 2, \ldots, n$.

The event $\{X = x_i\}$ of the sample space S_X can be written

$$\{X = x_i\} = \bigcup_{j=1}^{n} \{X = x_i, Y = y_j\}.$$

Table 4

	y_1	y_2	$\cdots$	y_n
x_1	$p(x_1, y_1)$	$p(x_1, y_2)$		$p(x_1, y_n)$
x_2	$p(x_2, y_1)$	$p(x_2, y_2)$		$p(x_2, y_n)$
$\vdots$				
x_m	$p(x_m, y_1)$	$p(x_m, y_2)$		$p(x_m, y_n)$

Note that the union is taken for $j = 1, 2, 3, \ldots, n$, and that it represents the occurrence of an event that is found by reading across a row in Table 4 for all entries in that row. As we know from proofs in Chapters 2 and 5, the probability of the union of elementary events is the sum of the individual probabilities:

$$(1) \qquad p_x(x_i) = \sum_{j=1}^{n} p(x_i, y_j).$$

Similarly, the event $\{Y = y_j\}$ is the union

$$\{Y = y_j\} = \bigcup_{i=1}^{m} \{X = x_i, \ Y = y_j\}.$$

In this case the union represents a column in Table 4. Again the probability of the union is the sum of the individual probabilities:

$$(2) \qquad _Y(y_j) = \sum_{i=1}^{m} p(x_i, y_j).$$

Since both

$$\sum_{i=1}^{m} p_X(x_i) = 1 \qquad \text{and} \qquad \sum_{j=1}^{n} p_Y(y_j) = 1,$$

it follows from (1) and (2) that

$$\sum_{i=1}^{m} \sum_{i=1}^{n} p(x_i, y_j) = \sum_{i=1}^{m} \left(\sum_{j=1}^{n} p(x_i, Y_j) \right) = \sum_{i=1}^{m} p_X(x_i) = 1.$$

By interchanging the order of summation, we also obtain

$$\sum_{i=1}^{n} \sum_{i=1}^{m} p(x_i, y_j) = \sum_{j=1}^{n} \left(\sum_{i=1}^{m} p(x_i, y_j) \right) = \sum_{j=1}^{n} p_Y(y_j) = 1.$$

Since the sums for X and Y by rows and columns can be conveniently recorded in the margins of the table for $p(\cdot, \cdot)$, they are called the *marginal probability point functions* of X and Y and coincide with the individual probability point functions for the sets S_X and S_Y.

Example 3

Suppose that two balls are drawn with replacement from an urn containing six balls, three of which are numbered 1, two are numbered 2, and one is numbered 3.   The sample space for pairs of numbers has nine elementary events.   Let X be the random variable whose value is the number on the first ball selected, and let Y be the random variable whose value is the sum of the numbers on the two balls selected.   The joint probability point function $p(\cdot, \cdot)$ and the marginal probability point functions are specified in Table 5.

Table 5

			Y			
X	2	3	4	5	6	$p_X(\cdot)$
1	$\frac{1}{4}$	$\frac{1}{6}$	$\frac{1}{12}$	0	0	$\frac{1}{2}$
2	0	$\frac{1}{6}$	$\frac{1}{9}$	$\frac{1}{18}$	0	$\frac{1}{3}$
3	0	0	$\frac{1}{12}$	$\frac{1}{18}$	$\frac{1}{36}$	$\frac{1}{6}$
$p_Y(\cdot)$	$\frac{1}{4}$	$\frac{1}{3}$	$\frac{5}{18}$	$\frac{1}{9}$	$\frac{1}{36}$	

∎

Example 4

In Example 1 of Section 5.1, two random variables, X and Y, were defined on the sample space S of the 16 outcomes from four tosses of a coin.   We let X denote the number of heads that occurred, and Y was defined to have the value $+1$ if the number of heads exceeds the number of tails, the value -1 if the number of heads was less than the number of tails, and the value 0 if the number of heads and the number of tails were the same.

The joint probability function for X and Y can be constructed by examining the events $\{X = i, \ Y = j\}$ of S for $i = 0, 1, 2, 3, 4$ and $j = -1, 0, 1$.   The results are given in Table 6.

Table 6

		y_j		
x_i	-1	0	1	$p_X(x_i)$
0	$\frac{1}{16}$	0	0	$\frac{1}{16}$
1	$\frac{4}{16}$	0	0	$\frac{4}{16}$
2	0	$\frac{6}{16}$	0	$\frac{6}{16}$
3	0	0	$\frac{4}{16}$	$\frac{4}{16}$
4	0	0	$\frac{1}{16}$	$\frac{1}{16}$
$p_Y(y_j)$	$\frac{5}{16}$	$\frac{6}{16}$	$\frac{5}{16}$	

∎

Up to now the domain of the functions used to define probability point functions and random variables were sets of real numbers, and the codomain was a set of real numbers. The symbol $p(\cdot, \cdot)$ is a real number that is associated with an ordered pair of numbers, one of the pair from S_X, the other from S_Y. Thus the domain of $p(\cdot, \cdot)$ is the set

$$S = \{(x_i, y_j) \mid x_i \in S_X, y_j \in S_Y\}.$$

S is called the *Cartesian product* of S_X and S_Y. We also can consider (x_i, y_j) as a point in the xy-plane and $p(x_i, y_j)$ a point on a line perpendicular to the xy-plane $p(x_i, y_j)$ units above the plane at (x_i, y_j). In this way we obtain a point $(x_i, y_j, p(x_i, y_j))$ in three-dimensional space.

The geometric representation of the probability point function for the random variables X and Y of Example 1 is given in Figure 1. At each of

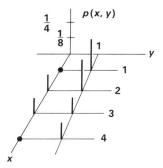

Figure 1

the points $(1, 0)$, $(1, 1)$, $(2, 0)$, $(2, 1)$, $(3, 0)$, $(3, 1)$, $(4, 0)$, and $(4, 1)$, the value of $p(x_i, y_j)$ is represented by the line segment whose length is proportional to the value.

The notation that generalizes $y = f(x)$ to functions of two variables is $z = g(x, y)$. A formal definition of a function of two variables follows.

Definition 2: Function of two variables

Let A and B be any two sets of real numbers and let

$$A \times B = \{(x, y) \mid x \in A, y \in B\}$$

be the Cartesian product of A and B. The correspondence (x, y) to z, where z is a real number, is a mathematical function of two variables x and y with domain $A \times B$ and codomain the real numbers. The value of the function at the point (x, y) is written as $z = f(x, y)$.

Example 5

Functions of two variables occur regularly in applied fields as well as in mathematics. In economics the relationship between consumption, denoted by z, and wages and profits, denoted by x and y, respectively, is expressed by

$$z = ax + by,$$

where a and b are positive real numbers. The domain of the function is the set of ordered pairs with both x and y positive real numbers, and the range is the set of positive z.

The natural frequency of vibrations of a taut wire can be expressed as a function of the length x and the diameter y of the wire by the expression

$$\frac{k}{xy},$$

where k is a positive real constant.   ∎

The generalization of the function concept to n variables is extremely useful in statistical applications as well as in providing models for physical phenomena. We shall return to this generalization as we need it.

Problems

1. A fair coin is tossed three independent times. Let X be the random variable whose value is 0 if the second toss is a tail and 1 if the second toss is a head. Let Y be the total number of heads. Construct the joint probability point function for X and Y.

2. An urn contains three red and two green balls. Two balls are drawn at random from the urn with replacement. Let X be the random variable whose value is 0 if the first ball is green and 1 if the first ball is red, and let Y be the random variable whose value is 0 if the second ball is green and 1 if the second ball is red. Construct the joint probability point function for X and Y.

3. Refer to Problem 2 and repeat the problem with the selection of the balls made without replacement.

4. Three balls are placed at random in three numbered jars. Let the number of empty jars be the random variable X and the number of balls in the first jar be the random variable Y. Find the joint probability function for X and Y.

5. Two fair dice are tossed. If X is the maximum of the two numbers and Y is the sum of the two numbers, find the joint probability function for X and Y.

6. The joint probability point function for X and Y is given by the function

$$p(x, y) = \tfrac{1}{32}(x^2 + y^2) \qquad x = 0, 1, 2, 3; \; y = 0, 1.$$

Determine the individual or marginal probability point functions for X and Y.

7. The joint probability point function for X and Y is given by the function

$$p(x, y) = \tfrac{1}{66}xy, \qquad x = 2, 4, 5; \; y = 1, 2, 3.$$

Determine the marginal probability point functions for X and Y.

8. One integer is selected from the set $\{1, 2, 3, 4, 5\}$. A second integer is selected from those integers greater than or equal to the first integer. Thus, if 4 is selected the first time, the second is chosen from $\{4, 5\}$. Let X and Y denote the numbers selected on the first and second draws, respectively. Find the joint and marginal probability point functions for X and Y.

9. Determine the marginal probability point functions of X and Y given the following joint probability point function:

		Y	
X	-1	0	1
-1	$\tfrac{1}{12}$	$\tfrac{1}{12}$	$\tfrac{2}{12}$
0	0	0	0
1	$\tfrac{2}{12}$	$\tfrac{2}{12}$	$\tfrac{4}{12}$

What is the probability that $X < 0$?

10. You make two throws of a die. Let X denote the number of times a 1 turns up and Y the number of times a 2 turns up. Describe the probability space on which X and Y are defined and determine the joint probability point function.

11. A die is tossed and if the number is less than 4, the die is tossed again. If the number is 4 or greater, three coins are tossed. Let X be the number that occurs on the first toss and Y the number of heads or the number on the die on the second toss. Find the joint probability point function for X and Y and find

$$P(X = 2, Y \leq 2).$$

12. Three cards are drawn without replacement from the 12 face cards of an ordinary deck of cards. Let X denote the number of black kings and Y the number of red queens selected. Find

$$P(X \geq 1 \text{ and } Y \geq 1).$$

13. An urn contains eight balls of which five are red and three are green. Four balls are drawn without replacement in a random fashion. What is the joint probability point function of X, the number of red balls drawn, and Y, the number of green balls drawn?

14. Prove that $A \times (B \cup C) = (A \times B) \cup (A \times C)$, where A, B, and C are arbitrary subsets of some universe.

15. The joint probability point function for X and Y is given by the following table:

			y_j		
x_i	-2	-1	1	3	4
-1	$\frac{4}{27}$	$\frac{1}{9}$	$\frac{1}{9}$	$\frac{2}{27}$	0
0	$\frac{2}{27}$	$\frac{2}{27}$	$\frac{1}{9}$	$\frac{1}{9}$	$\frac{1}{27}$
1	$\frac{1}{27}$	$\frac{2}{27}$	0	$\frac{1}{27}$	0

What is the probability that

(a) Y is positive? (b) XY is odd?
(c) X is positive and $Y^2 < 2$? (d) X is negative or Y is positive?

16. The same item is manufactured on two production lines. The capacity of one line is five items per day and the capacity of the other is three items per day. Let us assume that the output of each line is a random variable with the joint probability point function specified as follows:

				y_j		
x_i	0	1	2	3	4	5
0	0	0.01	0.03	0.05	0.07	0.09
1	0.01	0.02	0.04	0.05	0.06	0.08
2	0.01	0.03	0.05	0.05	0.05	0.06
3	0.01	0.02	0.04	0.06	0.06	0.05

What is the probability that

(a) $Y \geq 3$?
(b) $Y > X$?
(c) $X + Y \geq 3$?
(d) $Y - X = 1$?

17. The 25 students in a calculus class rated their instructor's overall performance as excellent, good, or poor. A two-way classification of these ratings with respect to the course grade the student received is given in the following table:

	Grade				
	A	B	C	D	F
Excellent	2	1	4	1	0
Good	1	3	6	0	1
Poor	0	2	2	2	0

Let X denote the ratings with the value 3 for excellent, 2 for good, 1 for poor, and Y the grades with 4 for A, 3 for B, 2 for C, 1 for D, and 0 for F. Determine the joint probability point function and the marginal probability point functions. Find the probability that $X > Y$.

18. Thirty percent of the population of a city is Protestant and 50 percent is Catholic. Let X be the number of Protestants and Y the number of Catholics that turn up in two random selections from the population of the city. Find the expected value of X and the expected value of Y.

7.2 Independent Random Variables

Random variables defined on the same sample space may be independent or dependent. By extending the notions of independent events (Section 4.3) and independent trials (Section 4.4), we can discuss the independence of random variables.

Definition 1: Independent random variables

Let X and Y be finite random variables defined on the same sample space S with $S_X = \{x_1, x_2, \ldots, x_m\}$ and $S_Y = \{y_1, y_2, \ldots, y_n\}$. The random variables X and Y are independent if

$$P(X = x_i, \ Y = y_j) = P(X = x_i)P(Y = y_j)$$

for all possible pairs of values of x_i and y_j. That is,

$$(1) \qquad\qquad p(x_i, y_j) = p_X(x_i)p_Y(y_j)$$

for the probability point functions.

If for even one pair of values (x_i, y_j) the equality in (1) does not hold, then X and Y cannot be independent.

Although the formal definition of independent random variables is given by the product of probabilities, it is helpful to think of independent random variables as having no influence on each other. In this connection, independence of joint random variables can be stated in terms of conditional probability. See Problems 12 and 13.

Example 1

An urn contains two white and three red balls. First, two balls are drawn one at a time without replacement, and then two balls are drawn one at a time with replacement. In both cases let X be the random variable with the value 0 if the first ball is white, and 1 if it is red; and let Y be the random variable with the value 0 if the second ball is white, and 1 if it is red. The joint probability point function and marginal probability point function for these cases are listed in Table 1. In (a)

Table 1

(a) $p(\cdot, \cdot)$ without Replacement

	y		
x	0	1	$p_X(\cdot)$
0	$\left(\frac{2}{5}\right)\left(\frac{1}{4}\right)$	$\left(\frac{2}{5}\right)\left(\frac{3}{4}\right)$	$\frac{2}{5}$
1	$\left(\frac{3}{5}\right)\left(\frac{2}{4}\right)$	$\left(\frac{3}{5}\right)\left(\frac{2}{4}\right)$	$\frac{3}{5}$
$p_Y(\cdot)$	$\frac{2}{5}$	$\frac{3}{5}$	

(b) $p(\cdot, \cdot)$ with Replacement

	y		
x	0	1	$p_X(\cdot)$
0	$(\frac{2}{5})(\frac{2}{5})$	$(\frac{2}{5})(\frac{3}{5})$	$\frac{2}{5}$
1	$(\frac{3}{5})(\frac{2}{5})$	$(\frac{3}{5})(\frac{3}{5})$	$\frac{3}{5}$
$p_Y(\cdot)$	$\frac{2}{5}$	$\frac{3}{5}$	

the values of X and Y are the same and the marginal probability point functions are exactly the same. When this is so, the random variables are said to be *identically distributed*. Moreover, X and Y are dependent because $p(1, 0) \neq p_X(1) \cdot p_X(0)$. In (b) X and Y are not only identically distributed, but are independent. ∎

Example 1 demonstrates the importance of the independence of random variables in relation to the joint probability point function. If one knows the joint probability point function, the marginal or individual probability point function can be calculated whether the random variables are independent or dependent. However, the converse is true only when the random variables are independent. Then the joint probability point function can be found by multiplying the pairs of individual probabilities.

Thus the assumption of independence is most important in the actual application of probability to statistical problems. Example 2 gives an indication of the way that the assumption of independent random variables arises naturally in application.

Example 2

Suppose that a machine is used for one task in the morning and another task in the afternoon. Let X be the number of breakdowns in the morning and Y the number in the afternoon, with the values and probability point function for X given by

x_i	0	1	2	3
$p_X(x_i)$	0.5	0.3	0.15	0.05

and the corresponding information for Y given by

y_j	0	1	2
$p_Y(y_j)$	0.7	0.2	0.1

If we assume that the random variables are independent, then the joint probability point function $p(\cdot, \cdot)$ can be computed using expression (1) in the definition of independent random variables, with the results given in Table 2.

Table 2

		y		
x	0	1	2	$p_X(\cdot)$
0	0.35	0.10	0.05	0.5
1	0.21	0.06	0.03	0.3
2	0.105	0.03	0.015	0.15
3	0.035	0.01	0.005	0.05
$p_Y(\cdot)$	0.70	0.20	0.10	

If X and Y denote the outcomes of two independent trials of a two-trial experiment, then naturally each of the outcomes of the pair (X, Y), considered as joint random variables, is independent. If $S_X = \{x_1, \ldots, x_m\}$ and $S_Y = \{y_1, \ldots, y_n\}$, then the common sample space is the set

$$S = \{(x_i, y_j) \mid i = 1, 2, \ldots, m, \text{ and } j = 1, \ldots, n\}.$$

The probability point function $P(\cdot)$ for S has the form

$$P((x_i, y_j)) = p_X(x_i) \cdot p_Y(y_j) \qquad \text{for } i = 1, 2, \ldots, m; j = 1, 2, \ldots, n.$$

Example 3

A rat runs a T-maze four times. Let X be the number of right turns in the first pair of runs and Y the number of right turns in the second pair of runs.

Both S_X and S_Y are the set $\{0, 1, 2\}$ with the identical probability point function

$$p(0) = \tfrac{1}{4}, \qquad p(1) = \tfrac{1}{2}, \qquad p(2) = \tfrac{1}{4}.$$

The joint probability table and point function is constructed using the definition of independence as follows:

		y		
x	0	1	2	$p_X(\cdot)$
0	$\frac{1}{16}$	$\frac{1}{8}$	$\frac{1}{16}$	$\frac{1}{4}$
1	$\frac{1}{8}$	$\frac{1}{4}$	$\frac{1}{8}$	$\frac{1}{2}$
2	$\frac{1}{16}$	$\frac{1}{8}$	$\frac{1}{16}$	$\frac{1}{4}$
$p_Y(\cdot)$	$\frac{1}{4}$	$\frac{1}{2}$	$\frac{1}{4}$	

In this case the joint probability can be checked by examining the elementary events of 4-tuples in the sample space of four runs. For instance,

$$P(X=1,\ Y=2) = P(\{(L,\ R,\ R,\ R),\ (R,\ L,\ R,\ R)\}) = \tfrac{2}{16} = \tfrac{1}{8}. \quad \blacksquare$$

Example 4 is another example of the importance of independence of random variables.

Example 4

A preventive serum for a disease is tested on a collection of 100 animals, 60 of which are inoculated prior to exposure to the disease and the remaining 40 of which are exposed but not inoculated. Twenty of the animals became diseased.

Let X have the value 0 if the animal is not diseased and 1 if diseased. Let Y have the value 0 if the animal was not inoculated and 1 if inoculated. If the results of the experiment are

		y	
x	0	1	$p_X(\cdot)$
0	$\frac{48}{100}$	$\frac{32}{100}$	$\frac{80}{100}$
1	$\frac{12}{100}$	$\frac{8}{100}$	$\frac{20}{100}$
$p_Y(\cdot)$	$\frac{60}{100}$	$\frac{40}{100}$	

then the random variables X and Y are independent since

$$p(x_i,\ y_j) = p_X(x_i)p_Y(y_j) \qquad \text{for } i, j = 1, 2.$$

Suppose the results of the test were slightly different from the above, as follows:

		y	
x	0	1	$p_X(\cdot)$
0	$\frac{50}{100}$	$\frac{30}{100}$	$\frac{80}{100}$
1	$\frac{10}{100}$	$\frac{10}{100}$	$\frac{20}{100}$
$p_Y(\cdot)$	$\frac{60}{100}$	$\frac{40}{100}$	

Even though the marginal probability point functions are the same, the random variables are not independent, because

$$p(1, 1) \neq p_X(1) \cdot p_Y(1). \quad \blacksquare$$

Problems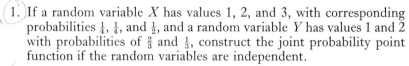

1. If a random variable X has values 1, 2, and 3, with corresponding probabilities $\frac{1}{4}$, $\frac{1}{4}$, and $\frac{1}{2}$, and a random variable Y has values 1 and 2 with probabilities of $\frac{2}{3}$ and $\frac{1}{3}$, construct the joint probability point function if the random variables are independent.

2. The joint probability point function for X and Y is

		y		
x	1	2	3	4
0	$\frac{1}{24}$	$\frac{1}{12}$	$\frac{1}{12}$	$\frac{1}{24}$
1	$\frac{1}{12}$	$\frac{1}{6}$	$\frac{1}{6}$	$\frac{1}{12}$
2	$\frac{1}{24}$	$\frac{1}{12}$	$\frac{1}{12}$	$\frac{1}{24}$

Are X and Y independent random variables?

3. A fair coin is tossed four times. Let X be the number of heads obtained in the first two tosses and Y the number of heads obtained in the last two tosses. Show that X and Y are independent.

4. Refer to Problem 3 and let Z be the random variable whose values are the total number of heads. Are Y and Z independent?

5. A fair coin is tossed three times. Let X be the random variable whose value is 0 if the first toss is a tail and 1 if the first toss is a head, and let Y be the random variable whose value is the algebraic difference of the number of tails from the number of heads. Are X and Y independent?

6. Three distinguishable balls are placed at random into three numbered cells. Let X be the random variable whose values are the number of occupied cells, Y the random variable whose values are the number of balls in the first cell, and Z the random variable whose values are the number of balls in the third cell. Are X and Y independent? Are Y and Z independent?

7. Suppose that in a game of coin tossing you bet \$1 on heads on the first throw and \$1 again on heads on the second throw. Let X be the random variable whose values are the gain on the first throw, and let Y be the random variable whose values are the gain on the second throw. Show that X and Y are independent.

8. Suppose that in the game of coin tossing of Problem 7 you bet $1 on heads on the first throw, as before.   If a head comes up, you will also bet $1 on the second toss; but if tails comes up on the first, you will bet $2 on tails in the second throw.   If X and Y are the random variables whose values are the gains on the first and second throws, respectively, determine if X and Y are dependent or independent.

9. Refer to Problem 8 of Section 7.1.   Are X and Y independent random variables?

10. In a class of 30 students, 20 are sophomores, 7 are juniors, and 3 are seniors.   At the same time, 18 have brown eyes, 10 have blue eyes, and 2 have green eyes.   Let X denote the color of eyes and Y the class.   If it is assumed that X and Y are independent, find the joint probability point function and the probability that a student selected at random is not a sophomore with brown eyes.

11. In a city election candidate A received 18 percent of the vote, candidate B received 31 percent, candidate C received 40 percent, and candidate D received 11 percent.   It is known that 25 percent of the voters are between 21 and 31 years of age, 30 percent are between 31 and 41, 30 percent are between 41 and 51, and 15 percent are over 51.   Suppose (unrealistically) that all eligible voters did vote and that age and candidate preference are independent random variables X and Y.   Find the joint probability point function for X and Y and determine the probability that a randomly selected voter was over 41 and preferred candidate D.

12. The conditional probability for joint random variables can be defined directly from the definition for events as

$$P(Y=y\,|\,X=x) = \frac{p(x, y)}{p_X(x)},$$

provided that $p_X(x) \neq 0$.   Let the joint probability point function for X and Y be given as follows:

	y			
x	1	2	3	4
0	$\frac{1}{4}$	$\frac{1}{16}$	$\frac{1}{8}$	0
1	$\frac{1}{16}$	0	$\frac{1}{16}$	$\frac{1}{8}$
2	$\frac{1}{8}$	$\frac{1}{8}$	0	$\frac{1}{16}$

(a) Find $P(Y=y\,|\,X=1)$ for $y=1, 2, 3, 4$.
(b) Show that $\sum_{y=1}^{4} P(Y=y\,|\,X=1) = 1$.
(c) Find $P(Y<3\,|\,X>0)$.

13. The joint probability function $p(x, y)$ of the random variables X and Y is given in the following table:

	y		
x	-1	0	1
-1	$\frac{1}{8}$	0	$\frac{1}{8}$
1	$\frac{1}{2}$	$\frac{1}{4}$	0

   (a) Make a table of values of the conditional probability for Y, given that $X = 1$.
   (b) Make a table of values of the conditional probability for X, given that $Y = -1$.

14. A number is selected from the set $\{3, 4, 5, \ldots, 10\}$ and then a number is selected from the set of positive integers less than the first number. Let X denote the first number selected and Y the second number. Find

   (a) $P(Y < 5 \mid X = x)$ for each value of X.
   (b) $F_Y(5)$, where F_Y is the probability distribution function for Y.

15. If X and Y are random variables defined on the same sample space, the joint probability point function can be represented in tabular form with the columns headed by values of Y and the rows by values of X. Show that if X and Y are independent, then the probabilities in each row may be obtained by multiplying the corresponding probabilities in any other row by the same number, where the number used for multiplying depends only on the rows involved and in no way depends on the values of the random variable Y.

16. Let X and Y be two independent random variables defined on the same sample space. Let $Z = f(X)$ and $W = g(Y)$, two new random variables functionally related to X and Y. Show that W and Z are independent random variables.

7.3 Sum and Product of Two Random Variables

If X and Y are random variables defined on the same sample space S, then combinations of X and Y, such as the sum $X + Y$ and the product XY, are also random variables. This is evident, since for each elementary event of the sample space S there is a corresponding real value for X and for Y.

In order to find the probability point function for $X + Y$ and XY, we shall use the joint probability point function for X and Y.

We begin with $Z = X + Y$ as the new random variable. The probability point function $p_Z(\cdot)$ is obtained by forming the sum of all those joint probabilities $p(x_i, y_j)$ for which $x_i + y_j = z_k$. Example 1 will help clarify the procedure.

Example 1

Let the joint probability point function for the random variables X and Y be given as shown in Table 1. Since $S_X = \{1, 2, 3, 4\}$, $S_Y = \{0, 1, 2\}$,

Table 1

			y	
x		0	1	2
1		0.15	0.05	0.05
2		0.1	0.05	0.1
3		0.1	0.05	0.1
4		0.1	0.05	0.1

and $Z = X + Y$, then $S_Z = \{1, 2, 3, 4, 5, 6\}$. If we wish to calculate the probability that $Z = 3$, we see that it is the sum of $P(X = 3, Y = 0)$, $P(X = 2, Y = 1)$, and $P(X = 1, Y = 2)$. Therefore, $P(Z = 3) = 0.1 + 0.05 + 0.05 = 0.2$. In a similar manner for each other value of Z, we construct the complete probability point function as

z_i	1	2	3	4	5	6
$p(z_i)$	0.15	0.15	0.2	0.2	0.2	0.1

Formally, the expression for probability point function for the sum of two random variables, $p_Z(\cdot)$, can be written as follows: If $S_X = \{x_1, x_2, \ldots, x_m\}$, $S_Y = \{y_1, y_2, \ldots, y_n\}$, and $S_Z = \{z_1, z_2, \ldots, z_t\}$, then

$$(1) \qquad p_Z(z_k) = \sum_{x_i + y_j = z_k} p(x_i, y_j) \qquad \text{for } k = 1, 2, \ldots, t,$$

where the sum is taken over all pairs (x_i, y_j) such that $x_i + y_j = z_k$.

We must show that this definition satisfies the requirements for a probability point function.

First we note that the numbers $p_Z(z_k)$ are nonnegative and less than 1, since they are sums of a joint probability function. Next we must show that

$$\sum_{k=1}^{t} p_Z(z_k) = 1.$$

By Equation (1),

(2)
$$\sum_{k=1}^{t} p_Z(z_k) = \sum_{k=1}^{t} \left(\sum_{x_i + y_j = z_k} p(x_i, y_j) \right).$$

Each of the $p(x_i, y_j)$ appears in the double sum when and only when $x_i + y_j = z_k$. Since each pair (x_i, y_j) gives rise to some z_k, every $p(x_i, y_j)$ is accounted for exactly once. Therefore,

$$\sum_{j=1}^{n} \sum_{i=1}^{m} p(x_i, y_j) = 1,$$

since it is the sum of a probability point function over its whole space.

By analogy to the foregoing example of the sum of random variables, the example of their product unfolds quickly. The probability point function for $W = XY$ is obtained as the sum of those joint probabilities for which the pairs of values of X and Y give the same value of W. In symbols,

(3)
$$p_W(w_k) = \sum_{x_i y_j = w_k} p(x_i, y_j),$$

where the sum is taken over all pairs (x_i, y_j) such that $x_i y_j = w_k$. It should be clear from the reasoning used previously that $p_W(w_k)$ is a proper probability function.

Example 2

Refer to the random variables X and Y in Example 1. If $W = XY$, then $S_W = \{0, 1, 2, 3, 4, 6, 8\}$, and the values for $p_W(\cdot)$ are calculated using equation (3). The complete probability point function is

w_i	0	1	2	3	4	6	8
$p_W(w_i)$	0.45	0.05	0.1	0.05	0.15	0.1	0.1

where, for example,

$$p_W(2) = p(2, 1) + p(1, 2) = 0.05 + 0.05 = 0.1. \quad \blacksquare$$

Knowing the probability point functions of random variables given by the sum and product of two random variables allows us to compute the expected value directly from the definition. We might ask, however, whether the expectation of the sum and product of two random variables is related to the expected values of the individual random variables. A very appealing result would be $E(X + Y) = E(X) + E(Y)$ and $E(XY) = E(X)E(Y)$. Unfortunately, the two situations are not quite the same, as we shall show. Consider the sum first.

Theorem 1

If X and Y are random variables defined on the same sample space with expected values $E(X)$ and $E(Y)$, then the expectation of the sum $X + Y$ is given by $E(X + Y) = E(X) + E(Y)$.

PROOF: Let $p_X(\cdot)$ be the probability point function for X, defined on $S_X = \{x_1, x_2, \ldots, x_m\}$, and let $p_Y(\cdot)$ be the probability point function for Y, defined on $S_Y = \{y_1, y_2, \ldots, y_n\}$. By definition

$$E(X) = \sum_{i=1}^{m} x_i p_X(x_i),$$

$$E(Y) = \sum_{j=1}^{n} y_j p_Y(y_j).$$

We can rewrite these sums in terms of the marginal probability point functions to obtain

$$E(X) = \sum_{i=1}^{m} x_i \left(\sum_{j=1}^{n} p(x_i, y_j) \right) = \sum_{i=1}^{m} \sum_{j=1}^{n} x_i p(x_i, y_j),$$

$$E(Y) = \sum_{i=1}^{n} y_j \left(\sum_{i=1}^{m} p(x_i, y_j) \right) = \sum_{j=1}^{n} \sum_{i=1}^{m} y_j p(x_i, y_j)$$

$$= \sum_{i=1}^{m} \sum_{j=1}^{n} y_j p(x_i, y_j).$$

Adding these equations, we obtain

$$(4) \qquad E(X) + E(Y) = \sum_{i=1}^{m} \sum_{j=1}^{n} (x_i p(x_i, y_j) + y_j p(x_i, y_j))$$

$$= \sum_{i=1}^{m} \sum_{j=1}^{n} (x_i + y_j) p(x_i, y_j).$$

Each term $x_i + y_j$ determines a value of the random variable $Z = X + Y$, which we call z_k. Collecting all those terms in the double sum for which $x_i + y_j = z_k$, we see that the sum of the corresponding probabilities is precisely $p_Z(z_k)$. Therefore, the right side of (4) can be represented as

$$\sum_{k=1}^{t} z_k p_Z(z_k),$$

which is precisely the expected value of $Z = X + Y$. Hence

$$E(X) + E(Y) = E(X + Y). \qquad \blacksquare$$

Example 3

There were 100 students enrolled in Calculus I and Humanities I. Let X be the grade points earned in Calculus I and Y the grade points earned in Humanities I. The joint probability point function for X and Y is given in Table 2.

Table 2

			y			
x	0	1	2	3	4	$p_X(\cdot)$
0	0.03	0.02	0	0	0	0.05
1	0.02	0.03	0.10	0	0	0.15
2	0	0.05	0.30	0.10	0	0.45
3	0	0	0.05	0.05	0.10	0.20
4	0	0	0.05	0.05	0.05	0.15
$p_Y(\cdot)$	0.05	0.10	0.50	0.20	0.15	

From the marginal probability point functions,

$$E(X) = \sum_{k=1}^{5} x_k p_X(x_k) = 2.25,$$

$$E(Y) = \sum_{k=1}^{5} y_k p_Y(y_k) = 2.30.$$

Therefore, using Theorem 1,

$$E(X + Y) = E(X) + E(Y) = 4.75.$$

The sample space for $Z = X + Y$ is $S_Z = \{0, 1, \ldots, 8\}$, and the probability point function can be calculated by adding the entries on the diagonals:

z_k	0	1	2	3	4	5	6	7	8
$p_Z(z_k)$	0.03	0.04	0.03	0.15	0.30	0.15	0.10	0.15	0.05

and, of course,

$$E(Z) = \sum_{k=1}^{9} z_k p_Z(z_k) = 4.75. \qquad \blacksquare$$

Our discussion of the sum of two random variables extends in a natural way to the sum of an arbitrary number of random variables. Let X_1, X_2, ..., X_n be n random variables defined on the same sample space S. For any elementary event of S, each random variable defines a real number, and the sum of these n real numbers defines a new random variable

$$Z = \sum_{k=1}^{n} X_k$$

on the sample space S. There is an immediate generalization of Theorem 1 to any finite sum of random variables.

Theorem 2

If X_1, X_2, ..., X_n are n random variables defined on the same sample space with expected values $E(X_k)$, $k = 1, 2, \ldots, n$, then

$$E\left(\sum_{k=1}^{n} X_k\right) = \sum_{k=1}^{n} E(X_k);$$

that is, the expected value of the sum is the sum of the expected values.

Since Theorem 1 establishes this theorem for $n = 2$, we have the first part of a mathematical induction argument. The details of the second part of the argument are left to the reader.

Example 4

The number of children in a family often can be considered to be a random phenomenon. Let the sample space be the set

$$S = \{0, 1, 2, \ldots, 8\},$$

where 8 is assigned to all families with eight or more children, and let us assume the following probabilities:

x_k	0	1	2	3	4	5	6	7	8
$p(x_k)$	0.05	0.15	0.26	0.32	0.10	0.06	0.03	0.01	0.02

If X denotes the random variable for this numerical-valued random phenomenon, then we find, by a direct calculation, that $E(X) = 2.67$.

Now define the random variable X_i, $i = 1, 2, \ldots, 50$, to be the results of 50 separate instances of the selection of a family and the recording of the number of children in the family. Each of the X_i is defined on the sample space S with the probability point function defined above. Therefore, $E(X_i) = 2.67$ for $i = 1, 2, \ldots, 50$, and if $Z = \sum_{i=1}^{50} X_i$, then

$$E(Z) = E\left(\sum_{i=1}^{50} X_i\right) = \sum_{i=1}^{50} E(X_i) = 50(2.67) = 133.5. \quad \blacksquare$$

The next theorem settles partially the problem of computing the expected value for the random variable $W = XY$. Notice, however, that we now need the additional hypothesis that X and Y are independent random variables.

Theorem 3

Let X and Y be random variables defined on the same sample space with expectations $E(X)$ and $E(Y)$. If X and Y are independent, then

$$E(XY) = E(X)E(Y).$$

PROOF: Let $S_X = \{x_1, \ldots, x_m\}$ with the probability point function $p_X(\cdot)$, and let $S_Y = \{y_1, \ldots, y_n\}$ with the probability point function $p_Y(\cdot)$. By definition

$$E(X) = \sum_{i=1}^{m} x_i p_X(x_i),$$

$$E(Y) = \sum_{j=1}^{n} y_j p_Y(y_j).$$

Now consider the product

$$E(X)E(Y) = \left(\sum_{i=1}^{m} x_i p_X(x_i) \right) \left(\sum_{j=1}^{n} y_j p_Y(y_j) \right),$$

which when multiplied out becomes

$$E(X)E(Y) = \sum_{i=1}^{m} \sum_{j=1}^{n} x_i y_j p_X(x_i) p_Y(y_j).$$

By hypothesis X and Y are independent random variables, so that

$$p_X(x_i) p_Y(y_j) = p(x_i, y_j).$$

Since this relationship is valid for all values of i and j, it follows that

$$(5) \qquad E(X)E(Y) = \sum_{i=1}^{m} \sum_{j=1}^{n} x_i y_j p(x_i, y_j).$$

The expression on the right side is precisely $E(XY)$. We can rearrange the right side of (5) using the same argument used on equation (3) to obtain

$$(6) \qquad E(X)E(Y) = \sum_{k=1}^{t} z_k p_Z(z_k),$$

where $Z = XY$. However, since the right side of (6) is just $E(Z) = E(XY)$, we have

$$E(X)E(Y) = E(XY). \qquad \blacksquare$$

Example 5

A number of subjects are observed in a learning experiment with a puzzle. Let X denote the elapsed time to complete the puzzle on the first trial, with 1 and 2 used to denote the intervals 0–10 minutes and 10–20 minutes, respectively. Let Y be the number of trials required to "learn the puzzle." Find $E(XY)$.

Suppose that the joint probability point function is that given in Table 3. Since the joint probability point function is the product of

Table 3

x	1	2	3	4	$p_X(\cdot)$
			y		
1	$\frac{3}{16}$	$\frac{3}{16}$	$\frac{1}{4}$	$\frac{1}{8}$	$\frac{3}{4}$
2	$\frac{1}{16}$	$\frac{1}{16}$	$\frac{1}{12}$	$\frac{1}{24}$	$\frac{1}{4}$
$p_Y(\cdot)$	$\frac{1}{4}$	$\frac{1}{4}$	$\frac{1}{3}$	$\frac{1}{6}$	

$p_X(x_i)$ and $p_Y(y_j)$ for all pairs x_i and y_j, X and Y are independent random variables and $E(XY)$ is the product of $E(X)$ and $E(Y)$. By the definition, $E(X) = \frac{5}{4}$ and $E(Y) = \frac{29}{12}$, so $E(XY) = \frac{145}{48}$ without further calculation. ∎

In the proof of Theorem 3 we used the analytical statement of independence to deduce a property of expectation. In Section 7.4 we shall use Theorem 3 to obtain an expression for the variance of the sum of two random variables in terms of their individual variances.

Problems

1. Suppose that X and Y have the following joint probability point function:

x	1	2	3
		y	
1	0.1	0.1	0
2	0.1	0.2	0.3
3	0.1	0.1	0

Find the probability point function of $X + Y$ and compute $E(X + Y)$.

2. Using the joint probability point function of Problem 1, find the probability point function for XY and compute $E(XY)$.

3. Using the joint probability point function in Problem 1, show that $E(XY) = E(X)E(Y)$ even though X and Y are dependent random variables.

4. If a cross of two varieties of peas yields yellow and green peas in the ratio $3 : 1$, what is the expected value of the total number of green peas in a planting of 100 seeds?

5. Seventy percent of the women who visit a Planned Parenthood Center have had no previous experience with any form of birth control. If 50 women are interviewed in a day, what is the expected value of the total number who have had some experience with birth control?

6. Let X and Y have the joint probability point function

		y
x	0	1
0	0	$\frac{1}{4}$
1	$\frac{1}{3}$	$\frac{5}{12}$

Compute $E(X + Y)$ and compare the result with $E(X) + E(Y)$.

7. Refer to the joint probability point function of Problem 6. Compute $E(XY)$ and compare the result with $E(X)E(Y)$.

8. Refer to the joint probability point function of Problem 6. Determine the probability point functions for $X + Y$ and XY.

9. Let X be the number of heads and Y the largest number of consecutive heads in five independent tosses of a fair coin, with $Y = 0$ if no heads appear.

(a) Compute $E(X)$ and $E(Y)$. (b) Compute $E(X + Y)$.
(c) Compute $E(XY)$.   (d) Are X and Y independent?

10. Of the registered voters in one city, 60 percent are registered Democratic, 40 percent have two or more cars, 75 percent are under 35, and 70 percent are married. Three people are selected at random and the number of positive responses in each category is recorded. What is the expected value of the total of these numbers?

11. A cube has one spot on each of four sides and two spots on each of the other two sides. Find the expected value of the total number of spots showing when 10 such cubes are tossed.

12. What is the expected weight of a box of 100 bolts, if the weight of an individual bolt is a random variable with an expected value of 1 ounce?

13. Sixty-five percent of the freshman class at a college came from public schools, and the remainder were graduates of private schools. Forty percent of the class took the placement examination in mathematics and $\frac{3}{4}$ of them received a satisfactory grade. Let X denote the type of high school with 0 for public and 1 for private. Let Y be 0 if a student did not take the mathematics examination, 1 if he did and failed, and 2 if he did and passed. Find $E(X+Y)$ and $E(XY)$.

14. Thirty percent of the population of a city are of Polish ancestry and 20 percent are of Irish ancestry. Let X be the number of people of Polish ancestry and Y the number of Irish ancestry selected in a random sample of two people. Find $E(X+Y)$ and $E(XY)$.

15. Five cards are drawn from an ordinary deck of playing cards. Let X denote the number of aces and Y denote the number of kings. Determine the joint probability point function and the expected value of $X+Y$.

16. A box contains three balls numbered 1, 2, and 3. First, a ball is drawn from the box; second, a fair coin is tossed the number of times shown on the ball. Let X be the number on the ball, and Y the number of heads that appear in the tosses of the coin. Determine $E(X)$, $E(Y)$, and $E(X+Y)$.

17. If X is a random variable with an expected value of 30 and Y is a random variable with an expected value of 50, what is the expected value of $X+Y$, and $X-Y$? Find the expected value of $aX+bY$, where a and b are any real numbers.

18. Let X and Y have the joint probability point function given in Problem 1. Using the approach that was used for $X+Y$, define the random variable $T=X/Y$, compute the probability point function for T, and find $E(T)$.

19. Let X and Y have the joint probability point function given in Problem 1. Proceeding in a manner similar to that used in the text for $X+Y$, define the random variables $Z=\text{maximum }(X, Y)$ and $W=\text{minimum }(X, Y)$, compute the probability point function for each, and find $E(Z)$ and $E(W)$.

20. Suppose that $p(x, y)=kxy$ at the points $(1, 1)$, $(2, 1)$, $(3, 1)$, and zero elsewhere.

(a) Evaluate the constant k.

(b) Determine $E(X + Y)$.

(c) Are X and Y independent?

21. Prove Theorem 2; that is, complete the mathematical induction argument.

22. Compute the expectation $E((X - E(X))(Y - E(Y)))$ under the assumption that X and Y are independent random variables.

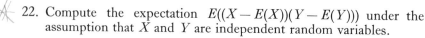

7.4 Variance of Sums of Random Variables

Having shown that $E(X + Y) = E(X) + E(Y)$, we would like to obtain a similar result for the variance of the sum of two random variables. We can do so, provided that X and Y are independent.

Theorem 1

Let X and Y be random variables defined on the same sample space with variances $V(X)$ and $V(Y)$. If X and Y are independent, then the variance of the sum is the sum of the variances; that is,

$$V(X + Y) = V(X) + V(Y).$$

PROOF: By definition, if $Z = X + Y$,

$$V(Z) = E(Z^2) - (E(Z))^2$$

can be written as

$$(1) \qquad V(X + Y) = E((X + Y)^2) - (E(X + Y))^2$$
$$= E(X^2 + 2XY + Y^2) - (E(X) - E(Y))^2$$

Applying Theorem 1 of Section 7.3 to the first term and squaring the second term of the right side of (1), we obtain

$$V(X + Y) = E(X^2) + E(2XY) + E(Y^2)$$
$$- (E(X))^2 - 2E(X)E(Y) + (E(Y))^2,$$

which we write as

$$(2) \qquad V(X + Y) = E(X^2) - (E(X))^2 + E(Y^2)$$
$$- (E(Y))^2 + E(2XY) - 2E(X)E(Y).$$

Since X and Y are independent, we know by Theorem 3 of Section 7.3 that

$$E(2XY) = 2E(X)E(Y).$$

Consequently, the last two terms of the right side of (2) add to zero. Since

$$V(X) = E(X^2) - (E(X))^2 \quad \text{and} \quad V(Y) = E(Y^2) - (E(Y))^2,$$

it follows that

$$V(X + Y) = V(X) + V(Y). \quad \blacksquare$$

Using the σ notation for the standard deviation, we also have

$$\sigma_{X+Y}^2 = \sigma_X^2 + \sigma_Y^2.$$

Example 1

Toss a fair die twice. Let X be the random variable whose values are the number of spots on the first toss, and let Y be the random variable whose values are the number of spots on the second toss. By a straightforward calculation, the variance $V(X) = \frac{35}{12}$. Since Y is identically distributed, it follows that the variance $V(Y) = \frac{35}{12}$. Furthermore, since these random variables are independent, $V(X + Y) = V(X) + V(Y) = \frac{35}{6}$. (Compare this relatively simple calculation with the one that must be performed to obtain the variance of the random variable that is the sum of the spots on a sample space of 36 elementary events for the toss of two dice.) $\quad \blacksquare$

Example 2

A learning experiment on the memorization of nonsense syllables was conducted in a psychology class of 30 students. Each student spent 1 minute memorizing nonsense syllables and then was tested to determine the number correctly learned. The results were

No. of syllables	3	4	5	6	7	8	9
No. of students	2	3	7	6	4	3	5

If X denotes the number of syllables learned for a randomly selected student, then

x_i	3	4	5	6	7	8	9
$p(x_i)$	$\frac{2}{30}$	$\frac{3}{30}$	$\frac{7}{30}$	$\frac{6}{30}$	$\frac{4}{30}$	$\frac{3}{30}$	$\frac{5}{30}$

and

$$E(X) = 6.2, \quad V(X) = 3.23.$$

The age distribution for the class, with age taken to the nearest year, is as follows:

Age	17	18	19	20	21
Number	1	6	20	1	2

Let Y denote the age of a student selected at random. Then, as above, the probability point function can be calculated with the expected value $E(Y) = 18.9$ and the variance $V(Y) = 0.6$.

Suppose that a new measure is formed by adding X and Y to form $Z = X + Y$. If we assume that X and Y are independent random variables, then for the new random variable $X + Y$ we have immediately that

$$E(X + Y) = 6.2 + 18.9 = 25.1,$$
$$V(X + Y) = 3.23 + 0.6 = 3.83.$$

Whether or not the assumption of independence is reasonable is a question that the psychologist must answer, and not a proper question here. ∎

In order to extend Theorem 1 to a sum of several random variables, we need the analogue of the definition of mutually independent events for mutually independent random variables.

Definition 1: Mutually independent random variables

The random variables $X_1, X_2, \ldots, X_n$, defined on the same sample space, are called mutually independent if and only if

$$P(X_1 = a_1, X_2 = a_2, \ldots, X_n = a_n)$$
$$= P(X_1 = a_1)P(X_2 = a_2) \cdots P(X_n = a_n)$$

for all possible values $a_1, a_2, \ldots, a_n$, and for all possible subsets of the set of n random variables.

The mutual independence of the random variables is the same as the mutual independence of the events $\{X_1 = a_1\}, \{X_2 = a_2\}, \ldots, \{X_n = a_n\}$ as defined in the postscript to Definition 2 of Section 4.3, which requires that all subsets of the set n events must also be independent. Consequently, any subset of mutually independent random variables are themselves mutually independent.

Now the additivity property of the variance can be easily generalized.

Theorem 2

Let $X_1, X_2, \ldots, X_n$ be n mutually independent random variables defined on the same sample space, with variances $V(X_i)$, $i = 1, 2, \ldots, n$.   Then

$$V\left(\sum_{i=1}^{n} X_i\right) = \sum_{i=1}^{n} V(X_i).$$

The proof of this theorem is another exercise in the use of the principle of mathematical induction and is left as an exercise for the reader.

Example 3

An experiment is conducted on the litter size of cocker spaniels.   Let X denote the litter size.   One assumes the probability point function for X to be as follows:

x_i	1	2	3	4 (or more)
$p(x_i)$	0.5	0.2	0.2	0.1

Let Z denote the random variable whose values are the total number of offspring produced in 75 matings.   We shall find the expected value and variance of Z by expressing Z as the sum of 75 random variables and then using Theorem 2 of this section and Theorem 2 of Section 7.3. Let X_i for $i = 1, 2, \ldots, 75$ denote the size of the ith litter.   Then $Z = \sum_{i=1}^{75} X_i$, since the value of the sum is the total of the litter sizes. For each i,

$$E(X_i) = \sum_{i=1}^{4} x_i p(x_i) = 1.9,$$

$$V(X_i) = \sum_{i=1}^{4} (x_i - 1.9)^2 p(x_i) = 1.09. \qquad \top = 1.04$$

Hence

$$E(Z) = E\left(\sum_{i=1}^{75} X_i\right) = \sum_{i=1}^{75} E(X_i) = 75(1.9) = 142.5.$$

Since we may assume that the X_i are independent random variables, we have

$$V(Z) = V\left(\sum_{i=1}^{75} X_i\right) = \sum_{i=1}^{75} V(X_i) = 75(1.09) = 81.75. \qquad \blacksquare$$

When two random variables are dependent, equation (2) provides the correct expression for the variance of the sum $X + Y$.   The expression

$$V(X-Y) = V(X) + V(Y)$$

$E(X)E(Y) - E(XY)$ is not zero as before. In fact, $E(XY) - E(X)E(Y)$ is called the covariance of X and Y and figures importantly in such topics as correlation analysis and regression analysis, which we have elected to omit from this book. The reader is urged to consult any book on mathematical statistics such as [1, Chap. 12] for further information on these topics.

Problems

1. Consider the following joint probability point function for X and Y:

		Y		
X	1	2	3	4
0	$\frac{1}{24}$	$\frac{2}{24}$	$\frac{2}{24}$	$\frac{1}{24}$
1	$\frac{2}{24}$	$\frac{4}{24}$	$\frac{4}{24}$	$\frac{2}{24}$
2	$\frac{1}{24}$	$\frac{2}{24}$	$\frac{2}{24}$	$\frac{1}{24}$

Find the probability point function for $X + Y$ and compute $V(X + Y)$.

2. Refer to Problem 1. Show that X and Y are independent random variables and that $V(X) + V(Y) = V(X + Y)$.

3. A die has one spot on each of four sides and two spots on each of the other two sides. Find the variance of the total number of spots showing when 10 such dice are tossed.

4. What is the variance of the sum of numbers that appear when six fair dice are rolled?

5. If a machine produces defective items with a probability of $\frac{1}{20}$, what is the variance of the number of defective items in a batch of size 500?

6. If a certain drug will effect a cure one time in four, what is the variance of the number of patients cured in a group of 200?

7. If X has an expected value of 40 and a standard deviation of 10, Y has an expected value of 35 and a standard deviation of 6, and X and Y are independent, find the expected value and the standard deviation of $X + Y$ and $X - Y$.

8. A machine makes small round disks of varying thicknesses. Let X be a random variable whose values are the thicknesses of the disks. Suppose that the standard deviation of the thicknesses is 0.003 inch.

If two disks are assembled one on top of the other (like checkers), what is the standard deviation of the heights of the finished assemblies?

9. Let X denote the number of trials needed to solve a puzzle. From previous testing, the probability point function

x_i	1	2	3	4	5 (or more)
$p(x_i)$	0.2	0.4	0.2	0.1	0.1

has been assumed for X. One hundred students are tested. Find the expected value and variance of the total number of trials.

10. A company produces gas furnaces and it has been found that the probability is 0.02 that the furnace will not last the full guarantee period. The cost to the company is $500 for each such unit. The probability that a part (or parts) will need to be replaced on other units is 0.15, with a cost of $25 per unit. The company sells 1000 furnaces. Let Z be the total amount spent to fulfill the guarantee. Find $E(Z)$ and $V(Z)$.

11. A college has found that the amount a new alumnus donates to the alumni fund the first year after graduation depends on the sex of the graduate. For a male, the amount X (in dollars) and the associated probabilities are

x_i	5	10	25	50	100
$p(x_i)$	0.20	0.25	0.5	0.10	0.05

For a female the amount Y (in dollars) and the associated probabilities are

y_i	5	10	25	50
$p(y_i)$	0.35	0.30	0.30	0.05

This year's graduating class has 120 men and 100 women. If Z is the total amount donated the first year after graduation, find $E(Z)$ and $V(Z)$.

12. In order to estimate their awareness of names in the current news, a standard test was given to the students in the sixth grade of Jones Elementary School. From previous tests the probability point

function of the random variable X, which denotes the number of correct identifications, has been found to be

x_i	0	1	2	3	4	5
$p(x_i)$	0.05	0.05	0.1	0.2	0.4	0.2

Let Z be the total score for 88 students. Find $E(Z)$ and $V(Z)$.

13. Refer to Problem 12. The heights of the 88 sixth graders were found to be

Height, inches	48	49	50	51	52	53
Number	10	24	30	16	4	4

Let X denote the height and Y the score of a randomly selected student. Find $V(X + Y)$ and $V(X - Y)$ under the assumption of the independence of X and Y.

14. The total income (to the nearest thousand) of 50 faculty households was found to be

Income, thousands of dollars	10	11	12	13	14	15
Number	12	5	8	10	7	8

The income from salaries alone was found to be

Income, thousands of dollars	9	10	11	12	13	14	15
Number	3	10	12	8	7	5	5

If Z denotes the total income, X the income from wages, and Y the income from other sources of a household selected at random, find $V(Y)$ under the assumption that X and Y are independent random variables.

15. Two coins and one die are tossed 10 times. Score 1 point for each head and each spot showing on each toss. Find the expected value and variance of the total score.

16. Ten numbers are selected at random and with replacement from the digits $0, 1, \ldots, 9$. If Z is the sum of those selected, find $E(Z)$ and $V(Z)$.

17. Let a and b be any real numbers and X and Y be independent random variables defined on the same space. Prove that

$$V(aX + bY) = a^2 V(X) + b^2 V(Y).$$

18. Let $X_1, X_2, \ldots, X_n$ be mutually independent random variables defined on the same sample space. Show that if k is any positive integer less than n and $Y_k = X_1 + X_2 + \cdots + X_k$, then Y_k and X_{k+1} are independent. (*Hint:* Make direct use of Definition 1.)

19. Using the results of Problem 18, prove Theorem 2 by mathematical induction.

20. Toss a fair coin n times and let $X_1, X_2, \ldots, X_n$ be the random variables whose values are 1 if the coin shows a head and 0 if the coin shows a tail, where X_1 is the result of the first toss, X_2 the result of the second toss, and so on. Calculate the variance of the sum of these n variables.

21. Let $X_1, X_2, \ldots, X_n$ denote n mutually independent random variables that are repetitions of the same Bernoulli trial with probability p of success. Show that the variance of the sum of the random variables is $np(1 - p) = npq$, where $p + q = 1$. The result is that of Theorem 5 of Section 6.4 for the binomial random variable with the parameters n and p.

22. Let X, Y, and Z be mutually independent random variables having the following probability point functions:

x_i	0	1
$p_X(x_i)$	$\frac{1}{4}$	$\frac{3}{4}$

y_i	-1	0	1
$p_Y(y_i)$	$\frac{1}{3}$	$\frac{1}{3}$	$\frac{1}{3}$

z	1	2	3	4
$p_Z(z_i)$	0.1	0.4	0.2	0.3

Construct the probability point function of $X + Y + Z$ and compute $V(X + Y + Z)$.

23. A boy saws boards in lengths of about 3 feet. Let X be a random variable whose values are the actual lengths of the boards. Suppose that $\sigma_X = 0.2$ inch. To check his precision the boy measures the

lengths of the boards and, of course, there is an error in his measurements. Let U be a random variable whose values are the measurements of the lengths of the boards, and suppose that $\sigma_U = 0.25$. Let Y be the random variable whose values are the differences between the actual and measured lengths (that is, the error in the measurement). What is the standard deviation of the errors in the measurement? Assume that X and Y are independent random variables.

7.5 Sample Random Variables— Sample Mean

Let us consider a dart game in which a dart is thrown at a board having three regions marked with the values -1, 0, and 1 (Figure 1).

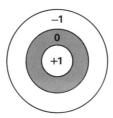

Figure 1

If X denotes the score after a person throws a dart one time, then X is a random variable that takes on the values $\{-1, 0, 1\}$. Now suppose that the same person throws the dart 50 times. Let X_1 be the random variable that denotes the score on the first throw, let X_2 correspond to the second throw, and so on for the 50 throws. Each of these 50 random variables is a duplication of the original random variable X and has the same probability point function as X. If we also assume that each throw has no influence on the others (except psychological ones, which we ignore), then there are 50 independent random variables with identical probability point functions. These are called sample random variables for a sample of size 50. These concepts are formalized in Definition 1.

Definition 1: Sample random variables

Let X be a random variable and let X_1, X_2, ..., X_n denote repeated observations (or measurements) of X. Each of these X_i is a random variable whose probability point function is the same as X, and the set $\{X_1, X_2, ..., X_n\}$ is a set of sample random variables of size n of the random variable X whose members are mutually independent.

Suppose that we are able to find a stable dart thrower for whom it is possible to determine a probability point function for his score X on one throw of the dart. Our data might look like

x_i	-1	0	1
$p(x_i)$	$\frac{1}{5}$	$\frac{3}{5}$	$\frac{1}{5}$

If three darts are thrown, there are 27 possible triples of different scores, which we can write as (x_1, x_2, x_3), where x_i is the outcome for the ith dart. We know the probability point function of X; therefore, the probability of each triple can be computed by taking the products of the probabilities for the individual scores. Thus the probability for the triple (x_1, x_2, x_3) is $p(x_1)p(x_2)p(x_3)$. In addition, let us record the total score for each triple $x_1 + x_2 + x_3$. The average score for each triple, therefore, is

$$(x_1 + x_2 + x_3)/3.$$

These calculations are listed in Table 1.

The three throws of the dart are three repetitions of the same random phenomenon, or equivalently a set of sample random variables of size 3. Let the mean, $\bar{X}$, be expressed as

$$\bar{X} = \tfrac{1}{3}(X_1 + X_2 + X_3),$$

where X_1, X_2, and X_3 are the sample random variables. The values that $\bar{X}$ may have are the "average" scores, and the corresponding probabilities can be computed by forming the sum of the probabilities in Table 1 that are associated with the same average score. For example, $\bar{X}$ will have the value $\frac{1}{3}$ for the triples $(-1, 1, 1)$, $(0, 0, 1)$, $(0, 1, 0)$, $(1, -1, 1)$, $(1, 0, 0)$, and $(1, 1, -1)$. Thus $P(\bar{X} = \frac{1}{3})$ is the sum of the corresponding probabilities:

$$\tfrac{1}{125} + \tfrac{9}{125} + \tfrac{9}{125} + \tfrac{1}{125} + \tfrac{9}{125} + \tfrac{1}{125} = \tfrac{30}{125}.$$

The complete set of values and the corresponding probabilities are

x_i	-1	$-\frac{2}{3}$	$-\frac{1}{3}$	0	$\frac{1}{3}$	$\frac{2}{3}$	1
$p_{\bar{X}}(x_i)$	$\frac{1}{125}$	$\frac{9}{125}$	$\frac{30}{125}$	$\frac{45}{125}$	$\frac{30}{125}$	$\frac{9}{125}$	$\frac{1}{125}$

Observe that $\bar{X}$ is a new random variable, which is defined on a new sample space, and that it has a new probability point function. A particular throw of three darts gives those values which are used in calculating $\bar{X}$, the average score. This example of the darts is somewhat artificial, because very little is usually known about the probability functions of X. Nevertheless, we shall see that it is still possible to take a sample and make assertions, predictions, and estimates about X and its probability function. We first define the sample mean in the obvious way.

Table 1

Triples of possible scores	Probability of the triple	Total score of triple	Average score
$(-1, -1, -1)$	$\frac{1}{125}$	-3	-1
$(-1, -1, 0)$	$\frac{3}{125}$	-2	$-\frac{2}{3}$
$(-1, -1, 1)$	$\frac{1}{125}$	-1	$-\frac{1}{3}$
$(-1, 0, -1)$	$\frac{3}{125}$	-2	$-\frac{2}{3}$
$(-1, 0, 0)$	$\frac{9}{125}$	-1	$-\frac{1}{3}$
$(-1, 0, 1)$	$\frac{3}{125}$	0	0
$(-1, 1, -1)$	$\frac{1}{125}$	-1	$-\frac{1}{3}$
$(-1, 1, 0)$	$\frac{3}{125}$	0	0
$(-1, 1, 1)$	$\frac{1}{125}$	1	$\frac{1}{3}$
$(0, -1, -1)$	$\frac{3}{125}$	-2	$-\frac{2}{3}$
$(0, -1, 0)$	$\frac{9}{125}$	-1	$-\frac{1}{3}$
$(0, -1, 1)$	$\frac{3}{125}$	0	0
$(0, 0, -1)$	$\frac{9}{125}$	-1	$-\frac{1}{3}$
$(0, 0, 0)$	$\frac{27}{125}$	0	0
$(0, 0, 1)$	$\frac{9}{125}$	1	$\frac{1}{3}$
$(0, 1, -1)$	$\frac{1}{125}$	0	0
$(0, 1, 0)$	$\frac{9}{125}$	1	$\frac{1}{3}$
$(0, 1, 1)$	$\frac{3}{125}$	2	$\frac{2}{3}$
$(1, -1, -1)$	$\frac{1}{125}$	-1	$-\frac{1}{3}$
$(1, -1, 0)$	$\frac{3}{125}$	0	0
$(1, -1, 1)$	$\frac{1}{125}$	1	$\frac{1}{3}$
$(1, 0, -1)$	$\frac{3}{125}$	0	0
$(1, 0, 0)$	$\frac{9}{125}$	1	$\frac{1}{3}$
$(1, 0, 1)$	$\frac{3}{125}$	2	$\frac{2}{3}$
$(1, 1, -1)$	$\frac{1}{125}$	1	$\frac{1}{3}$
$(1, 1, 0)$	$\frac{3}{125}$	2	$\frac{2}{3}$
$(1, 1, 1)$	$\frac{1}{125}$	3	1

Definition 2: Sample mean

Let $X_1, X_2, \ldots, X_n$ be a set of sample random variables of size n of the random variable X. Then the sample mean is the random variable

$$\bar{X} = \frac{1}{n}(X_1 + X_2 + \cdots + X_n) = \frac{1}{n} \sum_{i=1}^{n} X_i.$$

Now we have a random variable X, n "copies" X_k, $k = 1, \ldots, n$, and another random variable $\bar{X}$ formed from the sum of these random variables. Since the X_k are defined on the same sample space, we can appeal to the theorems of Sections 7.3 and 7.4 for a derivation of the expectation and the variance of $\bar{X}$. Remember that the X_k are mutually independent random variables.

Theorem 1

Let n be a positive integer, X a random variable for which expectation and variance exist, and $\bar{X}$ the sample mean for X for a sample of size n. Then

$$E(\bar{X}) = E(X),$$

$$V(\bar{X}) = \frac{1}{n} V(X).$$

PROOF: Let $Z_n = X_1 + \cdots + X_n$, so that $\bar{X} = (1/n)Z_n$. Using the previous results for the algebra of expectations, we can write the following:

$$E(\bar{X}) = \frac{1}{n} E(Z_n),$$

$$E(Z_n) = E\left(\sum_{i=1}^{n} X_i\right)$$

$$= E(X_1) + E(X_2) + \cdots + E(X_n)$$

$$= nE(X).$$

Each $E(X_k)$ for $k = 1, 2, \ldots, n$ has been replaced with $E(X)$, because the X_k have the same probability point function as X and, therefore, the same expected value. Hence,

$$E(\bar{X}) = \frac{1}{n} nE(X) = E(X).$$

The proof for the variance follows:

$$V(\bar{X}) = V\left(\frac{1}{n} Z_n\right) = \frac{1}{n^2} V(Z_n).$$

Because the X_k are mutually independent, we can write

$$\begin{aligned} V(Z_n) &= V(X_1 + \cdots + X_n) \\ &= V(X_1) + \cdots + V(X_n) \\ &= nV(X). \end{aligned}$$

Hence

$$\begin{aligned} V(\bar{X}) &= \left(\frac{1}{n}\right)^2 V(Z_n) \\ &= \left(\frac{1}{n}\right)^2 nV(X) \\ &= \frac{1}{n} V(X). \quad\blacksquare \end{aligned}$$

Example 1

In a psychological experiment using a maze with a reward, the probability that a rat will solve the maze and reach the food in a fixed length of time has been determined to be 0.3. Let X denote the Bernoulli random variable for this experiment with the values of 0, 1, and the probability $P(X=1) = 0.3$, $P(X=0) = 0.7$. The expected value $E(X) = 0.3$ and the variance $V(X) = 0.21$. Suppose that 50 trials of the experiment are run. The sample space for the sample mean $\bar{X}$ is the set $\{\frac{0}{50}, \frac{1}{50}, \frac{2}{50}, \ldots, \frac{49}{50}, 1\}$.

The probability that $\bar{X}$ take on the value $k/50$, where $k = 0, 1, 2, \ldots,$ 50, is the same as the probability that there are k successes in 50 Bernoulli trials, so we can write

$$P\left(\bar{X} = \frac{k}{50}\right) = \binom{50}{k}(0.3)^k(0.7)^{50-k}.$$

From Theorem 1, however, we know that the expected value and the variance of $\bar{X}$ are

$$E(\bar{X}) = 0.3 \quad \text{and} \quad V(\bar{X}) = \tfrac{1}{50}(0.21) = 0.0042. \quad\blacksquare$$

Example 2

From long use it has been found that a fixed dose of a drug causes an increase of the pulse rate in a random manner. Let X denote this increase in beats per minute with the following probability point function:

x_i	13	14	15	16	17	18	19	20
$p_X(x_i)$	0.05	0.1	0.3	0.25	0.1	0.1	0.05	0.05

The expected value is $E(X) = 16$ and the variance $V(X) = 2.9$.

For a sample of size 10, the sample mean $\bar{X}$ will have the parameters $E(\bar{X}) = 16$ and $V(\bar{X}) = 0.29$.

Suppose that the following results were recorded for ten patients: 13, 15, 16, 20, 16, 18, 15, 17, 15, and 18. The average of these data is 16.3, which is a value in the sample space of $\bar{X}$. Since $V(\bar{X})$ is small relative to the magnitude of $E(\bar{X})$, it is not surprising that the actual observed value of $\bar{X}$ is rather close to the expected value $E(\bar{X}) = 16$. ∎

This relationship of the expectation and variance of the sample mean to $E(X)$ and $V(X)$ can help us to understand actual statistical practices. In the first place, the new random variable $\bar{X}$ is a function of the sample random variables $X_1, X_2, \ldots, X_n$, and the value of $\bar{X}$ depends on the n-tuples of values of the sample random variables. The value of $\bar{X}$ that results from an actual n-tuple of sample values is a number that can be used as an estimate of $E(X)$. Therefore, the sample mean is called an *estimator* of $E(X)$ and is one example of an estimator for $E(X)$.

Since estimation theory is important in mathematical statistics, we should be familiar with some of the vocabulary. One of the most fundamental concepts is contained in Definition 3.

Definition 3: Statistic

Let $X_1, X_2, \ldots, X_n$ be a set of sample random variables of size n of the random variable X. Let h be a function whose domain is the n-tuples of values of the sample random variables and whose codomain is a set of real numbers. The random variable Y defined by this function is called a statistic and is denoted by

$$Y = h(X_1, X_2, X_3, \ldots, X_n).$$

The layman often uses the word statistic incorrectly to mean a piece of data. A person who lays down his life in defense of careless driving is not a statistic, but only an inert piece of data. On the other hand, the sample mean $\bar{X}$ is a statistic, because it satisfies Definition 3. The functional correspondence between $X_1, X_2, \ldots, X_n$ and the real numbers is given by

$$h(X_1, X_2, \ldots, X_n) = \frac{1}{n} \sum_{i=1}^{n} X_i.$$

We have just seen that the variance of $\bar{X}$ is related to $V(X)$ by

$$V(\bar{X}) = \frac{1}{n} V(X).$$

If we examine this relation closely, we observe that as the size of the sample is increased, $V(\bar{X})$ gets smaller. This follows because $V(X)$ is a fixed quantity and n increases as the sample size increases. Now, we recall that the variance in some sense measures the dispersion of the values of the random variable about the mean value, so for large values of n, the probability that an experimental (observed) value of $\bar{X}$ will be near $E(\bar{X}) = E(X)$ will be very high. This concept is extremely important in both probability theory and statistics, and we shall state this remark formally as a theorem in Chapter 10.

Example 3

In a biological experiment the elapsed time in seconds for a reaction after a given stimulus is considered a random phenomenon. Let X denote the elapsed time and assume that $E(X)$ and $V(X)$ exist.

In a series of 15 experiments, the following results were obtained:

4,   7,   5,   8,   4,   3,   3,   7,   8,   2,   4,   8,   8,   3,   3.

The average response time for this sample is found to be 5.1; that is, $\bar{X} = 5.1$. Since it is likely that a value of $\bar{X}$ is near $E(\bar{X})$ and since $E(\bar{X}) = E(X)$, the value of 5.1 can be taken as an estimate of $E(X)$. In another series of 15 experiments, the following results were obtained:

3,   6,   8,   7,   2,   7,   5,   4,   8,   4,   6,   5,   8,   2,   8.

In this case the average response time is 5.5, another of the values of $\bar{X}$ and another estimate of $E(X)$.   ∎

Problems

1. Let X be a random variable with a probability point function

x_i	0	1	2
$p_X(x_i)$	$\frac{1}{4}$	$\frac{1}{2}$	$\frac{1}{4}$

    (a) List the possible random samples of size 2 and determine the probability point function for $\bar{X}$ when $n = 2$.

    (b) Compute $E(\bar{X})$ and $V(\bar{X})$.

2. Let X be a random variable with a probability point function

x_i	−1	0	1	2
$p_X(x_i)$	0.1	0.5	0.3	0.1

    (a) List the possible random samples of size 2 and determine the probability point function for $\bar{X}$ when $n = 2$.

    (b) Compute $E(\bar{X})$ and $V(\bar{X})$.

3. There are ten chips in a bowl with one chip marked with a -1, five chips marked with a 0, and four chips marked with a 2. A chip is drawn at random. Let X be the number on the chip. Suppose that a random sample of size n is drawn with replacement. Compute $E(\bar{X})$ and $V(\bar{X})$ when

    (a) $n = 2$.
                    (b) $n = 10$.

4. A circular dial has ten equal sectors with two sections marked with a 1, three marked with a 2, four marked with a 3, and one marked with a 4. A pointer mounted on the dial is spun and the number of the sector in which the pointer stops is recorded. Let X be the random variable that denotes the result and $\bar{X}$ the sample mean for n spins. Find $E(\bar{X})$ and $V(\bar{X})$.

5. Nine city council members rated the political effectiveness of the mayor on a scale of 1, poor; 2, acceptable; and 3, excellent. Five ratings are 1's, three ratings are 2's, and one rating is a 3. Compute the sample average, $\bar{x}$, and the sample variance,

$$s^2 = \frac{1}{n} \sum_{i=1}^{n} (x_i - \bar{x})^2.$$

6. The following frequency distribution was obtained in a breeding experiment with mice:

No. in litter	1	2	3	4	5	6	7	8	9
Frequency	7	11	16	17	26	31	11	1	1

Compute the sample average, $\bar{x}$, and the sample variance,

$$s^2 = \frac{1}{n} \sum_{i=1}^{n} (x_i - \bar{x})^2.$$

7. The total nitrogen content in milligrams per cubic centimeter of rat blood plasma was determined for 60 rats of age 50 days as follows:

Nitrogen content	0.089	0.092	0.095	0.098	0.101	0.104	0.107
Frequency	2	8	13	17	10	7	3

What assertions, if any, can be made about X, the random variable that denotes the nitrogen content of rat blood plasma?

8. Fifty students are individually interviewed by a panel of student political science majors as to their political leanings. At the conclusion, each student was rated on a scale of 1, 2, 3, and 4, with 1 for liberal and 4 for conservative. If one assumes the uniform probability function for the assignment of the ratings, find the expected value and the variance of the sample mean of the ratings.

9. If the scores of students on a freshman achievement examination in mathematics show a standard deviation of 10, what will the standard deviation be for the sample mean of scores for a sample of size

    (a) 25?   (b) 100?

10. A deck of m cards is numbered 1, 2, ..., m. The deck is shuffled, placed face down, and the top card is turned face up. Before each card is turned, Art guesses what the number will be. If Art does not remember his guess from card to card, so that his guesses are independent and at random, find the expected value and the variance of the random variable whose value is the number of correct guesses.

11. A greenhouse has estimated that $\frac{2}{3}$ of the pansy seeds planted germinate. If 100 pansy seeds are planted, what is the expected number and the variance of the number that germinate? Suppose that 50 boxes of 100 seeds each are planted. Let $\bar{X}$ be the sample mean for the number that germinate in each box. Find $E(\bar{X})$ and $V(\bar{X})$.

12. The grade-point average for the fall term has an expected value of 2.6 and a standard deviation of 0.6. If samples of size 8 are taken from these measurements, what is the expected value and variance of the sample mean $\bar{X}$?

13. In one society it has been observed that the standard deviation of the number of cars per family is 1.5. A researcher desires the standard deviation of $\bar{X}$ to be 0.1, where $\bar{X}$ is his estimate of the mean number of cars per family. What size sample should he take?

14. If the standard deviation of the sample mean $\bar{X}$ in samples of size 9 is 4, what is the standard deviation of the population from which the sample is drawn?

15. A bowl contains an undetermined but large number of chips colored red and white, with the number of each color unknown. On 30 successive random draws with replacement, the following results were obtained:

W, W, R, R, R,   R, R, R, W, R,
R, W, R, W, W,   R, W, W, R, R,
R, R, R, R, W,   R, W, W, R, R.

What assertions can you make about the composition of the bowl (if any)?

16. In a large suburban community, the total income in thousands of dollars of a family unit was determined for 40 families with the following results:

Income, thousands of dollars	7	8	9	10	11	12	13	14	15	16
Frequency	4	6	5	5	6	4	3	4	1	2

What, if anything, can be said about the random variable X that denotes the total income of a family unit?

17. The scores of 20 students on a standardized psychology test, where the expected score is 70, were as follows:

61, 85, 76, 94, 67, 75, 68, 72, 65, 74,
70, 81, 79, 43, 74, 83, 90, 76, 81, 76.

Would you consider this sample an unusual one?

18. Another statistic of interest for a random sample of size n of a random variable X is the *minimum* of the sample: $K = \min(X_1, \ldots, X_n)$. If X is a random variable with a probability point function

x_i	-1	0	1
$p(x_i)$	$\frac{1}{5}$	$\frac{3}{5}$	$\frac{1}{5}$

determine the probability point function for K when the sample size is 3.

19. Refer to Problem 18. Still another statistic is the maximum of a sample: $L = \max(X_1, \ldots, X_n)$. Using the random variable of Problem 18, compute the probability point function for L when the sample size is 3.

20. Two samples of size n_1 and n_2, respectively, are drawn with replacement from a population specified by the probability point function of a random variable X. Let $\bar{X}_1$ and $\bar{X}_2$ denote the respective

sample means, and suppose that these means are independent random variables. Show that

$$E(\bar{X}_1 - \bar{X}_2) = 0,$$

$$V(\bar{X}_1 - \bar{X}_2) = V(X)\left(\frac{1}{n_1} + \frac{1}{n_2}\right).$$

21. If $X_1, X_2, \ldots, X_n$ are the sample random variables of a sample of size n for a Bernoulli random variable X with probability p of success, show that

$$E(\bar{X}) = p,$$

$$V(\bar{X}) = \frac{p(1-p)}{n}.$$

7.6 Chebyshev's Inequality and the Law of Large Numbers

Television-viewer surveys and voter-preference tabulations try to predict, from a small percentage of the people, the way a large population feels about a television program or a political candidate. There are many reasons why statisticians try to make predictions from a small sample. Often the cost of large surveys is prohibitive, or one may find that the population is so large that a complete survey is impossible.

A national election, for example, provides one of the most interesting tests of a statistician's ability to predict the outcome, because his predictions are irrefutably checked on election day. He may try to make his prediction on the basis of a "small" sample of 300 or a "large" sample of 3000. Intuitively a layman may feel that a sample of 3000 should give a much more accurate appraisal than the small sample, and later we shall provide some justification for this reasoning. However, even though professional pollsters can predict voting preferences within 1 or 2 percent, the actual vote can be so close that their prediction is wrong.

There are ways by which one can tell that if a sample of a certain size is chosen, the probability of being close to the true value is high. For example, suppose you wish to know how extensively a daily newspaper is read by the students of a particular school. You ask 50 students (chosen at random, of course) if they read the newspaper, and 28 answer yes. Then $\frac{28}{50} = 0.56$ is an estimate of the proportion of all students who read the newspaper. If you survey 100 students and 64 answer yes, the estimate is $\frac{64}{100} = 0.64$. Which estimate do you choose? Alternatively, suppose you wish to know the moisture content of corn stored in a bin. You take 10 small samples in a random fashion and determine the moisture content of the composite sample to be 13 percent. Later you repeat the procedure using 15 samples, and this time the moisture content is 11 percent. Which estimate do you choose?

The theorem that will provide a justification for a decision on these questions is known as the law of large numbers. There are several versions of the theorem, and the one that we will present is a special case. But first we shall prove a theorem known as *Chebyshev's inequality* (named after the Russian mathematician, P. L. Chebyshev, 1821–1894), which relates a random variable X, its mean $E(X)$, and its variance $V(X)$.

Theorem 1

Chebyshev's Inequality   Let X be a random variable with an expected value $E(X)$ and a variance $V(X)$. Let c be a positive number. Then the probability that a value of X occurs that differs from $E(X)$ by more than c is less than or equal to $V(X)/c^2$. Symbolically,

$$P(|X - E(X)| > c) \leq \frac{V(X)}{c^2}.$$

Theoretically, the probability of the event $\{|X - E(X)| > c\}$ can be computed, especially if the probability point function for X is known. If the probability point function is unknown, then Chebyshev's inequality states that no matter what the probability may be, it does not exceed the value of $V(X)/c^2$.

PROOF: Let us start with the definition of variance of a random variable, and from this set up an inequality for the probability that X deviates from $E(X)$ by more than c.

$$V(X) = E((X - E(X))^2)$$

$$= \sum_{k=1}^{n} (x_k - E(X))^2 p_X(x_k),$$

where as usual the set $\{x_1, x_2, \ldots, x_n\}$ is S_X. Since we are only interested in values of x_k for which $|x_k - E(X)| > c$, drop out all terms of the sum with $(x_k - E(X))^2 \leq c^2$ and we can write

$$V(X) \geq \sum_{k'} (x_{k'} - E(X))^2 p_X(x_{k'}),$$

where the symbol k' indicates that the summation is taken over those indices k for which $|x_k - E(X)| > c$. Each $(x_k - E(X))^2$ remaining in the sum is greater than c^2, so

$$V(X) \geq \sum_{k'} c^2 p_X(x_k) = c^2 \sum_{k'} p_X(x_k).$$

But

$$P(|X - E(X)| > c)$$

is precisely $\sum_{k'} p_X(x_k)$; therefore,

$$V(X) \geq c^2 P(|X - E(X)| > c),$$

$$\frac{V(X)}{c^2} \geq P(|X - E(X)| > c).$$

which is the desired inequality.   ∎

Example 1

The probability is $\frac{3}{5}$ that a convicted burglar released from prison will be arrested and convicted again for burglary. Let X denote the number of a group of 20 convicted burglars released from prison who ultimately are returned to prison. Since X can be considered as a binomial random variable with parameters $n = 20$ and $p = \frac{3}{5}$, we have

$$E(X) = 20 \cdot \tfrac{3}{5} = 12,$$

$$V(X) = 20(\tfrac{3}{5})(\tfrac{2}{5}) = 4.8.$$

According to Theorem 1, the probability that the number returned to prison differs from the expected value of 12 by more than 3 is no larger than

$$\frac{V(X)}{c^2} = \frac{4.8}{9} = 0.533.$$

The actual probability for the event

$$|X - 12| > 3$$

is given as follows:

$$P(|X - 12| > 3) = P(X < 9) + P(X > 15)$$

$$= \sum_{k=0}^{8} \binom{20}{k} \left(\frac{3}{5}\right)^k \left(\frac{2}{5}\right)^{20-k} + \sum_{k=16}^{20} \binom{20}{k} \left(\frac{3}{5}\right)^k \left(\frac{2}{5}\right)^{20-k}$$

$$= 0.0566 + 0.016$$

$$= 0.0726,$$

a value that is considerably less than 0.533.   Also, suppose $c = 1.0$ in the Chebyshev Inequality.   Then the statement becomes

$$P(|X - E(X)| > 1.0) \leq \frac{4.8}{(1.0)^2} = 4.8.$$

This is not incorrect, but neither is it very useful.   It serves to remind us, however, that the expression $V(X)/c^2$ is, in general, a crude estimate of the probability.   ∎

Example 2

The Chebyshev inequality can be used even if the probability distribution of X is not known.   Let X denote the score on a standardized mathematics test for which it is known (or assumed) that $E(X) = 50$ and $V(X) = 10$.   If n subjects take the test, then the sample mean $\bar{X}$ denotes the average test score.   From the results of Section 6.5 we know that

$$E(\bar{X}) = E(X) = 50,$$

$$V(\bar{X}) = \frac{1}{n} V(X) = \frac{10}{n}.$$

If a class of 100 completes the examination, then for c a positive number we can assert that

$$P(|\bar{X} - 50| > c) \leq \frac{1}{10c^2}.$$

This we obtain from Theorem 1 by substituting 50 for $E(\bar{X})$ and 1/10 for $V(\bar{X})$.   In particular, if $c = 2$, then

$$P(|\bar{X} - 50| > 2) \leq 0.025.$$

Since the event $|\bar{X} - 50| \leq 2$ is the complement of the event

$$|\bar{X} - 50| > 2,$$

we can write

$$P(|\bar{X} - 50| > 2) = 1 - P(|\bar{X} - 50| \leq 2),$$

so that

$$1 - P(|\bar{X} - 50| \leq 2) \leq 0.025.$$

Rearranging the inequality we have

$$P(|\bar{X} - 50| \leq 2) \geq 0.975.   \quad ∎$$

Let us return to the problem of justifying the statement "the larger the sample, the better the results."   Eventually we would like to use the

Chebyshev inequality, so we can start by choosing as the random variable the sample mean $\bar{X}$ for a sample of size n. Since $E(\bar{X}) = E(X)$ and $V(\bar{X}) = (1/n)V(X)$, Chebyshev's inequality gives

$$P(|\bar{X} - E(X)| > c) \leq \frac{(1/n)V(X)}{c^2} = \frac{V(X)}{nc^2}.$$

The right side of the inequality has two constants, $V(X)$ and c^2, and a variable n representing the sample size. Thus, as the sample size n is increased, the probability that a value of $\bar{X}$ deviates from the expected value $E(X)$ by more than c decreases to zero. We have just given the essential features of a proof of a special form of the law of large numbers, the mathematical statement that serves as the theoretical basis for the intuitive notion of probability as a measure of the relative frequency of occurrence. This idea will be explained further in the discussion following Example 3.

Theorem 2

Let X be a random variable with the expected value $E(X)$ and variance $V(X)$. If $\bar{X}$ is the sample mean for a set of random variables of size n, then the probability that $\bar{X}$ will deviate from $E(X)$ by more than c, for any positive number c, decreases to zero as the size of the sample increases.

The phrase "decreases to zero" is an intuitive form for the notion of limit of a sequence; see [18, Chap 8].

Example 3

The law of large numbers can be used to answer some practical questions. Let X be the income in thousands of dollars for a large group of people. Suppose that from other information about the group it is reasonable to assume that the variance of X is no greater than 2. How large a sample is required to maintain that the probability is at least 0.95 that the sample mean differs from the theoretical mean value of X by no more than 0.5? In other words, we would like to find the smallest value of n so that

$$P(|\bar{X} - E(X)| \leq 0.5) \geq 0.95.$$

Notice that the complement of the event $\{|\bar{X} - E(X)| > 0.5\}$ is $\{|\bar{X} - E(X)| \leq 0.5\}$. Therefore, by Theorem 3 of Section 2.5, we have

$$P(|\bar{X} - E(X)| \leq 0.5) = 1 - P(|\bar{X} - E(X)| > 0.5).$$

The original inequality can be rewritten as

$$1 - P(|\bar{X} - E(X)| > 0.5) \geq 0.95;$$

hence

$$0.05 \geq P(|\bar{X} - E(X)| > 0.5).$$

From Chebyshev's inequality we know that

$$P(|\bar{X} - E(X)| > 0.5) \leq \frac{2}{n(0.5)^2} = \frac{8}{n}.$$

Consequently, if we select n large enough so that $8/n \leq 0.05$, we have the desired value of n.

Clearly, for any $n \geq 160$ we can be assured that the probability is at least 0.95 that the value of $\bar{X}$ obtained will not differ from $E(X)$ by more than 0.5. ∎

Example 4

Refer to Example 2, where X denotes the score on a test for which $E(X) = 50$ and $V(X) = 10$ and $\bar{X}$ denotes the sample mean average score for n subjects. The statement of Theorem 2 enables us to attack such a question as this: How many scores are needed to assure us that the probability is at least 0.95 that $\bar{X}$ is within 2 of the expected value 50? In other words, what magnitude of n will guarantee the following statement to be true?

$$P(|\bar{X} - 50| \leq 2) \geq 0.95.$$

We must determine some value of n, so let us write Chebyshev's inequality with $c = 2$, n not specified, $E(\bar{X}) = 10$, and $V(\bar{X}) = 10/n$:

$$P(|\bar{X} - 50| > 2) \leq \frac{10}{n \cdot 2^2} = \frac{5}{2n}.$$

Using an argument entirely similar to that used in Example 3, we conclude that we want

$$P(|\bar{X} - 50| > 2) \leq 0.05.$$

For any value of n we know that the probability on the left is $\leq 5/2n$; thus if we select n so that

$$5/2n \leq 0.05,$$

we are done. Solving the inequality for n we find that for any $n \geq 50$ the conditions are satisfied. ∎

Example 5

Suppose the true weight of an object is w. In determining the weight, errors will occur, and suppose we call these random variable X, where the values of X, positive and negative, are taken to be the error (in milligrams). The apparent weight Y will be the sum of the error and the true weight; hence $Y = X + w$. Let us assume also that $E(X) = 0$ and $V(X) = 1$. Then

$$E(Y) = E(X + w) = w,$$
$$V(Y) = V(X) = 1,$$

and Y is a random variable with an expected value and a variance, so it is meaningful to ask how many weighings of the object are needed to ensure that

$$P(|\bar{Y} - w| \le 0.01) \ge 0.95.$$

Since $E(Y) = w$ and $V(Y) = 1$, $E(\bar{Y}) = w$ and $V(\bar{Y}) = 1/n$ for n weighings. Let $c = 0.01$ in Chebyshev's inequality:

$$P(|\bar{Y} - w| > 0.01) \le \frac{1}{n(0.01)^2}.$$

The statement $P(|\bar{Y} - w| \le 0.01) \ge 0.95$ can be rearranged as in the previous examples to read

$$P(|\bar{Y} - w| > 0.01) \le 0.05.$$

Therefore, we only have to solve the following inequality for n:

$$\frac{1}{n(0.01)^2} \le 0.05.$$

We find that any integer exceeding 20,000 will suffice. This number of weighings will guarantee with a probability of 0.95 that the average weighing does not deviate more than 0.01 from the true value. This is not to say that fewer weighings will not do the same thing; it is just that the theory guarantees that the number 20,000 will suffice. ■

In the case of Bernoulli random variables, the Chebyshev inequality is particularly revealing. Let X be a Bernoulli random variable with parameter p for success and let $X_1, X_2, \ldots, X_n$ be a set of sample random variables for X. The random variable $Y = \sum_{k=1}^{n} X_k$ is the number of successes in the n trials and is thus a binomial random variable with parameters n and p. The sample mean $\bar{X} = (1/n) \sum_{k=1}^{n} X_k$ is the ratio of the successes in n trials to the number of trials. Clearly,

$$E(\bar{X}) = E(X) = p,$$

$$V(\bar{X}) = \frac{1}{n} V(X) = \frac{1}{n} pq.$$

Using Chebyshev's inequality we obtain

(1) $$P(|\bar{X} - p| > c) \le \frac{pq}{nc^2}.$$

Now, let r denote the number of successes in n Bernoulli trials, and replace $\bar{X}$ with r/n. Since $pq < 1$, equation (1) can be written

$$P\left(\left| \frac{r}{n} - p \right| > c \right) \le \frac{1}{nc^2}.$$

Hence the probability that the difference between the frequency ratio of occurrence and the theoretical probability is larger than any positive constant c decreases to zero as the number of trials increases without bound. Since

$$P\left(\left|\frac{r}{n}-p\right|\leq c\right)=1-P\left(\left|\frac{r}{n}-p\right|>c\right),$$

we can say that

$$1\geq P\left(\left|\frac{r}{n}-p\right|\leq c\right)\geq 1-\frac{1}{nc^2}.$$

Hence for any constant, no matter how small, the probability that the frequency ratio and true probability differ by that small quantity c tends to 1 as n approaches infinity. Essentially, the last remark implies that the bigger the sample n, the better will be the approximation to p.

Example 6

A poll is conducted in a large high school to estimate the proportion of students that reads a daily newspaper regularly. How large should the sample size be to have a probability 0.8 that the ratio of the number who read a newspaper to the number polled differs from the true proportion p by at most 0.1?

Since a poll satisfies the requirements of a Bernoulli random variable, let X be a Bernoulli random variable with the values of 1 and 0, where 1 denotes success (that is, that person reads the paper), and let the associated probability p be the correct proportion of the students who read a daily newspaper regularly. The sample mean $\bar{X}$ is the ratio of the successes to the size of the sample. By Chebyshev's theorem we can write, as above,

$$P(|\bar{X}-p|>c)\leq\frac{1}{nc^2}.$$

By assumption, $c=0.1$, and we want a value of n to guarantee that

$$P(|\bar{X}-p|\leq 0.1)\geq 0.8$$

or

$$P(|\bar{X}-p|>0.1)\leq 0.2.$$

Solving,

$$\frac{1}{n(0.1)^2}\leq 0.2,$$

we find that n must exceed 500.

This is a rather large sample. However, using a different technique, we shall find that a much smaller sample will guarantee the same probability. Set $q = 1 - p$ and define a function f as

$$f(p) = p(1 - p) \qquad \text{for } 0 \leq p \leq 1.$$

The expression $p(1 - p)$ can be rewritten (by the technique of completing the square) as

$$f(p) = \tfrac{1}{4} - (p - \tfrac{1}{2})^2.$$

In this form it is apparent that the largest value that f can have occurs when $(p - \tfrac{1}{2}) = 0$. For other acceptable values of p, $(p - \tfrac{1}{2})^2$ is a positive number and $f(p)$ is less than $\tfrac{1}{4}$. We can therefore assert that

$$p(1 - p) \leq \tfrac{1}{4} \qquad \text{for } 0 \leq p \leq 1.$$

Therefore,

$$V(\bar{X}) = \frac{1}{n} V(X) = \frac{1}{n} p(p - 1) \leq \frac{1}{4n}.$$

Consequently, the new inequality to be examined is

$$\frac{1}{4n(0.1)^2} \leq 0.2.$$

The solution is $n = 125$, a more reasonable sample size indeed. ∎

Problems

1. In the long run, $\tfrac{2}{3}$ of the seeds of a certain variety germinate. Plant 100 seeds and use Chebyshev's inequality to obtain an estimate of the probability that the number germinating will differ from the expected number by more than 10.

2. From any previous experiments it has been determined that the probability is $\tfrac{3}{4}$ that a subject will complete a psychological test in the allotted time. The test is administered to 240 students. Use Chebyshev's inequality to obtain an estimate of the probability that the number completing the test will differ from the expected number by more than 20.

3. The proportion of television owners with color sets is to be estimated by a telephone survey. How large should the sample be to have a probability of 0.9 that the relative frequency obtained is within 0.05 of the true proportion?

4. Twenty-five coins are poured onto a table. Use Chebyshev's inequality to obtain an estimate of the probability that the number of heads will differ from the expected number by no more than 3.

5. The probability of one or more accidents in a large plant on any day is 0.05. Obtain an estimate of the probability that the number of accident-free work days in 100 days will differ from the expected number by more than 3.

6. Suppose X is a random variable with $E(X) = 0$ and $V(X) = 1$.

   (a) What is a maximum possible value of $P(|X| > 2)$?
   (b) What is the minimum possible value of $P(|X| \leq 2)$?
   (c) What value of k will guarantee that

   $$P(|X| \leq k) \geq 0.96?$$

7. The probability point function of the random variable X is given by

x_i	0	1	2	3
$p(x_i)$	$\frac{27}{64}$	$\frac{27}{64}$	$\frac{9}{64}$	$\frac{1}{64}$

   Find the exact probability of the event

   (a) $|X - E(X)| \leq \sigma$,
   (b) $|X - E(X)| \leq 2\sigma$,
   (c) $|X - E(X)| \leq 3\sigma$.

   and compare the result with the estimate from Chebyshev's inequality.

8. Suppose X assumes the values $0, 1, 2, \ldots, n$ and that $E(X) = 1$, $V(X) = 1$. Show that $P(X > 3) \leq \frac{1}{4}$.

9. If n shots are to be taken at a target, where the probability of hitting the target on any one show is $\frac{1}{4}$, use Chebyshev's inequality to find the value of n that is needed to have a probability of at least $\frac{1}{2}$ that the average number of hits for n shots will differ from $\frac{1}{4}$ by no more than $\frac{1}{4}$.

10. In a large elementary school system, it is planned to estimate the proportion of children with an IQ over 110 by the administering of an exhaustive test to a randomly selected set of students. How large should the sample be to be assured that the estimate is "accurate," that is, within 0.1 of the true proportion with a probability of 0.9?

11. You throw a single die, winning $4 if an odd number turns up, otherwise getting nothing. Suppose you play the game 10 times. Use Chebyschev's inequality to obtain an upper limit on the probability that your average winnings will differ from $E(\bar{X})$ by at least $1.

12. Use Chebyshev's inequality to determine how many times a fair coin must be tossed in order that the probability will be at least 0.90 that the ratio of the observed number of heads will lie between 0.4 and 0.6.

13. A large collection of voters is to be sampled to estimate the fraction that are Republican. Write an expression for an estimate of the probability that the relative frequency of Republicans differs from the true proportion by less than $n/100$, where n is the sample size.

14. A firm produces a large lot of portable radios, and an estimate of the total number in the lot that are defective is to be obtained by sampling. A random sample of radios is selected from the lot and each radio is thoroughly tested and inspected. The lot is large enough to assume that the sample is selected with replacement. Let p be the probability that a radio is defective and Y the random variable whose values are the total number defective in the sample. What size should the sample be to assure the investigator that the probability is less than 0.1 that the number of defectives differs from the expected number by more than $\frac{1}{10}$ of the sample size?

15. You are to determine the average height of the dandelions on the football field. The standard deviation σ of the height is not known, but after 50 measurements it seems reasonable to take the sample standard deviation of 2 cm for σ. With the assumption,

    (a) use Chebyshev's inequality to obtain a lower bound for the probability that the sample mean differs from the expected value of the height by less than 1 cm.
    (b) determine the number of measurements necessary to ensure a probability of at least 0.95 that the sample mean will differ from the expected value of the height by less than 0.5 cm.

16. The average length of stay in a hospital for patients over 60 years old is the subject of an investigation. The standard deviation of the length of stay is unknown; however, it is reasonable to accept a value of 1.5 days, the standard deviation of a sample of size 70, in its place. Under this assumption,

    (a) obtain a lower value for the probability the sample mean of the sample will differ from the true expected value by less than 0.5 days.
    (b) determine the size of the sample that would give a probability of at least 0.95 that the sample mean is within 0.5 of the true expected value.

17. If the probability of hitting a target is 0.6 and 50 shots are taken, use Chebyshev's inequality to determine a number h such that the probability that the ratio of the hits to the number of shots will differ from the true probability of a hit by not more than h.

18. Let X be a random variable with values $\{-1, +1\}$ and a probability of $\frac{1}{2}$ for each. Let $\bar{X}$ be the sample mean for n independent trials of the experiment connected to X. Write the Chebyshev inequality for $\bar{X}$ with c arbitrary.

19. If X is a binomial random variable with parameters n and p, write the Chebyshev's inequality for an arbitrary number c.

20. If X is the random variable whose values are the number of spots when one die is tossed, calculate

$$P(|X - E(X)| > b\sigma_x)$$

for $b = 1, 2$, and 3. Compare the probabilities with the estimates from Chebyshev's inequality.

21. Repeat Problem 11 when X is the total number of spots when two fair dice are tossed.

22. In Chebyshev's inequality let the constant c be written as $k\sigma$, where σ is the standard deviation of X, and so deduce that

$$P(|X - E(X)| \leq k\sigma) \geq 1 - \frac{1}{k^2}.$$

23. Using Chebyshev's inequality, determine a value of k that will guarantee that the probability is 0.90 that the deviation of X from the mean $E(X)$ is no more than $k\sigma$.

chapter eight

Countably Infinite Random Variables

8.1 Countably Infinite Sample Spaces

Although up to this point random phenomena referred to a finite number of outcomes, several references were made to the possibility of extending the basic probability notions to random phenomena with a countably infinite set of outcomes. For example, in the statements of the axioms for a probability space in Section 2.2, it was pointed out that they were valid for a countably infinite sample space. Axioms, of course, don't need proof. What then must we do to use the axioms for an infinite sample space? Observe that the sum of the probabilities over the sample space is 1. This means that the expression $\sum_{i=1}^{\infty} p_i = 1$ must be given some meaning. The key to success is being able to work with infinite series. Some knowledge of infinite series is all that is needed to extend the appropriate definitions and make the structure of the probability space available to the nonfinite case. For the most part, the notions and the concepts can be extended in a natural way by starting directly from the finite case.

In Section 1.2 the sample space was defined as

Definition 1: Sample Space

The collection of all possible outcomes of a random phenomenon is called the sample space.

In Section 1.3, an event and elementary event were defined as

Definition 2: Event

An event is a set of outcomes of the sample space.

Definition 3: Elementary Event

An elementary event is a set consisting of a single outcome of the sample space.

These statements apply without any significant change to the random phenomenon with a countably infinite set of outcomes. The notation we have been using for a finite sample S is now written as

$$S = \{e_1, e_2, e_3, \ldots, e_n, \ldots\}.$$

That is, three dots are added after e_n to indicate "and so on." The elementary events are E_i, $i = 1, 2, 3, \ldots$.

Example 1

For the random experiment of tossing a coin until a head occurs, the list of outcomes can be enumerated as follows:

$$e_1 = \text{H}$$
$$e_2 = \text{TH}$$
$$e_3 = \text{TTH}$$
$$e_4 = \text{TTTH}$$
$$\cdots$$
$$e_n = \underbrace{\text{TT} \ldots \text{TH}}_{n-1}$$
$$\cdots$$
$$e_* = \text{TTT} \ldots,$$

where the $*$ is used as a subscript to denote that outcome for which T occurs on every toss; that is, e_* consists of an infinite set of entries T. ∎

Example 2

Carl and Ralph agree to play chess until one of them wins two consecutive games. For this infinite sample space the outcomes can be enumerated as

$$e_1 = \text{AA}$$
$$e_2 = \text{BB}$$
$$e_3 = \text{ABB}$$
$$e_4 = \text{BAA}$$
$$e_5 = \text{ABAA}$$
$$e_6 = \text{BABB}$$
$$\cdots$$
$$e_* = \text{ABABAB} \ldots$$
$$e_{**} = \text{BABABA} \ldots.$$

In this example there are two outcomes of an infinite nature, and the symbols ∗ and ∗∗ are used to identify them.   ∎

As in the case for finite sample spaces, the set of all subsets of S comprises the set $\mathscr{E}$ of events and this set satisfies axioms B1 to B5 of Definition 1 of Section 2.1.   Of course $\mathscr{E}$ is an infinite set (it can even be shown that $\mathscr{E}$ contains a noncountable number of elements).

The probability function $P(\cdot)$ is defined in the same way as in the finite case; that is, $P(\cdot)$ is defined for the elementary events E_i, $i = 1, 2, 3, \ldots$, and is then extended to the arbitrary event A by adding the probabilities for the elementary events in A. Thus, if

$$P(E_i) = p_i, \qquad i = 1, 2, 3, \ldots,$$

where $0 \leq p_i \leq 1$ for all i, and

(1) $$\sum_{i=1}^{\infty} p_i = 1,$$

this defines the probability function for the probability space with sample space S.   If A is any event, then

(2) $$P(A) = \sum P(E_{i_j}),$$

where the sum is taken over all $e_{i_j} \in A$.

The expression (1) is an *infinite series* and (2) may or may not be one.   This marks the first important parting of ways of countably infinite sample spaces from finite sample spaces.   It may appear that the difference is merely the difference between adding a finite set of numbers and adding an infinite set of numbers.   Whereas every finite set of numbers gives a number under addition, some infinite sums get arbitrarily large, others get arbitrarily close to a fixed number, and some do neither.   We assume that the reader is familiar with the elements of the theory of infinite series.   (See, for example, [18, Chap. 8].)

Example 3

A coin is bent so that the probability that a head coming up is $\frac{3}{4}$.   The coin is tossed until a head does occur.   What is the probability that the coin was tossed less than 10 times?   What is the probability that an even number of tosses are made before a head comes up?

Let the sample space $S = \{e_1, e_2, e_3, \ldots\}$, where

$$e_1 = \text{H}$$
$$e_2 = \text{TH}$$
$$e_3 = \text{TTH}$$

$$\cdots$$

Since we may assume that these are repeated independent trials,

$$P(E_1) = \tfrac{3}{4}$$
$$P(E_2) = (\tfrac{1}{4})(\tfrac{3}{4})$$
$$P(E_3) = (\tfrac{1}{4})^2(\tfrac{3}{4}).$$

In general,

$$P(E_k) = (\tfrac{1}{4})^{k-1}(\tfrac{3}{4}).$$

If $P(E_k) = p_k = (\tfrac{1}{4})^{k-1}(\tfrac{3}{4})$, then $0 < p_k < 1$ for all k, and to show that these p_k define a probability function it remains only to show that $\sum_{k=1} p_k = 1$; that is,

$$\sum_{k=1}^{\infty} (\tfrac{1}{4})^{k-1}(\tfrac{3}{4}) = 1,$$

The infinite series is actually a geometric series $\sum_{k=1}^{\infty} ar^{k-1}$ with $a = \tfrac{3}{4}$ and $r = \tfrac{1}{4}$ which converges to the value $a/(1-r)$ provided $|r| < 1$. Thus,

$$\sum_{k=1}^{\infty} (\tfrac{1}{4})^{k-1}(\tfrac{3}{4}) = \frac{\tfrac{3}{4}}{1 - \tfrac{1}{4}} = 1$$

as desired. The first question can now be settled.

The event A that the coin was tossed less than 10 times is the subset

$$\{e_1, e_2, \ldots, e_9\}$$

of the sample space. $P(A)$ can be found by forming the sum of the probabilities of the elementary events whose union is A; that is,

$$P(A) = \sum_{k=1}^{9} P(E_k)$$

$$= \sum_{k=1}^{9} (\tfrac{1}{4})^{k-1}(\tfrac{3}{4}).$$

The last sum is the sum of a geometric progression with $r = \tfrac{1}{4}$, $a = \tfrac{3}{4}$, and $n = 9$. Thus,

$$P(A) = (\tfrac{3}{4}) \frac{1 - (\tfrac{1}{4})^9}{1 - \tfrac{1}{4}}$$

$$= 1 - (\tfrac{1}{4})^9.$$

The second equation is settled similarly. Following the same procedure for the event B, that the coin was tossed an even number of times, we note that

$$B = \{e_2, e_4, e_6, \ldots\}$$

and

$$P(B) = \sum_{m=1}^{\infty} P(E_{2m}).$$

Now one must find the sum of an infinite series,

$$P(B) = \sum_{m=1}^{\infty} (\tfrac{1}{4})^{2m-1}(\tfrac{3}{4}).$$

This is a geometric series with $a = \tfrac{3}{16}$ and $r = \tfrac{1}{16}$, so

$$P(B) = \frac{\tfrac{3}{16}}{1 - \tfrac{1}{16}}$$

$$= \tfrac{1}{5}. \quad \blacksquare$$

Problems

Describe the sample space for each of the random experiments defined in Problems 1–5.

1. A die is tossed until either a 2 or a 3 occurs.

2. A public-opinion pollster interviews people passing a corner, asking each for their voting preference. He stops when he finds the second Republican.

3. A couple has offspring until they produce one male and one female, then they stop.

4. A die is rolled until three 6's occur.

5. Three people, call them A, B, and C, alternately roll a die, with the first person to roll a 1 winning the game.

6. What is the probability that the number of tosses needed to get a 6 with a fair die is greater than six?

7. A sack contains 10 apples, 3 of which are spoiled. Suppose that an apple is drawn at random and returned to the sack if it is not spoiled. What is the probability that an odd number of draws are made?

8. A coin is to be tossed until two heads occur. What is the probability that more than 10 tosses are needed?

9. An experiment is to be performed repeatedly until there are two successful outcomes. If the probability of success on one trial is approximately 0.7, what is the probability more than five trials are required?

10. Two people, call them A and B, alternately toss a pair of dice, with the first person to toss a 10 winning the game. What are their respective probabilities of winning?

11. Two people, A and B, play a game repeatedly, with equal probability of winning, until the same person has won two successive games. What is the probability that they play an odd number of games?

12. Two people, A and B, play a game repeatedly, with a probability of $\frac{1}{3}$ that A wins and a probability of $\frac{2}{3}$ that B wins, until the same person has won two successive games. What is the probability that A wins the match?

13. Three people, call them A, B, and C, alternately toss a coin, with the first person to obtain a head winning the game. What are their respective probabilities of winning?

8.2   Countably Infinite Random Variables

Random variable, as defined in Chapter 5, applies with only a slight change to the countably infinite sample space. The domain of definition of the random variable is now an infinite set instead of a finite set of outcomes; otherwise, the definition in Chapter 5 is unchanged.

The two functions closely associated with a random variable, the probability point function and the probability distribution function, exist with a slight modification of the definition that provides for an infinite series instead of a finite sum. A probability point function is any function $p(\cdot)$ defined on the set

$$S_X = \{x_1, x_2, x_3, \ldots\}$$

with the properties

(a) $0 \leq p(x_n) \leq 1$ for all n,

(b) $\displaystyle\sum_{n=1}^{\infty} p(x_n) = 1.$

In (b) is the assertion that the infinite series converges to the value 1.

The probability distribution function is as before,

$$F(x) = \sum_{x_n < x} p(x_n).$$

Example 1

Consider the sample space whose elementary events are the outcomes when a fair die is tossed until the first occurrence of a 6. If Y denotes the roll of a 6 and N denotes the roll of other than a 6, the sample space can be described as

$$S = \{e_1, e_2, e_3, \ldots\},$$

where

$$e_1 = Y$$
$$e_2 = NY$$
$$e_3 = NNY$$
$$\cdots$$
$$e_n = \underbrace{NN \ldots N}_{n-1} Y.$$

$$\cdots$$

If X is defined to be the number of tosses needed to get a 6, then

$$S_X = \{1, 2, 3, \ldots\},$$

so

$$x_1 = 1, \; x_2 = 2, \; \ldots, \; x_n = n, \; \ldots.$$

The probability point function $p(\cdot)$ is

$$p(x_n) = (\tfrac{5}{6})^{n-1}(\tfrac{1}{6}),$$

since the first $n-1$ tosses must be something other than a 6 and the last toss a 6. Obviously, each $p(x_n) > 0$ and

$$\sum_{n=1}^{\infty} p(x_n) = \sum_{n=1}^{\infty} (\tfrac{5}{6})^{n-1}(\tfrac{1}{6}).$$

This is a geometric series with $a = \tfrac{1}{6}$ and $r = \tfrac{5}{6}$. The sum of a geometric series is $a/(1-r)$, so the value is

$$(\tfrac{1}{6})/(1 - \tfrac{5}{6}) = 1,$$

as expected.

For the probability distribution function $F(x)$, since $p(x) = 0$ except at at the points $x_n = n$, $n = 1, 2, 3, \ldots$.

$$
\begin{aligned}
F(x) &= 0 & x &< 1 \\
&= p(x_1) & 1 &\le x < 2 \\
&= p(x_1) + p(x_2) & 2 &\le x < 3 \\
&\quad \cdots \\
&= \sum_{k=1}^{m} p(x_k) & m &\le x < m+1
\end{aligned}
$$

$$\cdots$$

But

$$\sum_{k=1}^{m} p(x_k) = \sum_{k=1}^{m} (\tfrac{5}{6})^{k-1}(\tfrac{1}{6})$$

$$= (\tfrac{1}{6}) \frac{1 - (\tfrac{5}{6})^m}{1 - \tfrac{5}{6}}$$

$$= 1 - (\tfrac{5}{6})^m,$$

where the formula for the sum of the terms of a finite geometric progression was used to calculate the right side. ▮

The expected value and the variance of a finite random variable are finite sums. For countably infinite random variables we obtain infinite series:

$$E(X) = \sum_{n=1}^{\infty} x_n p(x_n),$$

$$V(X) = \sum_{n=1}^{\infty} [x_n - E(X)]^2 p(x_n).$$

Of course, $E(X)$ and $V(X)$ do not exist unless these infinite series converge. If $E(X)$ and $V(X)$ do exist, all their algebraic properties are also valid, such as $E(X + Y) = E(X) + E(Y)$ and $E(aX + b) = aE(X) + b$, $V(X + Y) = V(X) + V(Y)$ if X and Y are independent.

Example 2

Refer to the random variable X in Example 1, where

$$S_X = \{1, 2, 3, \ldots, n, \ldots\},$$
$$p(x_k) = (\tfrac{5}{6})^{k-1}(\tfrac{1}{6}) \qquad k = 1, 2, 3, \ldots.$$

By definition

(1) $$E(X) = \sum_{n=1}^{\infty} x_n p(x_n)$$

$$= \sum_{n=1}^{\infty} n(\tfrac{5}{6})^{n-1} \tfrac{1}{6}.$$

(2) $$V(X) = \sum_{n=1}^{\infty} [x_n - E(X)]^2 p(x_n)$$

$$= \sum [n - E(X)]^2 (\tfrac{5}{6})^{n-1}(\tfrac{1}{6}).$$

That the series (1) and (2) converge can be established relatively easily using the ratio test for convergence. In fact, the values $E(X) = \tfrac{6}{5}$ and $V(X) = \tfrac{6}{25}$ are obtainable from the general case of the geometric random variable. ▮

In Chapter 4, the characterization of random phenomenon as independent repeated trials with two outcomes, called Bernoulli trials, requires the number of trials to be finite.   Suppose that the number of trials is allowed to be infinite, but otherwise each trial is independent with two outcomes.   For these Bernoulli trials, as before, let α denote the probability for success on each trial and let $\beta = 1 - \alpha$ denote the probability for failure.

A model of such an "infinite" experiment is accomplished by repeating the experiment until the first success occurs.   Let X be the random variable whose values are the number of repetitions up to and including the first success.   Then

$$S_X = \{1, 2, 3, \ldots\}$$

with $x_k = k$, and

$$p(x_k) = \beta^{k-1}\alpha, \qquad k = 1, 2, \ldots,$$

since there must be $k - 1$ failures before the first success.   Because the trials are independent, the probabilities can be multiplied, for independence of the trials is assumed as we did in developing the binomial probability point function.

Obviously $p(x_k)$ is positive and less than 1 for all k, so if

$$\sum_{k=1}^{\infty} p(x_k) = 1,$$

$p(k) = q^{k-1}p$ is indeed a probability point function.   But

$$\sum_{k=1}^{\infty} \beta^{k-1}\alpha$$

is a geometric series with $a = \alpha$ and $r = \beta$, and since the value of such a series is $a/(1 - r)$,

$$\sum_{k=1}^{\infty} \beta^{k-1}\alpha = \frac{\alpha}{1 - \beta} = 1.$$

Theorem 1

If a random experiment consists of Bernoulli trials with α the probability of success on each trial, then the probability point function for the sample space of the number of repetitions up to and including the first success is

$$p(k) = (1 - \alpha)^{k-1}\alpha.$$

The function $p(\cdot)$ is called the *geometric probability point function.*

The probability distribution function is readily obtainable as the sum of a geometric progression, as follows:

$$F(x) = 0 \qquad\qquad x < 1$$
$$= p(1) \qquad\qquad 1 \le x < 2$$
$$= p(1) + p(2) \qquad 2 \le x < 3$$
$$\cdots$$
$$= \sum_{k=1}^{m} p(k) \qquad\qquad m \le x < m+1$$
$$\cdots$$

Now

$$\sum_{k=1}^{m} p(k) = \sum_{k=1}^{m} \beta^{k-1}\alpha$$
$$= \alpha\,\frac{1-\beta^m}{1-\beta}.$$
$$= 1 - \beta^m.$$

Thus

$$F(x) = \begin{cases} 0 & x < 1 \\ 1 - \beta^m & m \le x < m+1, \quad m = 1, 2, \ldots. \end{cases}$$

Definition 1: Geometric random variable

A random variable X with a geometric probability point function, where α is the probability of success, is called a *geometric random variable* with parameter α.

Expressions for $E(X)$ and $V(X)$ for a geometric random variable X are given in the next theorem.

Theorem 2

If X is a geometric random variable of parameter α, then

(a) $$E(X) = \frac{1}{\alpha},$$

(b) $$V(X) = \frac{1-\alpha}{\alpha^2}.$$

PROOF: In the proof of this theorem we use the derivative of a function, which therefore requires some knowledge of calculus. Since X is a geometric random variable with parameter α,

$$p(k) = \beta^{k-1}\alpha.$$

From the definition of the expected value,

$$E(X) = \sum_{k=1}^{\infty} x_k p(x_k),$$

that is,

$$E(X) = \sum_{k=1}^{\infty} k\beta^{k-1}\alpha,$$

and this last sum is to be evaluated. Rewrite the sum as

$$\alpha \sum_{k=1}^{\infty} k\beta^{k-1}.$$

Now consider the geometric series $\sum_{k=1}^{\infty} \beta^k$ as a function of β; that is,

$$f(\beta) = \sum_{k=1}^{\infty} \beta^k.$$

The value of the infinite series is $\beta/(1-\beta)$, so

$$f(\beta) = \frac{q}{1-q}.$$

Since $k\beta^{k-1} = D_\beta \beta^k$, the relationship between $f(\beta)$ and $\sum_{k-1}^{\infty} k\beta^{k-1}$ can be written as

$$\sum_{k=1}^{\infty} k\beta^{k-1} = \sum_{k=1}^{\infty} D_\beta \beta^k = D_\beta \left(\sum_{k=1}^{\infty} \beta^k \right) = D_\beta(f(\beta))$$

$$= D_\beta \frac{\beta}{1-\beta}.$$

(The interchange of D_q and $\sum$ is a legitimate operation for power series.) Therefore,

$$\sum_{k=1}^{\infty} k\beta^{k-1} = \frac{1}{(1-\beta)^2},$$

$$E(X) = \alpha \left(\frac{1}{(1-\beta)^2} \right) = \frac{1}{\alpha}.$$

To find $V(X)$ let us first find $E(X^2)$. Now

$$E(X^2) = \sum_{k=1}^{\infty} k^2 p(k)$$

or

$$E(X^2) = \sum_{k=1}^{\infty} k^2 \beta^{k-1}\alpha = \alpha \sum_{k=1}^{\infty} k^2 \beta^{k-1}.$$

From the above, $\sum_{k=1}^{\infty} k\beta^{k-1} = 1/(1-\beta)^2$, and if we multiply both sides of this expression by q, we have

$$\sum_{k=1}^{\infty} k\beta^k = \frac{\beta}{(1-\beta)^2}.$$

Now $D_\beta k\beta^k = k^2\beta^{k-1}$, so

$$\sum_{k=1}^{\infty} k^2\beta^{k-1} = \sum_{k=1}^{\infty} D_\beta k\beta^k = D_\beta \frac{\beta}{(1-\beta)^2}$$

$$= \frac{1+\beta}{(1-\beta)^3}.$$

Therefore,

$$E(X^2) = \alpha\frac{1+\beta}{(1-\beta)^3} = \frac{1+\beta}{\alpha^2},$$

Using $V(X) = E(X^2) - (E(X))^2$, we have

$$V(X) = \frac{1+\beta}{\alpha^2} - \frac{1}{\alpha^2}$$

$$= \frac{\beta}{\alpha^2}. \quad\blacksquare$$

In Example 2, the values of $E(X)$ and $V(X)$ that were given are easily obtained from Theorem 2, where the probability of success in the random experiment is $\frac{1}{6}$. The results $\frac{6}{5}$ and $\frac{6}{25}$ are obvious.

Example 3

The probability that an experiment will succeed without destroying the apparatus is 0.4. What is the probability that three or fewer experiments will fail before a successful run? What is the expected number of experiments?

Let X be the geometric random variable whose values are the number of trials up to and including the first success. Then the parameter $\alpha = 0.4$, and according to Theorem 2, $E(X) = 2.5$ with $V(X) = 3.75$. The probability of three or fewer failures is

$$P(X \le 4) = \sum_{k=1}^{4} (0.6)^{k-1}(0.4) = 0.8704. \quad\blacksquare$$

Problems

1. A countably infinite random variable X has the probability point function

$$p(k) = (\tfrac{1}{3})^{k-1}(\tfrac{2}{3}), \qquad k = 1, 2, 3, \ldots.$$

Find

(a) $P(X > 7)$ (b) $P(X$ is a multiple of 3$)$

2. An urn contains five red and three white marbles. Let X denote the number of draws with replacement required until a red marble is drawn. What is the probability that fewer than 10 draws are needed?

3. If X is a countably infinite random variable with the probability point function

$$p(0) = \tfrac{6}{16},$$
$$p(1) = \tfrac{5}{16},$$
$$p(2k) = (\tfrac{1}{4})(\tfrac{1}{5})^{k-1}, \qquad k = 1, 2, \ldots,$$

show that $p(\cdot)$ is a proper probability point function.

4. The probability of a successful rocket firing is considered to be 0.9. What is the probability that more than two attempts are necessary?

5. The cost of an experiment is \$500, but if the experiment fails, there is an additional cost of \$200 to reset the apparatus for the next experiment. Assume that the trials of the experiment are Bernoulli trials with a probability of 0.4 of success on any trial. What is the expected value of the overall cost of the experiment?

6. A countably infinite random variable Z has the probability point function

$$p(k) = 1/k(k+1), \qquad k = 1, 2, 3, \ldots.$$

Find $P(Z > 33)$. Does Z have an expected value?

7. Let X be a countably infinite random variable with the probability point function

$$p(0) = \tfrac{1}{2},$$
$$p(2k) = 1/4k(k-1), \qquad k = 2, 3, \ldots,$$
$$p(-2k) = 1/4k(k-1), \qquad k = 2, 3, \ldots.$$

Show that $p(\cdot)$ is a proper probability point function and find $E(X)$. Does $V(X)$ exist?

8. Let Z be the random variable whose values are the number of failures before the first success, where the probability of success on each trial is α. Find $E(Z)$.

9. A single die is tossed repeatedly until either a 1 or a 6 comes up. You win the game if you get a 6 before you get a 1. What is the probability that you win?

10. Let Y be a random variable with the probability point function

$$p(k) = (\tfrac{1}{3})^k, \qquad k = 1, 2, \ldots,$$
$$p(-k) = (\tfrac{1}{3})^k, \qquad k = 1, 2, \ldots.$$

(a) Verify that $p(\cdot)$ is a probability point function.
(b) Find $P(|X| > 8)$.

11. Let the random variable X denote the number of times a light switch operates before having to be discarded. The probability point function is

$$p(k) = c(\tfrac{1}{4})^k \qquad k = 0, 1, 2, \ldots.$$

(a) Find the value of c for $p(\cdot)$ to be a probability point function.
(b) Find $P(X > 10)$ and $P(X > m)$, where m is any positive integer.

12. Refer to Problem 11. Using the definition of conditional probability given in Section 4.1, show that for the random variable X of Problem 11, $P(X > m + n \mid X > m) = P(X > n)$, where m and n are any positive integers.

13. John and Neal play golf regularly with the probability $\tfrac{3}{4}$ that John will win. Neal vows to play until he has won twice. Let X denote the number of times they play until Neal wins twice. Find the probability point function for X. Use the identity $\sum_{k=1}^{\infty} kr^{k-1} = 1/(1-r)^2$ for $|r| < 1$ to show that your function is indeed a proper probability point function.

8.3 The Poisson Probability Point Function

In Chapter 4 we developed the binomial random variable in order to study random phenomena called Bernoulli trials, in which it was possible to count the number of successes in a definite number of trials as well as the number of times the event did not occur. In other words, the random experiment culminated in either success or failure.

There are many natural random phenomena where the number of times the event occurs can be counted easily but where it makes no sense to count failures or nonoccurrence. For example, the number of lightning flashes in a 10-minute interval can be counted, but you cannot count the number of times the lightning does not flash. The number of floods on the Mississippi River in a decade is another example, as is the number of airplanes in line waiting to take off at a major airport, or the number of nonsense syllables learned in a unit of learning time. These random phenomena are all examples of the occurrence of an isolated event in a continuum of time.

The random phenomena where the isolated event occurs within some area or volume is of the same type and will be considered along with the occurrences in time. The natural sample space for these random phenomena is the countably infinite space $S = \{0, 1, 2, \ldots\}$. It turns out that the probability point function of a random variable for the sample space is the Poisson probability point function. We shall now derive the Poisson probability point function as a limiting case of the binomial probability point function.

Let X be a random variable that takes on the values 0, 1, 2, ... so that the sample space $S_X = \{0, 1, 2, \ldots\}$. We shall now find the probability point function for X on S_X.

Suppose Y_n is a binomial random variable with parameters n and p. We can interpret Y_n as the number of successes in n Bernoulli trials with probability p of success on each trial. Suppose that as the number of trials increase, the probability p of success on each trial changes in such a way that the product $n \cdot p$ is a constant λ. This means that for each n, the probability p depends on n and λ in such a way that $p_n \cdot n = \lambda$. The symbol p_n denotes the probability for the value n. For each value of n, the probability that $Y_n = k$, where k is some positive integer less than n, as given by the binomial probability point function, is

$$p(k) = P(Y_n = k) = \binom{n}{k} p_n (1 - p_n)^{n-k}.$$

Let us find the limit of this expression as $\{p_n\}$ converges to zero. We rewrite $p(k)$ as

$$p(k) = \frac{n!}{k!(n-k)!} p_n^k (1 - p_n)^{n-k}.$$

But $p_n = \lambda/n$; hence

$$p(k) = \frac{n(n-1)\cdots(n-k+1)}{k!} \left(\frac{\lambda}{n}\right)^k \left(1 - \frac{\lambda}{n}\right)^{n-k}.$$

Therefore,

$$p(k) = \frac{\lambda^k}{k!} \frac{n(n-1)\cdots(n-k+1)}{n^k} \left(1 - \frac{\lambda}{n}\right)^{n-k},$$

which we rewrite as

$$p(k) = \frac{\lambda^k}{k!} 1\left(1 - \frac{1}{n}\right)\left(1 - \frac{2}{n}\right) \cdots \left(1 - \frac{k-1}{n}\right)\left(1 - \frac{\lambda}{n}\right)^n \left(1 - \frac{\lambda}{n}\right)^{-k}.$$

Now, k is a fixed integer, so each of the factors $(1 - 1/n)$, $(1 - 2/n)$, ..., $[1 - (k-1)/n]$ approaches 1 as $n \to \infty$. The factor $(1 - \lambda/n)^{-k}$ also approaches 1 as $n \to \infty$. However,

$$\lim \left(1 - \frac{\lambda}{n}\right)^n = e^{-\lambda},$$

for if we put $t = -\lambda/n$, then

$$\left(1 - \frac{\lambda}{n}\right)^n = (1 + t)^{-\lambda/t} = [(1 + t)^{1/t}]^{-\lambda}.$$

Since $t \to 0$ as $n \to \infty$, we have (see [18], Section 13.4)

$$\lim_{t \to 0} [(1+t)^{1/t}]^{-\lambda} = e^{-\lambda}.$$

Thus the limit of $p(k)$ as $n \to \infty$ is

$$p(k) = P(X = k) = \frac{\lambda^k}{k!} e^{-\lambda}, \qquad k = 0, 1, 2, \ldots.$$

For each k, $p(k)$ is positive, because $\lambda > 0$ and $e^{-\lambda} > 0$, but before we can claim that $p(k) = (\lambda^k/k!)e^{-\lambda}$ is a valid probability point function we must show that the infinite series $\sum_{k=0}^{\infty} p(k)$ has the value 1. In order to do this we need the value of the infinite series $\sum_{k=0}^{\infty} x^k/k!$, where x is an arbitrary real number. It can be shown using the theory of Taylor's series that $\sum_{k=0}^{\infty} x^k/k! = e^x$. If we assume this fact, then

$$\sum_{k=0}^{\infty} p(k) = \sum_{k=0}^{\infty} \frac{\lambda^k}{k!} e^{-\lambda}$$

$$= e^{-\lambda} \sum_{k=0}^{\infty} \frac{\lambda^k}{k!}$$

$$= e^{-\lambda} \cdot e^{\lambda}$$

$$= 1.$$

Taylor's series will not be discussed in this book.

Definition 1: Poisson random variable

If X is a random variable with sample space $S_X = \{0, 1, 2, \ldots\}$ and probability point function

$$p(k) = \frac{\lambda^k}{k!} e^{-\lambda}, \qquad k = 0, 1, 2, \ldots,$$

where λ is a positive constant, then X is a Poisson random variable.

The following typical problem will help to justify the assertion that the Poisson probability function obtains from the binomial probability function. It will also identify the type of random phenomena for which the Poisson model is appropriate.

Suppose one observes and notes the number of telephone calls that arrive in 1 hour at a certain exchange; that is, one records the occurrence of a number of similar events in an interval of time. If we divide the time interval of 1 hour into a large number n of nonoverlapping time subintervals, each of the same length, then for each interval we can consider the occurrence of at least one call to be a success in a sequence of n Bernoulli trials. If we assign a probability to the occurrence of at least one call in the small interval, then we can compute the probability that a certain number of the intervals had calls. More than one call can occur in a small interval, so this does not give us the probability for the total number of calls. However, if we subdivide the 1-hour time interval even further, so that it is reasonable to assume that the probability of two or more calls in a small interval is negligible, we can then define success in the Bernoulli trial to be the occurrence of exactly one call, and failure as the nonoccurrence of a call. With this assumption on the probability, and assuming independence between the small time intervals, the total number of calls can be considered a binomial random variable with n very large. As the number n of small intervals increases, the probability that exactly one call occurs in the small interval decreases. Ultimately, we get the Poisson random variable, whose values are the total number of calls in the time interval of 1 hour. In this respect, the Poisson random variable represents the occurrences of an event in a continuum (that is, random occurrence in time, in area, or in space).

The next theorem contains the vital facts of the Poisson probability point function.

Theorem 1

If X is a Poisson random variable with parameter λ, then

$$E(X) = \lambda,$$
$$V(X) = \lambda.$$

PROOF: From the definition of $E(X)$,

$$E(X) = \sum_{k=0}^{\infty} kp(k)$$

$$= \sum_{k=0}^{\infty} k \frac{\lambda^k}{k!} e^{-\lambda}$$

$$= e^{-\lambda} \sum_{k=1}^{\infty} \frac{\lambda^k}{(k-1)!},$$

where the k has been divided from numerator and denominator, and the first term, which is zero, has been dropped. λ is a common factor of a convergent infinite series, so

$$E(X) = \lambda e^{-\lambda} \sum_{k=1}^{\infty} \frac{\lambda^{k-1}}{(k-1)!}.$$

The infinite series is now just $\sum_{s=0}^{\infty} \lambda^s/s!$, obtained by replacing $k-1$ by s. As mentioned above, it can be shown that this infinite series has the value e^λ; therefore,

$$E(X) = \lambda e^{-\lambda} e^\lambda = \lambda.$$

To get the variance we use the expression

$$V(X) = E(X^2) - [E(X)]^2;$$

we only have to calculate $E(X^2)$.

Now,

$$E(X^2) = \sum_{k=0}^{\infty} k^2 p(k)$$

$$= \sum_{k=0}^{\infty} k^2 \frac{\lambda^k}{k!} e^{-\lambda}$$

$$= \lambda e^{-\lambda} \sum_{k=1}^{\infty} k \frac{\lambda^{k-1}}{(k-1)!},$$

where the common factors of λ and $e^{-\lambda}$ are factored out and the k cancelled between the numerator and denominator. In the last sum replace $k-1$ by s. Thus

$$E(X^2) = \lambda e^{-\lambda} \sum_{k=1}^{\infty} k \frac{\lambda^{k-1}}{(k-1)!} = \lambda e^{-\lambda} \sum_{s=0}^{\infty} (s+1) \frac{\lambda^s}{s!}.$$

Using the algebra of convergent infinite series, we have

$$E(X^2) = \lambda e^{-\lambda} \left(\sum_{s=0}^{\infty} s \frac{\lambda^s}{s!} + \sum_{s=0}^{\infty} \frac{\lambda^s}{s!} \right).$$

Clearly,

$$\sum_{s=0}^{\infty} \frac{s\lambda^s}{s!} = \sum_{s=1}^{\infty} \frac{\lambda^s}{(s-1)!} = \lambda \sum_{s=1}^{\infty} \frac{\lambda^{s-1}}{(s-1)!} = \lambda e^\lambda,$$

$$\sum_{s=0}^{\infty} \frac{\lambda^s}{s!} = e^\lambda.$$

Therefore,

$$E(X^2) = \lambda e^{-\lambda} (\lambda e^\lambda + e^\lambda)$$
$$= \lambda^2 + \lambda.$$

Hence

$$V(X) = \lambda^2 + \lambda - (\lambda)^2 = \lambda. \quad \blacksquare$$

The constant λ therefore is the expected value of X as well as the variance of X. This is consistent with the earlier definition of $\lambda = np$, where n and p

are the parameters for a binomial probability point function. The parameter λ may be interpreted as the mean rate at which the events occur per
unit of time, area, or volume, or simply as the mean rate of occurrence of
events in the given amount of time, area, or volume. If X is a Poisson random
variable denoting the number of events that occur in a time interval t, and
λ is the mean rate of occurrence per unit of time, then the probability point
function for X has the parameter λt, so

$$P(X = k) = \frac{(\lambda t)^k}{k!} e^{-\lambda t}.$$

Example 1

A hole-in-one golf tournament is staged with the first player to score a
hole-in-one the winner. The probability a player can score a hole-in-
one is assumed to be $1/5000$ (unlikely) and it is anticipated that there
will be 3000 attempts. If we assume that these are $n = 3000$ Bernoulli
trials with the probability of success on one trial $p = 1/5000$, then the
probability of no winner is

$$p(0) = \binom{3000}{0}\left(\frac{1}{5000}\right)^0 \left(\frac{4999}{5000}\right)^{3000} = \left(\frac{4999}{5000}\right)^{3000}.$$

If we use the Poisson approximation,

$$\lambda = (3000)\frac{1}{5000} = 0.6.$$

Thus, using Table 6,

$$p(0) = \frac{(0.6)^0}{0!} e^{-0.6} = e^{-0.6} \doteq 0.55. \qquad \blacksquare$$

Example 2

A car rental agency usually rents 50 cars per day, and the requests for
convertibles average 3 in 100. One day there are three convertibles
available. If there are 50 transactions that day, what is the probability
that more than 3 will request convertibles?

The random phenomenon here is the occurrence of customers requiring special cars, and, from the data, this occurs at the rate of 3 per
100 customers. Hence $\lambda = 1.5$. If we assume a Poisson model and let
X be a Poisson random variable with $\lambda = 1.5$, then

$$P(X > 3) = \sum_{k=4}^{\infty} p(k) = \sum_{k=4}^{\infty} \frac{(1.5)^k}{k!} e^{-1.5}$$

$$= 1 - \sum_{k=0}^{3} \frac{(1.5)^k}{k!} e^{-1.5}.$$

From Table 3 we find that the value is approximately 0.066. $\qquad \blacksquare$

Example 3

Suppose that it is known that a certain kind of bacteria is distributed in water at the rate of two bacteria per cubic centimeter of water. If we assume that this phenomenon can be approximated by a Poisson model, what is the probability that a sample of 2 cubic centimeters will contain at least two bacteria?

Since the rate of occurrence is two bacteria per cubic centimeter, the rate for a 2-cubic-centimeter sample will be $2 \cdot 2 = 4$. Let X be the number of bacteria in a sample of 2 cubic centimeters with $\lambda = 4$. The required probability is

$$P(X \geq 2) = 1 - P(X \leq 1),$$

where

$$P(X \leq 1) = \sum_{k=0}^{1} \frac{4^k}{k!} e^{-4}$$

$$= e^{-4} + \frac{4}{1!} e^{-4} = 5e^{-4}.$$

Therefore,

$$P(X \geq 2) = 1 - 5e^{-4}. \quad \blacksquare$$

Example 4

At a supermarket, customers arrive at a checkout counter at the rate of 30 per hour. What is the probability that 5 or fewer customers will arrive in any 20-minute period?

The arrival of a customer at a checkout counter is the occurrence of random event in the continuum of time so that a Poisson model is appropriate. From the data, the mean rate at which the arrivals occur is 30 per hour, which is equivalent to 10 per each unit of 20-minutes duration. Thus we put $\lambda = 10$ and evaluate

$$P(X \leq 5),$$

where X is a Poisson random variable. Hence

$$P(X \leq 5) = \sum_{k=0}^{5} e^{-10} \frac{(10)^k}{k!},$$

and from Table 3,

$$P(X \leq 5) = 0.0671. \quad \blacksquare$$

Example 5

Weather records show that of the 30 days in September, on the average 3 days are rainy. What is the probability that the next September will have at most 2 rainy days?

First, let us assume that the days of September are 30 repeated independent trials of a random experiment with a probability $p = \frac{3}{30} = 0.1$ of success, so that X, the number of rainy days, is a binomial random variable. Then

$$P(X \leq 2) = \sum_{k=0}^{2} \binom{30}{k} (0.1)^k (0.9)^{30-k},$$

and from some detailed calculations,

$$P(X \leq 2) = 0.411.$$

If X is taken to be a Poisson random variable with $\lambda = 30(0.1) = 3$, then from Table 3,

$$P(X \leq 2) = 0.423,$$

so this approximation leaves a little to be desired. ∎

Problems

1. The number X of customers arriving at a gas station each hour during the day can be considered a Poisson random variable with parameter $\lambda = 5$.

   (a) $P(X > 5)$
   (b) $P(X < 3)$
   (c) $P(X = 3)$
   (d) $P(X = 0)$

2. A machine produces defective items with a probability of 0.1. If a sample of size 10 is drawn at random from the output, what is the probability that there will be no more than 1 defective? Write the binomial probability expression and the approximation using the Poisson probability.

3. An insurance company has insured 10,000 people against a particular kind of accident. The rates have been set on the assumption that on the average 2 persons in 1000 will have this accident each year. What is the probability that more than 25 of the insured will collect on their policy in a given year?

4. A flask contains 500 cubic centimeters of vaccine drawn from a vat that contains on the average 5 live viruses per 1000 cubic centimeters. What is the probability that the flask contains (a) no live viruses? (b) more than 1 live virus?

5. The number of customers that arrive at a cafeteria for the noon meal in any 1-minute interval can be considered to be a Poisson random variable with an expected value of 3. What is the probability that

   (a) more than 5 customers arrive between 12 noon and 12:01?
   (b) no customer arrives between 12:03 and 12:06?

6. The number of sales of lawnmowers in any week has been found to be a Poisson random variable with parameter 4. What is the probability that

   (a) there is one sale each week for 3 weeks?
   (b) there are more than two sales in 1 week?

7. The number X of orders per week for television sets at a particular retail store was found to have a Poisson probability function with a parameter $\lambda = 6$. Find

   (a) $P(1 < X < 4)$ (b) $P(X \geq 4)$
   (c) $P(X < 4)$ (d) $P(X = 4)$

8. Refer to Problem 7. How many television sets should the dealer have on hand at the beginning of the week so that the probability that he will be able to supply all orders that week is at least 0.90?

9. The number X of telephone calls that come to a switchboard may be considered a Poisson random variable with an average rate of occurrence of 8 calls per minute. If the switchboard can handle at most 12 calls per minute, write an expression for the probability that the switchboard will receive more calls than it can handle.

10. For a large fleet of milk trucks the average number that are inoperative on any given day is 2. If there are two standby trucks available, what is the probability that

    (a) no standby truck will be needed?
    (b) the number of standby trucks is inadequate?

11. In general, 3 percent of the people that make reservations for a particular airline flight do not show up. The plane has a capacity of 98 passengers and the policy of the airline is to sell reserved seats to 100 people. What is the probability that every person who shows up for the flight will find a seat?

12. If bank tellers make bookkeeping errors at the rate of $\frac{3}{4}$ an error per page of entries, what is the probability that in three pages there will be 2 or more errors?

13. Suppose the number of particles emitted from a radioactive source can be considered a Poisson random variable with a mean rate of one particle every 2 seconds. The source was observed for four time intervals of 6 seconds each. What is the probability that in at least one of the four time intervals, three or more particles were emitted?

14. A digital computer has been found to average about one transistor failure per 10 hours. What is the probability for the successful completion of a program that takes 3 hours to perform, if the computer stops when three or more transistors are inoperative?

15. If the mean rate of occurrence of accidents in a factory is two accidents per week, what is the probability that there will be two or fewer accidents during the time interval of

    (a) 1 week?   (b) 2 weeks?

16. Refer to Problem 15. In a 10-week period, what is the probability that there will be 3 or fewer weeks during which there are two or fewer accidents in that week?

17. If a Poisson random variable X has a parameter λ, what is the probability point function for the random variable $Y = 2X + 4$? What is the expected value and the variance of Y?

18. State and discuss the appropriateness of assumptions that lead to a model in which the number of times lightning will strike in Poweshiek County in Iowa in a given month is a Poisson random variable.

19. Discuss as in Problem 18 a model in which the number of accidents at a specific intersection in a given week is a Poisson random variable.

20. At a certain college there are 4000 men students, and it has been found that the probability that a male student selected at random on a given day will need a hospital bed is 1/2500. How many beds should the campus infirmary have so that the probability a student will be turned away is less than 0.01? [That is, find k so that $P(X > k)$ is less than 0.01, where X is the number of students requiring beds.]

21. The number of misprints on a page has been found to be governed by a Poisson probability function with an average of one misprint per page. What is the probability of more than one misprint per page? Write an expression for the probability that there will be no more than 10 pages in a 500-page book with more than one misprint per page.

22. It has been observed that the average number of customers that enter a gift shop in an hour is 20. What is the probability that during a 3-minute interval either no one will enter the shop or at least 2 persons will enter the shop? Write an expression for the probability that during a set of 20 3-minute intervals, 11 or more of the intervals have the property that either no one or at least 2 persons entered during the interval.

23. Let X be a Poisson random variable with parameter λ. Write an expression for the probability that X assumes an even value.

24. Let X be a Poisson random variable with parameter λ. Using the definition of expected value compute (a) $E(X(X-1))$, and (b) $E(X(X-1)\cdots(X-(n-1)))$.

25. Prove: If X and Y are independent Poisson random variables, X with parameter λ_1 and Y with parameter λ_2, then $X + Y$ is a Poisson random variable with parameter $\lambda = \lambda_1 + \lambda_2$.

chapter nine

Continuous Random Variables

9.1 **A Sample Space of Real Numbers**

9.2 **Distribution and Density Functions**

9.3 **Functions of Random Variables**

9.4 **Expectation and Variance**

9.1 A Sample Space of Real Numbers

We are now in a position to step up our ideas of probability theory from discrete sample spaces to sample spaces of intervals of real numbers, or, as they are usually called, continuous sample spaces. Through Chapter 7 we confined our attention to the finite sample space, and in Chapter 8 the probability concepts were extended to discrete sample spaces with an infinite set of outcomes. The countably infinite sample space cannot be extended to the continuous sample space quite as easily, for there are several rather sophisticated adjustments that must be made. On the other hand, many of the terms can be used almost without modification. As an illustration of the kind of shift that is involved, consider the following.

Although the sample space is still the set of outcomes of a random phenomenon, let us for the moment consider only sample spaces with numerical-valued outcomes, so that the term random variable can be used in the description. Let X be a random variable whose values are the heights of the students enrolled at a college. This physical characteristic is usually considered to be a random phenomenon. If each measurement is recorded to the nearest inch, then a reasonable sample space would be the set of integers from 1 to 180. The sample space is finite, the set of events is the set of all subsets of the sample space and a probability function $P(\cdot)$ that assigns to each integer k, $k = 1, 2, \ldots, 180$ a nonnegative real number such that the sum is 1 completes the definition of the probability space.

The probability question: What is the probability that X exceeds 60 but not 72; that is, $P(60 < X \leq 72) = ?$ As it was done in Chapter 5,

$$P(60 < X \le 72) = F_X(72) - F_X(60).$$

Suppose for some reason or another the measurements were made to the nearest $\frac{1}{4}$ inch. This change in the recording of the measurements quadruples the number of elementary events, necessitating a redefinition of the probability function $P(\cdot)$, but nothing conceptually has been changed.

Now, extend the sample space to include *all* real numbers between 0 and 180 inches, that is, the interval (0, 180). The random variable X still denotes the height of a student enrolled at the college and the question

$$P(60 < X \le 72) = ?$$

still makes good sense, maybe. The interval sample space includes the previous sample space; however, how are we to find the probability associated with an interval of real numbers—a noncountably infinite set? More precisely, how is $P(\cdot)$ to be defined? Is an event in this sample space any subset of real numbers from the interval (0,180)? Suppose we could define $P(\cdot)$ for all possible intervals (a, b) with $0 < a < b < 180$. Will this noncountably infinite set of sets have the properties that are required by axioms B1–B5 of Definition 1 in Section 2.1; that is, is this set of sets a Borel field of sets?

The tough part of the transition from discrete to continuous probability spaces is the definition and description of the set of events on which $P(\cdot)$ is defined. Very roughly, the Borel field of sets that works for the real line is the smallest set of subsets that includes all intervals and satisfies axioms B1–B5. We refer the interested student to more elaborate treatises for a discussion of this deep concept. Once the existence of the Borel set is accepted, there will be little difficulty in transferring the ideas already defined for discrete probability spaces to continuous probability spaces. Among the inducements for studying the continuous probability space is that it has more flexibility in applications and is far more manageable computationally than the discrete space because the concepts and techniques of the calculus are available.

The definition for sample space given in Definition 1 of Section 1.2 makes sense for the continuous case.

Definition 1 : Sample space

The collection of all possible outcomes of a random phenomenon is called the sample space.

For those random phenomena where the collection of outcomes is noncountably infinite, the sample space will be intervals or a half-line of real numbers. For example, the time that elapses between the instant a light bulb is turned on and the instant it burns out has a sample space of $S = \{t \mid t$ a positive real number$\}$, which is a half-line.

The actual time of arrival of a commuter train scheduled to arrive at 8:24 A.M. seems to vary randomly from 8:22 A.M. to 8:29 A.M., so that the interval $-2 \le t \le 5$ would be a suitable sample space.

In a study to determine some causes for bank failures in the depression of the 1930s, the number of failures in a county can be represented by a discrete sample space, but the *proportion* of banks that failed is more easily represented by a continuous sample space.

According to Definition 1 of Section 1.3, an event is a subset of the sample space; however, not every subset of real numbers can be regarded as an event. The set of subsets that can be events must of course satisfy the five axioms listed in Definition 1 of Section 2.2. As a way of talking about such sets, the set of subsets that can be events is called a *Borel field* of sets, and any member of the Borel field of sets is called a Borel set of real numbers.† Since we shall only be concerned with a Borel field of sets, we shall omit a discussion of sets that are not Borel sets.

The third part of the probability-space definition is the probability function $P(\cdot)$ defined on the set of events and satisfying axioms PA1, PA2, and PA3 of Section 2.2. Nothing new needs to be said about $P(\cdot)$ for the noncountably infinite sample space other than that it is defined for every Borel set of real numbers.

Random variable as given in Definition 1 of Section 5.1 is easily extended to the continuous case, but some modification is required.

Definition 2: Random variable

A random variable is a function X whose domain is a sample space S complete with a probability function $P(\cdot)$ defined in a collection of subsets of S that satisfy axioms B1–B5, and whose codomain is the set of real numbers, with the provision that for every Borel set A of real numbers, the subset $\{y \mid y \in S, \ X(y) \in A\}$ of S belongs to the domain of definition of $P(\cdot)$.

Observe that although a random variable is still a function from a sample space to the real number, there is a provision for the noncountably infinite sample space to guarantee that the probability that an observed value of X does lie in some specific set A of real numbers can be found. That is, the

† These terms honor E. Borel, who lived in the latter part of the nineteenth century and who made fundamental contributions to probability theory as well as the theory of integrals.

definition of X requires that for any Borel set A of real numbers, $P(X \text{ in } A)$ has an answer in terms of the original sample space.

Since we are more interested in the range set S_X of the random variable than in the sample space where X is defined, a probability function $P_X(\cdot)$ for the set of events of S_X is defined as follows: Let A be any Borel set in the Borel field of sets of S_X and define $P_X(A) = P(\{y \,|\, y \in S, X(y) \in A\})$. Now $P_X(\cdot)$ is a well-defined function on the subsets of the sample space of real numbers S_X, and as soon as we establish that $P_X(\cdot)$ is a probability function, then S_X and $P_X(\cdot)$ form a probability space. This we do later.

Example 1

Suppose a point Q is chosen on the circumference of a circle of radius 1. The sample space S is the set of all points on the circle. Let T denote the set of events where T consists of all the Borel subsets of S. The function $P(\cdot)$ is defined on T in such a way that arcs of equal length have equal probability. Let R be a fixed point on the circle. Define the random variable X to be the length of the shorter arc from R to Q.

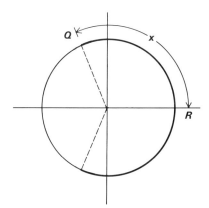

Figure 1

Thus $S_X = \{x \,|\, 0 \le x \le \pi\}$ and $P_X(X \le a)$ for $0 \le a \le \pi$ is the probability that the point Q will be selected on the arc of length $2a$ centered at point R. Hence

$$P_X(X \le a) = \frac{2a}{2\pi} = \frac{a}{\pi}, \qquad 0 \le a \le \pi.$$

If $a \ge \pi$, then $P_X(X \le a) = 1$, while if $a \le 0$, then $P_X(X \le a) = 0$. ∎

Example 2

A patient is selected at random from the population of patients at a large hospital. Define the random variable X to be the blood pressure of the patient selected. Assume that the probability that X lies between two real numbers a and b is equal to the area under the graph of

$$f(x) = \begin{cases} \dfrac{x - 90}{1350}, & 90 \leq x \leq 120, \\[2mm] \dfrac{180 - x}{2700}, & 120 \leq x \leq 180, \\[2mm] 0, & \text{otherwise.} \end{cases}$$

Determine $P_X(X \leq a)$.

Since $f(x)$ is zero for $x < 90$ and $x > 180$, S_X can be taken to be the interval $(90, 180)$. $P_X(X \leq a)$, for $a < 90$ is 0, but for any value of $a: 90 < a < 180$, $P_X(X \leq a) = \int_{90}^{a} f(x)\, dx$. If $90 < a < 120$,

$$\int_{90}^{a} f(x)\, dx = \int_{90}^{a} \frac{x - 90}{1350}\, dx$$

$$= \frac{(x - 90)^2}{2700} \bigg|_{90}^{a}$$

$$= \frac{(a - 90)^2}{2700}.$$

If $120 < a < 180$, then

$$\int_{90}^{a} f(x)\, dx = \int_{90}^{120} \frac{x - 90}{1350}\, dx + \int_{120}^{a} \frac{180 - x}{2700}\, dx$$

$$= \frac{900}{2700} + \frac{-(180 - x)^2}{5400} \bigg|_{120}^{a}$$

$$= \frac{1}{3} + \frac{3600}{5400} - \frac{(180 - a)^2}{5400}$$

$$= 1 - \frac{(180 - a)^2}{5400}.$$

For $180 < a$, $P(X \leq a) = 1$. In summary,

$$P(X \leq a) = \begin{cases} 0, & a \leq 90, \\[2mm] \dfrac{(a - 90)^2}{2700}, & 90 < a \leq 120, \\[2mm] 1 - \dfrac{(180 - a)^2}{5400}, & 120 < a \leq 180, \\[2mm] 1, & 180 < a. \quad \blacksquare \end{cases}$$

Problems

1. Let a point Q be chosen on a line segment of length 4. Define the random variable X to be the length of the left segment. Assume that equal lengths have equal probability and determine S_X and $P_X(X \leq a)$.

2. Let a point Q be chosen on the perimeter of a rectangle of length 4 and width 2. If R is one of the vertices, define the random variable X to be the shorter distance from Q to R measured along the perimeter of the rectangle. Assume that equal lengths have equal probability and determine S_X and $P_X(X \leq a)$.

3. A dart is thrown at a circular target of radius 2. Define the random variable X to be the distance from the point where the dart hits the center of the circle. Assume that equal areas have equal probability and determine S_X and $P_X(X \leq a)$.

4. A point Q is selected on the interval from 0 to 2. Define the random variable X to be the distance from Q to the origin. However, assume that the selection of a point in the interval 0 to 1 is twice as likely as the selection of a point in the interval 1 to 2. Determine S_X and $P_X(X \leq a)$.

5. A point Q is chosen in the rectangle $0 < x < 2$, $0 < y < 3$. Define the random variable Z to be product xy of the coordinates of point P. Assume that equal areas have equal probability and determine S_X and $P_Z(Z < 1)$.

6. A child is lost in a region approximately the shape of an equilateral triangle whose sides have length 2. Assume that the child wanders aimlessly so that equal areas have equal probability. Define the random variable X to be the perpendicular distance from the child to one side of the triangle. Determine S_X and $P_X(X \leq a)$.

7. A point Q is chosen on the perimeter of a rectangle of width 2 and length 4. Define the random variable X to be the distance from Q to the point R, the center of the rectangle, and assume that equal lengths have equal probability. Determine S_X and find $P_X(X \leq 2)$.

8. A dandelion is selected from a large field. Define the random variable X to be the length of the stem. Suppose the lengths vary from 1 inch to 10 inches and that the probability of a length between 4 and 7 inches is 4 times as great as the rest combined. The probability of a stem with a length less than 4 is equal to the probability of a stem greater than 7. Furthermore, the lengths between 1 and 4 are of equal probability, the lengths between 4 and 7 are of equal probability, and the lengths between 7 and 10 are

of equal probability.   Determine S_X and find $P_X(X \leq a)$ for any real number a.

9. Suppose a point A is chosen on a line segment $[0, 1]$, and define the random variable X to be the length of the segment from the origin to Q.   Assume that the probability that X lies between two real numbers is equal to the area under the graph of

$$f(x) = \begin{cases} 2x, & 0 < x < 1, \\ 0, & \text{otherwise.} \end{cases}$$

Determine $P_X(X \leq a)$ and $P_X(X > \frac{1}{2})$.

10. Suppose a man Q is selected from the male student population at a university and define the random variable X to be the height of the individual selected.   Assume that the probability that X lies between two real numbers is equal to the area under the graph of

$$f(x) = \begin{cases} x - 5, & 5 < x < 6, \\ -x + 7, & 6 < x < 7, \\ 0, & \text{otherwise.} \end{cases}$$

Determine $P_X(X \leq a)$ and $P_X(X > 6.5)$.

9.2 Distribution and Density Functions

In Section 5.2 the probability point function and the probability distribution function were defined for a finite random variable X.   Since the distribution function was defined in terms of the probability that X lies in the infinite interval $(-\infty, x]$, the same definition can be applied without change to the random variable on a noncountably infinite sample space.

Definition 1: Probability distribution function

The function $F(\cdot)$ whose value for each real number x is given by

$$F(x) = P_X(X \leq x)$$

is called the probability distribution function for the random variable X.

The real number x defines an event of S_X, the set $A = \{y \mid y \in S_X, y \leq x\}$. By extending S_X to all real numbers if necessary, the set A becomes the interval $(-\infty, x]$, and the event A has a probability number $F(x)$ attached to it.

Consequently, the domain of F is all real numbers, while the real numbers between 0 and 1 inclusive are its codomain. If $F(x)$ is a continuous function of x, then F is said to define a *continuous probability distribution* and X is called a *continuous random variable*.

The connection between the probability of the events $\{a < X \le b\}$ and the probability distribution function $F(\cdot)$ stated in Theorem 1 of Section 5.2 holds without any change in that proof. Hence

(1) $$P(a < X \le b) = F(b) - F(a).$$

Before we discuss what properties a function must have in order to qualify as a probability distribution function, we define the analogue of the probability point function for a continuous random variable.

Definition 2: Probability density function

A nonnegative function f defined for all real numbers x such that

$$F(x) = \int_{-\infty}^{x} f(t)\, dt$$

is called a probability density function for the random variable X whose probability function is $F(\cdot)$.

The function f must have enough properties so that the improper integral exists, and it is usual to have f defined and continuous for all but a finite number of points.

Since $F(b) = \int_{-\infty}^{b} f(t)\, dt$ and $F(a) = \int_{-\infty}^{a} f(t)\, dt$, the probability expressed in (1) can be written in terms of the integral as follows:

$$P(a < X \le b) = \int_{-\infty}^{b} f(t)\, dt - \int_{-\infty}^{a} f(t)\, dt = \int_{a}^{b} f(t)\, dt.$$

Since $f \ge 0$, we can also say that $P(a < X \le b)$ is the area under the curve f over the interval $(a, b]$. Consequently, any function that is (a) defined and continuous for all but a finite number of points, (b) nonnegative for all x, and (c) has $\int_{-\infty}^{\infty} f(t)\, dt = 1$. qualifies as a probability density function for some random variable X, and defines a unique probability distribution function. Moreover, whenever f is continuous, we know by the first fundamental theorem of calculus (see [18, p. 394]) that $F'(x) = f(x)$.

Example 1

The function

$$f(x) = \begin{cases} x, & 0 \le x \le 1, \\ 2 - x, & 1 \le x \le 2, \\ 0, & \text{otherwise,} \end{cases}$$

qualifies as a probability density function. The function is non-negative and continuous for all x, and, in fact, is 0 for $x \leq 0$ and for $x \geq 2$.

The integral $\int_{-\infty}^{\infty} f(t)\, dt$, which can be reduced to the integral $\int_{0}^{2} f(t)\, dt$ since f is zero outside $[0, 2]$, can be evaluated by calculating the two integrals $\int_{0}^{1} t\, dt$ and $\int_{1}^{2} (2 - t)\, dt$. The sum of the two integrals is 1, so f is indeed a probability density function. The graph of f is given in Figure 1.

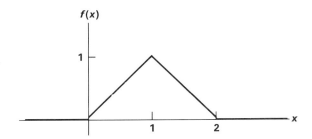

$f(x)$

1

1   2   x

Figure 1

The probability distribution function can be computed as follows:

For $x < 0$: $\qquad F(x) = \int_{-\infty}^{x} dt = 0.$

For $0 \leq x \leq 1$: $\quad F(x) = \int_{-\infty}^{x} f(t)\, dt = \int_{0}^{x} t\, dt = \dfrac{x^2}{2}.$

For $1 < x \leq 2$: $\quad F(x) = \int_{-\infty}^{x} f(t)\, dt$

$$= \int_{0}^{1} t\, dt + \int_{1}^{x} (2 - t)\, dt$$

$$= \tfrac{1}{2} + \left(2x - \dfrac{x^2}{2}\right) - \tfrac{3}{2}$$

$$= 2x - \dfrac{x^2}{2} - 1.$$

For $2 < x$, $F(x) = 1$. Hence

$$F(x) = \begin{cases} 0, & x \leq 0, \\[2mm] \dfrac{x^2}{2}, & 0 < x \leq 1, \\[2mm] 2x - \dfrac{x^2}{2} - 1, & 1 < x \leq 2, \\[2mm] 1, & x > 2. \end{cases}$$

The graph of $F(x)$ is given in Figure 2.   ▮

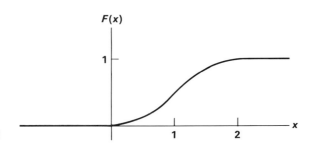

Figure 2

Example 2

Let the length of life of a television component (in thousands of hours) be a random variable X with a probability density

$$f(x) = \begin{cases} k(4 - x^2), & 0 \le x \le 2, \\ 0, & \text{otherwise.} \end{cases}$$

Determine the value of the constant k, find the probability distribution function, and compute the probability $X > 1.5$.

For f to be a probability density function we must determine k so that $\int_{-\infty}^{\infty} f(t) \, dt = 1$.   Since f is zero for $x < 0$ and for $x > 2$, this means that

$$\int_0^2 k(4 - t^2) \, dt = 1.$$

However,

$$\int_0^2 k(4 - t^2) \, dt = k\left(4t - \frac{t^3}{3}\right)\Big|_0^2$$

$$= k \cdot \frac{16}{3}.$$

Certainly $k = \frac{3}{16}$ makes f a probability density function.   The graph of f is given in Figure 3.

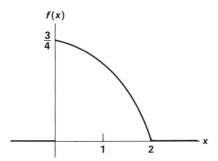

Figure 3

The probability distribution function F is 0 for $x < 0$, while for $0 \le x \le 2$,

$$F(x) = \int_0^x \frac{3}{16}(4 - t^2)\, dt = \frac{3}{16}\left(4t - \frac{t^3}{3}\right)\Big|_0^x$$

$$= \frac{3}{16}x\left(4 - \frac{x^2}{3}\right).$$

If $x > 2$, then $F(x) = 1$.   Therefore,

$$F(x) = \begin{cases} 0, & x < 0, \\[2mm] \dfrac{3}{16}x\left(4 - \dfrac{x^2}{3}\right), & 0 \le x \le 2, \\[2mm] 1, & 2 < x. \end{cases}$$

To find $P(X > 1.5)$ we can use the probability distribution function and the complement to write

$$P(X > 1.5) = 1 - F(1.5)$$
$$= 1 - 0.914$$
$$= 0.086.$$

Alternatively, as the area under the probability density function curve for the interval $(1.5, 2)$, $P(X > 1.5) = \frac{3}{16}\int_{1.5}^2 (4 - t^2)\, dt$, which should evaluate to 0.086.   ∎

The expression $F(b) - F(a)$ represents the probability associated with the interval (a, b), while the quotient $[F(b) - F(a)]/(b - a)$ is the probability per unit length of interval.   If f is continuous at the point a, then we know that

$$\lim_{b \to a} \frac{F(b) - F(a)}{b - a} = F'(a) = f(a)$$

from the calculus.   In this sense $f(a)$ is the instantaneous probability per unit length of interval, or the density of the probability at $x = a$ and is actually the rate of change of the probability at $x = a$.

In the expression $P_X(a < x \le b) = F(b) - F(a)$, suppose we let a approach b; that is,

$$\lim_{a \to b^-} P_X(a < x \le b) = F(b) - \lim_{a \to b^-} F(a).$$

The left side becomes $P_X(X = b)$.   Since $F(x)$ is assumed continuous, $\lim_{a \to b^-} F(a) = F(b)$ and we conclude that $P_X(X = b) = 0$.   This does not mean that X cannot take on the value b. It does mean that the probability of picking at random a specific point is zero.

The probability assigned to the closed interval $a \leq X \leq b$ is therefore the same as the probability for an open interval $a < X < b$, when the probability distribution function is continuous. Indeed,

$$P_X(a < X < b) = \int_a^b f(x) \, dx = F(b) - F(a).$$

The probability for the events $\{a \leq X \leq b\}$ and $\{a \leq X < b\}$ is also $F(b) - F(a)$ when $F(\cdot)$ is continuous.

Not every function can be a probability distribution function. A distribution function must certainly have the following properties:

1. $0 \leq F(x) \leq 1$, for all x, since F is the probability of an event.
2. F must be nondecreasing; that is, if $a < b$, then $F(a) \leq F(b)$.

Statement 2 is a consequence of a theorem in calculus that relates the derivative to increasing–decreasing functions. (See [18, p. 466].) Since $F'(x) = f(x)$ and $f(x) \geq 0$ for all x, the function F is nondecreasing.

As x increases to ∞, the limit of $F(x)$ must exist, because F is a bounded monotone function.

3. $\lim_{x \to \infty} F(x) = 1$.

Similarly, as x approaches $-\infty$,

4. $\lim_{x \to -\infty} F(x) = 0$.

In advanced probability theory it is shown that these properties and an additional requirement that $F(x)$ be continuous from the right characterize those functions that can be distribution functions.

Example 3

Suppose that the time in minutes that a man has to wait at a certain subway station for a train is found to be a random phenomenon, with a probability function specified by the probability distribution function:

$$F(x) = \begin{cases} 0, & x \leq 0, \\ \frac{1}{2}x, & 0 \leq x \leq 1, \\ \frac{1}{2}, & 1 \leq x \leq 2, \\ \frac{1}{4}x, & 2 \leq x \leq 4, \\ 1, & x \geq 4. \end{cases}$$

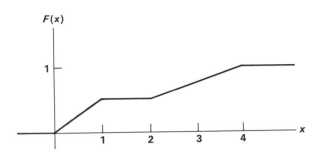

Figure 4

The graph of $F(x)$ is given in Figure 4. $F(x)$ is continuous for all x, so

$$f(x) = \begin{cases} 0, & x < 0, \\ \frac{1}{2}, & 0 < x < 1, \\ 0, & 1 < x < 2, \\ \frac{1}{4}, & 2 < x < 4, \\ 0, & x > 4. \end{cases}$$

The graph of $f(x)$ is given in Figure 5.

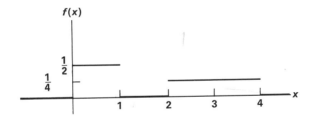

Figure 5

The probability that the man will have to wait for a train more than 3 minutes is

$$P(X > 3) = 1 - F(3) = 1 - \tfrac{3}{4} = \tfrac{1}{4}.$$

The probability that he will have to wait between 1 and 3 minutes is

$$P(1 < X < 3) = F(3) - F(1) = \tfrac{3}{4} - \tfrac{1}{2} = \tfrac{1}{4}. \quad \blacksquare$$

Problems

In Problems 1–6 verify that the given function is a probability density function. Compute the probability distribution function and sketch its graph.

1. $f(x) = \begin{cases} 2x - 4, & 2 < x < 3, \\ 0, & \text{otherwise} \end{cases}$

2. $f(x) = \begin{cases} e^{-x}, & x \geq 0, \\ 0, & \text{otherwise} \end{cases}$

3. $f(x) = \begin{cases} \frac{1}{4}|x|, & |x| < 2, \\ 0, & \text{otherwise} \end{cases}$

4. $f(x) = \dfrac{1}{\pi(1 + x^2)}$

5. $f(x) = \begin{cases} -\frac{1}{9}(x - 3), & 0 < x < 3, \\ \frac{1}{4}(x - 3), & 3 < x < 5, \\ 0, & \text{otherwise} \end{cases}$

6. $f(x) = \frac{1}{2}e^{-|x|}$

7. If the probability density function for the random variable X is
$$f(x) = \begin{cases} Ax(1 - x), & 0 < x < 1, \\ 0, & \text{otherwise,} \end{cases}$$
   (a) determine the value of A that makes $f(\cdot)$ a probability density function.
   (b) compute the probability distribution function.
   (c) find $P_X(\frac{1}{4} < X < \frac{1}{2})$.

8. The quantity of refined ore in tons that a plant produces in a day is found to be a random variable X with a probability density function
$$f(x) = \begin{cases} Ax, & 0 < x < 5, \\ A(10 - x), & 5 < x < 10, \\ 0, & \text{otherwise.} \end{cases}$$
   (a) Determine the value of A that makes f a probability density function.
   (b) Compute the probability distribution function.
   (c) Find $P_X(X > 5)$.

9. The fraction of alcohol in a certain compound is considered to be a random variable X with a probability density function
$$f(x) = \begin{cases} Ax^3(1 - x), & 0 < x < 1, \\ 0, & \text{otherwise.} \end{cases}$$
   (a) Determine the value of A that makes f a probability density function.
   (b) Compute the probability distribution function.
   (c) Find $P_X(X \leq \frac{1}{3})$.

10. The length of life of a light bulb measured in hours may be considered to be a random variable X with a probability density function
$$f(x) = \begin{cases} \dfrac{A}{x^3}, & 1500 < x < 2500, \\ 0, & \text{otherwise.} \end{cases}$$

(a) Determine the value of A that makes f a probability density function.
(b) Compute the probability distribution function.
(c) Find $P_X(X > 2000)$.

11. Verify that for the arbitrary constants c and k the function f is a probability density function.

$$f(x) = \begin{cases} \dfrac{ck^c}{x^{c+1}}, & x > k, \\ 0, & \text{otherwise.} \end{cases}$$

In Problems 12–15 sketch the given probability distribution function and find the corresponding probability density function.

12.
$$F(x) = \begin{cases} 0, & x \le 0, \\ \dfrac{x}{5}, & 0 < x < 5, \\ 1, & 5 \le x \end{cases}$$

13.
$$F(x) = \begin{cases} 0, & x < -1, \\ \frac{1}{2}x^3 + \frac{1}{2}, & -1 \le x \le 1, \\ 1, & 1 < x \end{cases}$$

14.
$$F(x) = \begin{cases} e^{2x}, & x \le 0, \\ 1, & 0 < x \end{cases}$$

15.
$$F(x) = \begin{cases} 0, & x \le 0, \\ 1 - e^{-3x}, & 0 < x \end{cases}$$

16. Let X have the probability density function

$$f(x) = \begin{cases} \frac{3}{4}(1 - x^2), & -1 \le x \le 1, \\ 0, & \text{otherwise.} \end{cases}$$

If two independent observations of X are made, what is the probability at least one is greater than $\frac{1}{4}$?

17. The definition of conditional probability presented in Chapter 4 has meaning for continuous random variables as well. Let X be a continuous random variable with the probability density function

$$f(x) = \begin{cases} 6x(1 - x), & 0 < x < 1, \\ 0, & \text{otherwise.} \end{cases}$$

Let A be the event $\{\frac{1}{3} < X < \frac{2}{3}\}$ and B the event $\{X < \frac{1}{2}\}$. Calculate $P(B|A)$.

18. Suppose the life in hours of a light bulb is a random variable X with a probability density function

$$(x)f = \begin{cases} 6e^{-6x}, & 0 < x, \\ 0, & \text{otherwise.} \end{cases}$$

Compute $P_X(m < X \leq m + 1)$ for m an arbitrary integer.

19. A random variable X has a probability density function

$$f(x) = \begin{cases} 4x - 2, & 0 < x < 1, \\ 0, & \text{otherwise.} \end{cases}$$

Determine a number a such that $P(X < a) = 2P(X > a)$.

20. The life in hours of a certain computer component is a continuous random variable X with a probability density function

$$f(x) = \begin{cases} \dfrac{100}{x^2}, & x > 100, \\ 0, & \text{otherwise.} \end{cases}$$

If there are five such components installed in the machine, what is the probability that at most one unit will have to be replaced after 150 hours of service?

21. The quantity of fuel oil measured in thousands of gallons required to operate a factory each day can be considered to be a random variable X with a probability density function

$$f(x) = \begin{cases} 1, & \frac{1}{2} < x < 1, \\ \frac{1}{2}, & 2 < x < 3, \\ 0, & \text{otherwise.} \end{cases}$$

(a) Sketch the probability density function.
(b) Determine the probability distribution function.
(c) Sketch the probability distribution function.
(d) Find $P(\frac{3}{4} < X < \frac{5}{2})$.

9.3 Functions of Random Variables

In a study of the length of stay of patients in a hospital it was established that there is a relationship between the age of a patient and the number of days he is in the hospital. Suppose the probability density and distribution functions are known for the random variable X, the age of people in a geographical region. What are the probability density and distribution functions for the random variable that denotes length of stay?

To solve this problem we must first obtain empirically or otherwise a relation between length of stay Y and age X. For example, $Y = \frac{2}{5}X + 1$ might serve as a model. Then we need to invoke the ideas of Section 6.3, where we discussed functions of random variables and how to assign probability numbers to the new random variable. The problem simply stated is, if X is a random variable, form the new random variable $Y = g(X)$ and find $P_Y(\cdot)$ in terms of $P_X(\cdot)$. The ideas of Section 6.3 go over to the continuous random variable, particularly when the function g is a differentiable function. The main concern is to express the probability distribution function $F(\cdot)$ for the new random variable $Y = g(X)$ in terms of the probability distribution function for X, and then, by differentiation, obtain the probability density function for Y in terms of the probability density function for X. Some examples will clarify the procedure.

A particularly important, but special, case is the linear function $g(X) = ax + b$, where a and b are fixed real numbers.

Example 1

Suppose X has a probability density function $f_X(\cdot)$ defined as follows:

$$f_X(x) = \begin{cases} 3x^2, & 0 \leq x \leq 1, \\ 0, & \text{otherwise}, \end{cases}$$

and let $Y = g(X) = 2X + 3$. Let $F_Y(\cdot)$ denote the probability distribution function for Y such that

$$F_Y(y) = P_Y(Y \leq y).$$

The event $\{Y \leq y\}$ in the range set S_Y gives rise to the event $\{2X + 3 \leq y\}$ in the range set S_X. We therefore write

$$F_Y(y) = P_Y(Y \leq y) = P_X\left(X \leq \frac{y - 3}{2}\right).$$

From the definition of probability distribution function, the last probability is $F_X[(y - 3)/2]$ for all y. In integral form we can write

$$F_Y(y) = \int_{-\infty}^{(y-3)/2} f_X(t) \, dt.$$

Since $f_X(t) = 3t^2$ is zero for $t < 0$ and $t > 1$, the integrand will be zero for all values of y that do not satisfy $0 \leq (y-3)/2 \leq 1$. Solving this inequality we have $3 \leq y \leq 5$. Consequently,

$$F_Y(y) = \int_0^{(y-3)/2} 3t^2 \, dt = t^3 \Big|_0^{(y-3)/2} = \frac{(y-3)^3}{8} \qquad \text{for } 3 \leq y \leq 5.$$

If $y < 3$, then $F_Y(y) = 0$, and if $y > 5$, then $(y-3)/2 > 1$ and $f_X(t) = 0$. Hence $F_Y(y) = \int_0^1 3t^2 \, dt = 1$ for $y > 5$. Thus

$$F_Y(y) = \begin{cases} 0, & y < 3, \\ \dfrac{(y-3)^3}{8}, & 3 \leq y \leq 5, \\ 1, & y > 5. \end{cases}$$

To find the probability density function $f_Y(y)$, differentiate $F_Y(\cdot)$:

$$f_Y(y) = \begin{cases} 0, & y < 3, \\ \dfrac{3(y-3)^2}{8}, & 3 < y < 5, \\ 0, & y > 5. \end{cases}$$

The same result can be attained in an alternative way if we utilize the composite function relationship of $F_X(\cdot)$ and the inverse function for $g(\cdot)$, which is an increasing function. The function $g(x) = 2x + 3$ has an inverse, which we denote by $g^{-1}(x) = (x-3)/2$. Thus since the event $\{g(X) < y\}$ is the same as the event $\{X < g^{-1}(y)\}$, we can write

$$F_Y(y) = F_X(g^{-1}(y)).$$

To find the probability density function for Y, differentiate the right side using the chain rule:

$$f_Y(y) = f_X(g^{-1}(y)) \cdot \frac{d}{dy}(g^{-1}(y)).$$

The function $f_X(x)$ is different from zero only for $0 < x < 1$, so $f_X(g^{-1}(y))$ will be different from zero only for $0 < (y-3)/2 < 1$. Hence

$$f_Y(y) = \begin{cases} 3\left(\dfrac{y-3}{2}\right)^2 \cdot \dfrac{1}{2}, & 0 < \dfrac{y-3}{2} < 1, \\ 0, & \text{otherwise.} \end{cases}$$

Since the inequality $0 < (y-3)/2 < 1$ is equivalent to the inequality $3 < y < 5$, this result coincides with the previous calculations. The graphs of $f_X(\cdot)$ and $f_Y(\cdot)$ are given in Figures 1 and 2. ▮

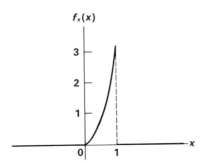

Figure 1

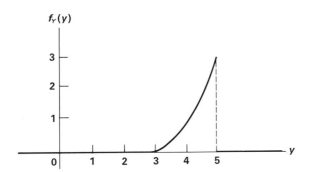

Figure 2

In the preceding example two important facts were tacitly assumed. The first was the existence of the inverse of the function g and the second was the equivalence of the event $\{Y \le y\}$ in the form $\{2X + 3 \le y\}$ and the event $\{X \le (y-3)/2\}$. These two assumptions are connected with the existence of the inverse function. In order to ensure that the inverse function exists, it is sufficient to require that g be differentiable and either increasing or decreasing. If g is increasing, then the event $\{Y \le y\}$ is equal to the event $\{X \le g^{-1}(y)\}$, while if g is decreasing, $\{Y \le y\}$ is equivalent to the event $\{X \ge g^{-1}(y)\}$. To guarantee that one of these two cases does occur, we require that $g'(x)$ is either always positive or always negative. All of this is stated formally in the next theorem.

Theorem 1

Let X be a continuous random variable with a probability density function $f_X(\cdot)$ that is positive for $a < x < b$. If $g(x)$ is a differentiable function, with either $g'(x) > 0$ or $g'(x) < 0$ for $a < x < b$, then the random variable $Y = g(X)$ has a probability density function $f_Y(\cdot)$ given by

$$f_Y(y) = f_X(g^{-1}(y)) \left| \frac{d}{dy} g^{-1}(y) \right|,$$

$$g(a) < y < g(b) \qquad \text{if } g \text{ is increasing,}$$
$$g(b) < y < g(a) \qquad \text{if } g \text{ is decreasing.}$$

PROOF: First, assume that $g(x)$ is increasing. Then the event $\{Y \le y\}$ is equal to the event $\{X \le g^{-1}(y)\}$, and we can write

$$\begin{aligned}
F_Y(y) &= P_Y(Y \le y) \\
&= P_X(g(X) \le y) \\
&= P_X(X \le g^{-1}(y)) \\
&= F_X(g^{-1}(y)).
\end{aligned}$$

Differentiating both sides of the last line above with respect to y, we obtain

$$f_Y(y) = f_X(g^{-1}(y)) \frac{d}{dy} (g^{-1}(y)).$$

Now suppose that $g(x)$ is decreasing. The event $\{Y \le y\}$ is equal to the event $\{X \ge g^{-1}(y)\}$, and so we can write

$$\begin{aligned}
F_Y(y) &= P_Y(Y \le y) \\
&= P_X(g)(X \le y) \\
&= P_X(X \ge g^{-1}(y)) \\
&= 1 - F_X(g^{-1}(y)).
\end{aligned}$$

Differentiating both sides of the last line with respect to y, we obtain

$$f_Y(y) = -f_X(g^{-1}(y)) \frac{d}{dy} g^{-1}(y).$$

Since $g'(x)$ is negative,

$$\frac{d}{dy} g^{-1}(y) = \frac{1}{g'(x)} < 0$$

$$-\frac{d}{dy} g^{-1}(y) > 0.$$

Consequently, the proof is complete.   ∎

Example 2

Suppose X is the random variable defined in Example 1 with the probability density function $f_X(\cdot)$ given by

$$f_X(x) = \begin{cases} 3x^2, & 0 < x < 1, \\ 0, & \text{otherwise.} \end{cases}$$

Let $Y = g(X) = e^{-X}$. Thus $g(x) = e^{-x}$ and, to use Theorem 1, we must find the inverse function $g^{-1}(x)$. However, if $y = e^{-x}$, $\ln y\dagger = -x$, or $x = -\ln y$. Therefore,

$$g^{-1}(y) = -\ln y \qquad \text{and} \qquad \left| \frac{d}{dy} \ln y \right| = \frac{1}{y}.$$

Hence

$$f_Y(y) = \begin{cases} 3(-\ln y)^2 \dfrac{1}{y}, & 0 < -\ln y < 1, \\ 0, & \text{otherwise.} \end{cases}$$

Since $g(x) = e^{-x}$, $g'(x) = -e^{-x}$, which is always negative; thus g is decreasing. Therefore, the set of values of y for which $f_Y(\cdot)$ is positive is the interval $g(1) < y < g(0)$, which can be written as $e^{-1} < y < 1$. If $e^{-1} < y$, then $-1 < \ln y$, and if $y < 1$, then $\ln y < 0$. Hence $e^{-1} < y < 1$ implies that $-1 < \ln y < 0$; that is, $0 < -\ln y < 1$. ∎

Example 3

If X is a continuous random variable with probability density function $f_X(x)$, find the probability density function for $Y = X^2$.

The function $g(x) = x^2$ does not have an inverse g^{-1} valid for all x, so we cannot use Theorem 1 as it stands. However, we can proceed from the definition of the probability distribution function as before; that is,

$$F_Y(y) = P(Y \leq y).$$

Since $Y \geq 0$ for all values, if $y < 0$, then $F_Y(y) = 0$, so we can assume that $y > 0$. Thus

$$P(Y \leq y) = P(X^2 \leq y)$$
$$= P(-\sqrt{y} \leq X \leq +\sqrt{y}).$$

Hence

(1) $\qquad F_Y(y) = F_X(\sqrt{y}) - F_X(-\sqrt{y}) \qquad$ for $y > 0$.

To find the probability density function of Y, differentiate (1) using the chain rule:

$$f_Y(y) = f_X(\sqrt{y}) \tfrac{1}{2} y^{-1/2} - f_X(-\sqrt{y})(-\tfrac{1}{2} y^{-1/2})$$
$$= \tfrac{1}{2} y^{-1/2}(f_X(\sqrt{y}) + f_X(-\sqrt{y})).$$

If $y \leq 0$, then $f_Y(y) = 0$. ∎

† $\ln y$ denotes the natural logarithm of y.

Problems

1. If X is a continuous random variable with the probability density function

$$f(x) = \begin{cases} 6x(1-x), & 0 < x < 1, \\ 0, & \text{otherwise}, \end{cases}$$

and $Y = 3X - 2$, determine $f_Y(\cdot)$ and $F_Y(\cdot)$.

2. Refer to Problem 1. Find the probability density function for the random variable $Y = X^2$.

3. If X is a continuous random variable with a probability density function

$$f(x) = \begin{cases} \frac{1}{2}, & 1 < x < 3, \\ 0, & \text{otherwise}, \end{cases}$$

find $f_Y(\cdot)$ and $F_Y(\cdot)$ for $Y = \frac{1}{2}X + 3$.

4. Refer to Problem 3. Find the probability density function for $Y = e^X$.

5. If X is a continuous random variable with a probability density function

$$f(x) = \begin{cases} e^{-x}, & x > 0, \\ 0, & \text{otherwise}, \end{cases}$$

find the probability density function for $Y = X^3$.

6. If X is a continuous random variable with a probability density function

$$f(x) = \begin{cases} 2x, & 0 < x < 1, \\ 0, & \text{otherwise}, \end{cases}$$

find the probability density function for $Y = \sqrt{X}$.

7. The temperature, measured in degrees Fahrenheit, of a certain object is a randon variable X with a probability density function

$$f(x) = \begin{cases} \frac{1}{100}(x - 40), & 40 < x < 50, \\ -\frac{1}{100}(x - 60), & 50 < x < 60, \\ 0, & \text{otherwise}. \end{cases}$$

Find the probability density function for the random variable

$Y = \frac{5}{9}(X - 32)$, which is the same temperature measured in degrees centigrade.

8. If X is a continuous random variable with a probability density function

$$f(x) = \begin{cases} \frac{1}{2}, & -1 < x < 1, \\ 0, & \text{otherwise,} \end{cases}$$

find the probability density function for $Y = -X + 2$.

9. If X is a continuous random variable with a probability density function

$$f(x) = \begin{cases} 1, & 0 < x < 1, \\ 0, & \text{otherwise,} \end{cases}$$

find the probability density function for $Y = X^2 + 1$.

10. Refer to Problem 9. Find the probability density function for the random variable $Y = 1/(X + 1)$.

11. If X is a continuous random variable with a probability density function

$$f(x) = \begin{cases} \frac{1}{2}, & -1 < x < 1, \\ 0, & \text{otherwise,} \end{cases}$$

find the probability density function for $Y = X^2$. [*Hint*: Since $g(x) = x^2$ is neither increasing or decreasing over $(-1, 1)$, use the technique of Example 1.]

12. Refer to Problem 11. Find the probability density function for $Y = |X|$.

13. If the radius of a sphere is a random variable with a probability density function

$$f(x) = \begin{cases} 3x^2, & 0 < x < 1, \\ 0, & \text{otherwise,} \end{cases}$$

find the probability density function for the volume $V = \frac{4}{3}\pi X^3$.

14. Prove: If X is a continuous random variable with probability density function $f_X(\cdot)$ and $Y = aX + b$, where a and b are real numbers with $a > 0$, then

$$f_Y(y) = \frac{1}{a} f_X\left(\frac{y - b}{a}\right).$$

15. Let X be a random variable with a probability density function

$$f(x) = \begin{cases} 1, & 0 < x < 1, \\ 0, & \text{otherwise.} \end{cases}$$

Choose a value for X and construct a chord X units from the center of a circle of radius 1 and let Y be the length of the chord. Find $P(Y < \sqrt{3})$.

16. Refer to Problem 15. Let Z be a random variable with a probability density function

$$f(z) = \begin{cases} \dfrac{1}{\pi}, & 0 < z < \pi, \\ 0, & \text{otherwise.} \end{cases}$$

Choose a value for Z and construct a chord subtending a central angle of Z radians and let Y be the length of the chord. Find $P(Y < \sqrt{3})$.

Note: The different answers to Problems 15 and 16 point up some of the difficulty with the term random in the question "What is the probability that a random chord drawn in a circle of radius 1 is less than $\sqrt{3}$?" This is known as Bertrand's paradox.

17. If X is a continuous random variable with a probability density function

$$f(x) = \begin{cases} e^{-x}, & x > 0, \\ 0, & \text{otherwise,} \end{cases}$$

find the probability density function for $Y = \ln X$.

9.4 Expectation and Variance

Nearly everything stated in Sections 6.2 and 6.4 regarding the expectation and variance of a discrete random variable applies to the continuous random variable.

Definition 1: Mathematical expectation

Let X be a random variable with probability density function $f(\cdot)$. The mathematical expectation of X is defined to be

$$E(X) = \int_{-\infty}^{\infty} xf(x)\,dx$$

provided that the integral $\int_{-\infty}^{\infty} |x| f(x)\,dx$ converges.

Example 1

Suppose X is a continuous random variable with a probability density function $f(\cdot)$:

$$f(x) = \begin{cases} 6x(1-x), & 0 < x < 1, \\ 0, & \text{otherwise.} \end{cases}$$

The expectation of X is

$$E(X) = \int_{-\infty}^{\infty} xf(x)\,dx$$

$$= \int_{0}^{1} x(6x(1-x))\,dx$$

$$= 6\left(\frac{x^3}{3}\right) - \left(\frac{x^4}{4}\right)\Big|_{0}^{1} = \tfrac{1}{2}. \quad \blacksquare$$

Example 2

Let c and k be positive constants. Then the function f is a probability density function for a random variable X:

$$f(x) = \begin{cases} \dfrac{ck^c}{x^{c+1}}, & x \geq k, \\ 0, & \text{otherwise.} \end{cases}$$

The mathematical expectation of X is

$$E(X) = \int_{-\infty}^{\infty} xf(x)\,dx$$

$$= \int_{k}^{\infty} x\,\frac{ck^c}{x^{c+1}}\,dx$$

$$= ck^c\,\frac{x^{-c+1}}{-c+1}\Big|_{k}^{\infty} \quad (\text{if } c \neq 1)$$

$$= \frac{ck}{c-1}, \quad \text{provided } c > 1.$$

If c satisfies the inequality $0 < c \leq 1$, then the integral does not converge, and the mathematical expectation for X does not exist. This probability law is known as the *Pareto distribution* and is useful in the construction of models for the study of such economic variables as income. $\quad \blacksquare$

If a probability density function is interpreted as the density of mass of a physical unit of mass distributed continuously on a thin rod, then there is a corresponding interpretation for $E(X)$. The value of $f(x)$ at any point is call~d the *mass density* at the point and can be considered as the instantaneous mass. The product $xf(x)$ is a function whose values denote the moment about the zero point and $\int_{-\infty}^{\infty} xf(x)\,dx$ is called the first moment of the unit mass distribution. If a unit mass is located at a position $E(X)$ units from the zero point, the moment of this mass system will be exactly that of the original system. The number $E(X)$ is often called the *center of mass*.

In Theorem 1 of Section 6.3, the expectation of the random variable Y, defined as a function of the random variable X by $Y = g(X)$, is given in terms of the probability function for X. A corresponding theorem can be stated for continuous random variables; however, the proof will be omitted, for it is rather intricate.

Theorem 1

Let X be a continuous random variable with a probability density function $f(\cdot)$ and let $Y = g(X)$, where g is a continuous function. Then if $E(Y)$ exists.

$$E(Y) = E(g(X)) = \int_{-\infty}^{\infty} g(x)f(x)\,dx.$$

One need not use this theorem to find $E(Y)$. One could first find $f_Y(y)$ and compute $E(Y)$ directly from the definition of $E(Y)$. The content of Theorem 1 is the assertion that $E(Y)$ can be obtained by using the function $g(x)$ with the probability density function for X—thus it is unnecessary to find the probability density function for Y in order to obtain $E(Y)$.

Example 3

Let X be a random variable with the probability density function f defined as follows:

$$f(x) = \begin{cases} \frac{1}{243}(x-10)^2, & 1 \le x \le 10, \\ 0, & \text{otherwise.} \end{cases}$$

(X might denote the number of pounds a person loses on a 2-week diet.) Let $Y = X^2$ and find $E(Y)$ using Theorem 1.

Since $g(x) = x^2$,

$$E(Y) = \int_{-\infty}^{\infty} x^2 f(x)\,dx$$

$$= \int_{1}^{10} x^2 \frac{1}{243}(x-10)^2\,dx$$

$$= \frac{1}{243} \int_{1}^{10} (x^4 - 20x^3 + 100x^2)\,dx$$

$$= \frac{1}{243}\left(\frac{x^5}{5} - 5x^4 + 100\frac{x^3}{3}\right)\Big|_{1}^{10} = 13.7. \quad \blacksquare$$

With this theorem, the properties for the expectation of continuous random variables follow directly from the properties of the integral.

Theorem 2

If X is a continuous random variable for which the expected value $E(X)$ exists, and a and b are arbitrary real numbers, then $E(aX + b) = aE(X) + b$.

PROOF: By Theorem 1, if $Y = aX + b$, then

$$E(Y) = E(aX + b) = \int_{-\infty}^{\infty} (ax + b)f(x)\, dx,$$

and the integral can be rewritten as

$$E(aX + b) = a \int_{-\infty}^{\infty} xf(x)\, dx + b \int_{-\infty}^{\infty} f(x)\, dx.$$

Since

$$E(X) = \int_{-\infty}^{\infty} xf(x)\, dx \quad \text{and} \quad \int_{-\infty}^{\infty} f(x)\, dx = 1,$$

$$E(aX + b) = aE(X) + b. \quad \blacksquare$$

Example 4

A businessman must buy a fixed quantity of a perishable product from a supplier a month's supply at a time. Over a period of time he has found it reasonable to assume that the quantity he can sell in a month is a random variable X with a probability density function

$$f(x) = \begin{cases} \frac{1}{3}, & 0 < x < 3, \\ 0, & \text{otherwise.} \end{cases}$$

Suppose that for each unit of the product sold he realizes a profit of $30, but for each unit not sold there is a loss of $10. What quantity should be ordered to maximize the expected profit?

Let us denote the profit by the random variable Y and find an expression for Y in terms of X. The function $g(X)$ will also depend on the quantity available for sale, so let z denote this variable. If $X > z$, all units are sold and $Y = 30X$. If $X < z$, $z - X$ are not sold; then $Y = 30X - 10(z - X)$. Thus

$$Y = \begin{cases} 30, & x \geq z, \\ 40X - 10z, & X < z. \end{cases}$$

Even though it takes two statements to define the function $Y = g(X)$, Theorem 1 is still applicable, so that we may write

$$E(Y) = \int_0^3 g(x)f(x)\,dx.$$

To evaluate the integral, we must take account of the value of z. Hence

$$E(Y) = \int_0^z (40x - 10z)\tfrac{1}{3}\,dx + \int_z^3 30z\tfrac{1}{3}\,dx.$$

Evaluating each of these integrals we find that

$$E(Y) = 30z - \tfrac{20}{3}z^2,$$

with the restriction that $0 \le z \le 3$. The expected value of Y is a polynomial of degree 2, and to find the maximum we find the critical points. Taking the derivative

$$D_z E(Y) = 30 - \tfrac{40}{3}z$$

and solving for z, we see that the only critical value is at $z = \tfrac{9}{4}$. For this value $E(Y)$ is maximum on the interval. ∎

The variance for a continuous random variable follows analogously.

Definition 2 : Variance

Let X be a continuous random variable with probability density function f. The variance of X is defined to be

$$V(X) = E((X - E(X))^2) = \int_{-\infty}^{\infty} (x - E(X))^2 f(x)\,dx.$$

The positive square root of the variance is called the standard deviation of X and denoted by σ_X.

The computation of the variance can be simplified using Theorem 1 of Section 6.4. This theorem can be used on continuous random variables because the proof depended only on an algebraic property of expectation. The same remark applies to Theorem 2 of Section 6.4, and the theorems are restated here for convenience.

Theorem 3

If X is a continuous random variable with variance $V(X)$, then

$$V(X) = E(X^2) - (E(X))^2.$$

Theorem 4

If X is a continuous random variable and a and b are arbitrary real numbers, then

$$V(aX + b) = a^2 V(X).$$

Example 5

In Example 1, X had the probability density function

$$f(x) = \begin{cases} 6x(1 - x), & 0 < x < 1, \\ 0, & \text{otherwise,} \end{cases}$$

and we obtained the result $E(X) = \frac{1}{2}$.

Since

$$E(X^2) = \int_{-\infty}^{\infty} x^2 f(x)\, dx$$

$$= \int_0^1 6x(1 - x)x^2\, dx$$

$$= 6 \left. \frac{x^4}{4} - \frac{x^5}{5} \right|_0^1$$

$$= \tfrac{3}{10}$$

the variance $V(X) = \tfrac{3}{10} - (\tfrac{1}{2})^2 = \tfrac{11}{40}$. ∎

Example 6

Whenever the expected value of a random variable is zero, the variance is just $E(X^2)$, according to Theorem 3. This occurs whenever the probability density function is an even function and the limits are symmetric [for example, $\int_{-a}^{a} xf(x)\, dx = 0$].

Let the probability density function for the random variable X be given by

$$f(x) = \begin{cases} \frac{3}{32}(4 - x^2), & -2 < x < 2, \\ 0, & \text{otherwise.} \end{cases}$$

Since $f(-x) = f(x)$ for all x, f is an even function and the graph of f is symmetric about the y-axis. The expected value $E(X)$ is the integral

$$E(X) = \int_{-2}^{2} x \tfrac{3}{32}(4 - x^2)\, dx.$$

The integrand $h(x) = \tfrac{3}{32}x(4 - x^2)$ is now an odd function with $h(-x) = -h(x)$ for all x. This means that the area between the curve of $h(x)$

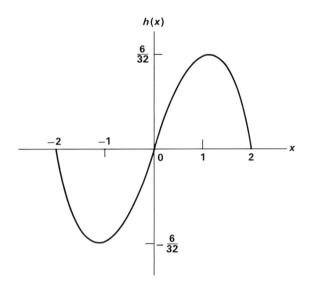

Figure 1

and the x-axis for the interval form -2 to 0 is the negative of the area from 0 to 2, and so $E(X) = 0$. The graph of $h(x)$ is given in Figure 1.

The variance $V(X)$ is $E(X^2)$, which can be found by evaluating the integral

$$E(X^2) = \int_{-2}^{2} x^2 \cdot \tfrac{3}{32}(4 - x^2)\, dx$$

$$= \tfrac{3}{32}\left(\tfrac{4}{3}x^3 - \tfrac{1}{5}x^5\right)\Big|_{-2}^{2}$$

$$= \tfrac{4}{5}.$$

Thus $E(X) = 0$ and $V(X) = \tfrac{4}{5}$. ∎

Problems

In Problems 1–6 find the expectation and the variance of the random variable with the given probability density function.

1.
$$f(x) = \begin{cases} 2x - 4, & 2 < x < 3, \\ 0, & \text{otherwise} \end{cases}$$

2.
$$f(x) = \begin{cases} \tfrac{1}{2}|x|, & |x| < 2, \\ 0, & \text{otherwise} \end{cases}$$

3.
$$f(x) = \begin{cases} -\frac{1}{9}(x-3), & 0 < x < 3, \\ \frac{1}{4}(x-3), & 3 < x < 5, \\ 0, & \text{otherwise} \end{cases}$$

4.
$$(x) = \begin{cases} \frac{1}{2}, & -1 < x < 1, \\ 0, & \text{otherwise} \end{cases}$$

5.
$$f(x) = \begin{cases} e^{-x}, & 0 < x \\ 0, & \text{otherwise} \end{cases}$$

6.
$$f(x) = \frac{1}{2} e^{-|x|}$$

7. If X is a continuous random variable with probability density function

$$f(x) = \begin{cases} 6x(1-x), & 0 < x < 1, \\ 0, & \text{otherwise,} \end{cases}$$

and $Y = 3X - 2$, find $E(Y)$ and $V(Y)$.

8. Refer to Problem 7. Find the expectation and variance for the random variable $Y = X^2$.

9. If X is a continuous random variable with probability density function

$$f(x) = \begin{cases} 1, & 0 < x < 1, \\ 0, & \text{otherwise,} \end{cases}$$

find the expectation and variance for the random variable $Y = X^2 + 1$.

10. Refer to Problem 9. Find $E(Y)$ and $V(Y)$ for $Y = 1/(X+1)$.

11. If X is a continuous random variable with the probability density function

$$f(x) = \begin{cases} 1, & \frac{1}{2} < x < 1, \\ \frac{1}{2}, & 2 < x < 3, \\ 0, & \text{otherwise,} \end{cases}$$

find $E(X)$ and $V(X)$.

12. Not every random variable has an expectation. If X has the probability density function

$$f(x) = \frac{1}{\pi} \frac{1}{1+x^2},$$

show that $E(X)$ does not exist.

13. If X is a continuous random variable with a probability density function

$$f(x) = \begin{cases} 2x, & 0 < x < 1, \\ 0, & \text{otherwise,} \end{cases}$$

find $E(Y)$ and $V(Y)$ for $Y = \sqrt{X}$.

14. A random variable X has a probability density function

$$f(x) = \begin{cases} 2xe^{-x^2}, & x > 0, \\ 0, & \text{otherwise.} \end{cases}$$

If $Y = X^2$, compute $E(Y)$

(a) using Theorem 1 and integration by parts.
(b) finding $f_Y(\cdot)$ first and then using the definition of $E(Y)$.

15. The length of life of an electronic device measured in hundreds of hours may be considered to be a random variable X with a probability density function

$$f(x) = \begin{cases} e^{-x}, & x > 0, \\ 0, & \text{otherwise.} \end{cases}$$

The cost to manufacture one item is \$3 and the item sells for \$7, unless the device fails before 90 hours, in which case a total refund is guaranteed. Find the expected value of the profit for each item.

16. A chemical is formed by the interaction of two compounds, and the resulting chemical contains a variable percentage of sulfur, which is considered a random variable X. The probability density function for X is

$$f(x) = \begin{cases} \dfrac{6}{10^6} x(100 - x), & 0 < x < 100, \\ 0, & \text{otherwise.} \end{cases}$$

If the net profit $Y = a + bX$, where a and b are real constants, find the expected value of the profit.

17. Suppose that a manufacturer produces a certain type of fertilizer which if not used within a certain time must be discarded. Let X be the random variable that denotes the number of units of fertilizer purchased from the manufacturer in 1 year and suppose that the probability density function for X is

$$f(x) = \begin{cases} \frac{1}{2}, & 2 < x < 4, \\ 0, & \text{otherwise.} \end{cases}$$

For each unit sold, a profit of \$300 is earned; for each unit not sold there is a loss of \$100. If four units of fertilizer are produced, what is the expected profit?

18. Refer to Problem 17. Suppose the manufacturer decides at the beginning of the year to produce a units of fertilizer, where a is a positive real number. What value of a will give the maximum expected profit? (*Hint*: Express the expected value of the profit as a function of a and find the maximum value of a.)

19. A circular dart target is divided into three regions by three concentric circles of radii 1, 2, and 3, respectively. If a dart lands in the inner circle the score is 5, in the next ring 3 points, in the outer ring, 1 point, and 0 if it misses the target altogether. Suppose the distance of a hit from the center of the target is considered a random variable X with a probability density function

$$(x) = \begin{cases} xe^{-x}, & x > 0, \\ 0, & \text{otherwise.} \end{cases}$$

What is the expected value of the score from five throws?

20. Show that the expected value of a random variable X with a probability density function $f(x)$ that is symmetric about $x = a$ is $E(X) = a$, provided that $E(X)$ exists, of course.

21. If X is a random variable with $E(X) = 10$ and $V(X) = 5$, for what values of a and b does $Y = aX + b$ have an expectation of 0 and a variance of 1?

chapter ten

Some Special Continuous Probability Density Functions

10.1 Uniform Probability Function

If a random phenomenon assumes numerical values on some interval $[a, b]$, we can assign a probability function so that the probability of an event on intervals of equal length are equal. This would correspond to the discrete case where the same probability is assigned to each elementary event. To find the value that should be assigned to an elementary event in a finite sample space, we used the reciprocal of the number of elements. The corresponding situation in the continuous case would be that the probability density function $f(x)$ is equal to a positive constant c on some interval and requires that $P(x_1 < X < x_1 + h) = P(x_2 < X < x_2 + h)$, provided that the end points are in the interval; that is, the probability for intervals of equal length is the same. In particular, if the sample space is $[a, b]$, we must choose the constant c so that

$$\int_a^b c \, dx = 1.$$

Clearly, the choice for c must be $1/(b-a)$.   Hence

$$P(x_1 < X < x_1 + h) = \int_{x_1}^{x_1 + h} \frac{1}{b-a} \, dx = \frac{h}{b-a}$$

$$= \int_{x_2}^{x_2 + h} \frac{1}{b-a} \, dx = P(x_2 < X < x_2 + h).$$

We call this probability function the uniform probability distribution and define it formally.

Definition 1: Uniform probability function

Let X be a continuous random variable with a probability density function specified by

$$f(x) = \begin{cases} \dfrac{1}{b-a}, & a < x < b, \\ 0, & \text{otherwise.} \end{cases}$$

We say that X is a uniform random variable over the interval $[a, b]$ and that $f(\cdot)$ is the uniform probability density function for the interval $[a, b]$.

If A is any subinterval of $[a, b]$, then

$$P(A) = \frac{\text{length of } A}{\text{length of } S} = \frac{\text{length of } A}{b-a},$$

where S is the sample space $\{x \mid a < x < b\}$.   Expression (1) is very similar to the assignment of a probability number to a finite sample space with equally likely elementary events, where the word size replaces length.

The intuitive notion of picking a point at random on an interval is formalized by means of the uniform probability density function.   If one selects a point on the interval $[a, b]$ by some random device, this is equivalent to assigning to the random variable X which denotes the coordinate of the point, the uniform probability density function for $[a, b]$.

The probability distribution function, the expectation, and the variance for the uniform probability function follow easily from the appropriate definitions.

The probability distribution function is

$$F(x) = \begin{cases} 0, & x \le a, \\ \dfrac{x-a}{b-a}, & a < x < b, \\ 1, & b \le x. \end{cases}$$

The graphs of the probability density function and the probability distribution function are given in Figure 1.

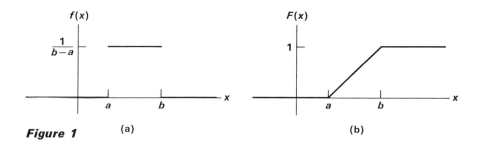

Figure 1   (a)   (b)

Theorem 1

If X is a uniform random variable on the interval $[a, b]$, then

$$E(X) = \frac{a \mid b}{2},$$

$$V(X) = \frac{(a-b)^2}{12}.$$

PROOF: By direct application of the definition of expectation, we may write

$$E(X) = \int_{-\infty}^{\infty} xf(x)\, dx = \int_{a}^{b} x\, \frac{1}{b-a}\, dx$$

$$= \frac{x^2}{2}\, \frac{1}{b-a}\Big|_{a}^{b} = \frac{a+b}{2}.$$

To find the variance, we first calculate $E(X^2)$:

$$E(X^2) = \int_{-\infty}^{\infty} x^2 f(x)\, dx = \int_{a}^{b} x^2\, \frac{1}{b-a}\, dx;$$

hence

$$E(X^2) = \frac{x^3}{3}\, \frac{1}{b-a}\Big|_{a}^{b} = \frac{b^2 + ab + a^2}{3}.$$

Therefore,

$$V(X) = E(X^2) - [E(X)]^2$$

$$= \frac{b^2 + ab + a^2}{3} - \left(\frac{a+b}{2}\right)^2$$

$$= \tfrac{1}{12}(a-b)^2. \quad \blacksquare$$

Example 1

A point is chosen at random on the line segment $[-2, 2]$. What is the probability that the point selected lies between 1 and 2?

Let X be the uniform random variable over the interval $[-2, 2]$. Thus the probability density function is given by

$$f(x) = \begin{cases} \frac{1}{4}, & -2 < x < 2, \\ 0, & \text{otherwise.} \end{cases}$$

The probability distribution function is given by

$$F(x) = \begin{cases} 0, & x \leq -2, \\ \dfrac{x+2}{4}, & -2 < x < 2, \\ 1, & x \geq 2. \end{cases}$$

Therefore, $P(1 < X < 2) = F(2) - F(1) = 1 - \frac{3}{4} = \frac{1}{4}$. Observe that this equals the length of the interval $(1, 2)$ divided by the length of the interval $(-2, 2)$. ∎

Example 2

The time it takes a man to travel from his home to a train station varies from 15 to 20 minutes. The commuter train he is trying to catch leaves the station promptly at 7:48 A.M. If he leaves home promply at 7:30 A.M., find the probability that he will catch his train.

Let us assume that the time required to travel from home to station can be described by a uniform random variable over the interval $[15, 20]$. If X denotes the time of arrival of the man at the station, then X is uniform over the interval 7:45 A.M. to 7:50 A.M., or the interval 0 to 5 with 7:45 A.M. corresponding to 0. The probability the man catches the train is

$$P(X \leq 3) = F(3) = \frac{3-0}{5} = \frac{3}{5}. \quad ∎$$

Example 3

A circle of radius 3 is divided into eight equal sectors marked as in Figure 2, with a random spinner set at the center. What is the probability that the tip of the spinner stops on a sector labeled -1, 0, and 1, respectively, and that on two successive spins the sum of the results will be zero?

Let X be the random variable denoting the point of the tip of the spinner, so that X is uniform over the circumference of a circle of radius 3,

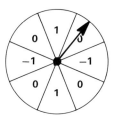

Figure 2

that is, the interval $[0, 6\pi]$.   Let Y be the randon variable that takes on the numbers -1, 0, and 1, so that $S_Y = \{-1, 0, 1\}$.   Then

$$P(Y = -1) = \frac{2\frac{3}{4}\pi}{6\pi} = \frac{1}{4},$$

$$P(Y = 0) \quad = \frac{4\frac{3}{4}\pi}{6\pi} = \frac{1}{2},$$

$$P(Y = 1) \quad = \frac{2\frac{3}{4}\pi}{6\pi} = \frac{1}{4}.$$

If we let Y_i, $i = 1, 2$, be the result of the ith spin, then $Y_1 + Y_2 = 0$ when $Y_1 = -1$, $Y_2 = 1$, or $Y_1 = 1$, $Y_2 = -1$, or $Y_1 = Y_2 = 0$. Hence $P(Y_1 + Y_2 = 0) = P(Y_1 = -1, \quad Y_2 = 1) + P(Y_1 = 1, \quad Y_2 = -1) + P(Y_1 = 0, Y_2 = 0) = \frac{1}{4} \cdot \frac{1}{4} + \frac{1}{4} \cdot \frac{1}{4} + \frac{1}{2} \cdot \frac{1}{2} = \frac{3}{8}.$ ∎

Problems

1. A point is chosen at random on the interval $[-1, 2]$.   What is the probability that the chosen point has a positive coordinate?

2. A piece of string of length 10 is cut at a random point into two pieces.   What is the probability that the ratio of the length of the shorter piece to the length of the longer piece is less than $\frac{1}{3}$?

3. A clock with a second hand is stopped at random.   What is the probability that the second hand is between 4 and 7?

4. Buses arrive at a bus stop at 8:20, 8:06, 8:14, and 8:20.   Your arrival at the bus stop is uniform over the interval 8:00 to 8:20. What is the probability that you will have to wait less than 1 minute?

5. A point is chosen at random on the circumference of a circle of radius 1, with a fixed point designated.   What is the probability that the shorter arc between the variable point and the fixed point is less than one third of the longer arc?

6. The signal light at the intersection of Main and Broad is green for 3 minutes and red for 2 minutes, starting on the hour with green for Main. If your arrival at the intersection on Main Street is uniform between 8:27 A.M. and 8:35 A.M., what is the probability you will have to stop?

7. A point r is selected at random on the interval $(0, 1)$ and the circle with its center at $(0, 0)$ and radius r is drawn. What is the probability that the area is greater than 1?

8. If X is a uniform random variable over the interval $[1, 4]$, find the probability density function for $Y = 3X + 1$.

9. If X is a uniform random variable over the interval $[2, 3]$, find the probability density function for $Y = \ln X$.

10. If X is a uniform random variable over the interval $[-a, a]$, determine a value of a, if possible, for each of the following.

    (a) $P(X > 1) = \frac{1}{4}$ (b) $P(X < \frac{1}{2}) = \frac{1}{2}$
    (c) $P(|X| < 1) = \frac{1}{3}$ (d) $P(X < \frac{1}{2}) = \frac{2}{3}$

11. Refer to Problem 10. If X is a uniform random variable over the interval $[0, a]$, answer the same questions.

12. If Z is uniform random variable over the interval $[0, 6]$, what is the probability that the roots of the quadratic equation $x^2 + Zx + Z = 0$ are real?

13. If a point is chosen at random in the interval $[1,4]$ five independent times, what is the probability that no more than two of the selections have a coordinate between 2 and 3?

14. Prove: If X is a uniform random variable over the interval $[a, b]$, then for any subinterval $[c, d]$ where $a \le c < d \le b$, $P(c < X < d)$ is the same for all subintervals having the same length.

15. If X is a uniform random variable over $[a, b]$ with $E(X) = \alpha$ and $V(X) = \beta$, find a and b.

16. Let X be a random variable with expectation μ and standard deviation σ. If Y is a uniform random variable over the interval $(\mu - \sigma\sqrt{3}, \mu + \sigma\sqrt{3})$, show that $E(X) = \mu$ and $V(X) = \sigma^2$ that is, Y has the same expectation and variance as X.

17. A point is chosen at random on the perimeter of a square of side 2. If Y is a random variable that denotes the distance from the point to the center of the square, find the probability density and distribution functions for Y.

18. Seventy numbers are chosen independently and at random between 1 and 3, and their average is computed. What are the expected value and the variance of the average?

10.2 Normal Probability Function

Perhaps the most important and useful probability function is the normal distribution function. Not only does it arise naturally in many applications, but it also serves as a highly accurate approximation to other distributions, such as the binomial distribution (see Example 2, Section 10.4). The probability density function for the sample mean $\bar{X}$, for example, turns out to have approximately a normal probability function independent of the probability density function for X. Indeed, important theoretical probability arguments hinge on a judicious application of the normal probability function.

Definition 1: Normal probability function

Let X be a continuous random variable with a probability density function specified by

(1)
$$f(x) = \frac{1}{\sqrt{2\pi}\sigma} \exp\left[-\frac{1}{2}\left(\frac{x-\mu}{\sigma}\right)^2\right]$$

for $-\infty < x < \infty$ with the parameters μ and σ satisfying $-\infty < \mu < \infty$ and $\sigma > 0$. We say that X is a normal random variable with parameters μ and σ and that $f(x)$ is the normal probability density function $N(\mu, \sigma)$.

Showing that $f(x)$ is an acceptable probability density function requires some work. In particular, we would need to know something about double integrals to prove that $\int_{-\infty}^{\infty} f(x)\,dx = 1$. We must therefore omit the proof and shall assume that the function in (1) is indeed a probability function. That is,

(2)
$$\int_{-\infty}^{\infty} f(x)\,dx = \int_{-\infty}^{\infty} \frac{1}{\sqrt{2\pi}\,\sigma} \exp\left[-\frac{1}{2}\left(\frac{x-\mu}{\sigma}\right)^2\right] dx = 1.$$

The probability distribution function $F(x)$ is defined in the usual manner.

$$F(x) = \frac{1}{\sqrt{2\pi}\,\sigma} \int_{-\infty}^{x} \exp\left[-\frac{1}{2}\left(\frac{t-\mu}{\sigma}\right)^2\right] dt.$$

Unfortunately, there is no known antiderivative for the density function $f(x)$, so the integral cannot be evaluated except for the two cases $x = \infty$, for which $F(\infty) = 1$, and $x = \mu$, for which $F(\mu) = \frac{1}{2}$. If a value of $F(x)$ is desired for any other value of x, there are tables of approximate values. (see Table 5 in the Appendix).

The graph of the normal density function is the familiar bell-shaped graph symmetric around the point $x = \mu$, with $f(\mu) = 1/(\sqrt{2\pi}\,\sigma)$. The derivative is

$$f'(x) = -\frac{1}{\sqrt{2\pi}\sigma}\left(\exp\left[-\frac{1}{2}\left(\frac{x-\mu}{\sigma}\right)^2\right]\right)\frac{x-\mu}{\sigma},$$

which is zero only at $x = \mu$. For $x < \mu, f'(x) > 0$, while for $x > \mu, f'(x) < 0$, so the function has a maximum at $x = \mu$.

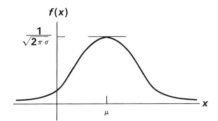

Figure 1

Theorem 1

If X is a normal random variable with parameters μ and σ, then

$$E(X) = \mu \qquad \text{and} \qquad V(X) = \sigma^2.$$

To verify that the expected value of X is μ and that the variance is σ^2 is an exercise in the evaluation of integrals using the techniques of integration by parts and substitution. The details can be found in [18, p. 612].

A normal random variable X is completely specified when and only when the values of the expectation μ and the standard deviation σ are given.

In Section 6.4 we showed that the standardized random variable $Z = (X - \mu)/\sigma$, where $E(X) = \mu$ and $V(X) = \sigma^2$, has mean 0 and standard deviation 1. Thus if X is a normal random variable with $\mu = 0$, $\sigma = 1$, the density and distribution functions are given by

$$\phi(x) = \frac{1}{2\pi}\, e^{-(1/2)x^2},$$

$$\Phi(x) = \int_{-\infty}^{x} \phi(t)\, dt.$$

The symbols ϕ and Φ are frequently used to denote the density and distribution functions of the standardized normal random variable.

Tables for both $\phi(x)$ and $\Phi(x)$ are given in Tables 4 and 5 of the Appendix. If X is a standardized normal random variable, then

$$P(a < X < b) = \frac{1}{\sqrt{2\pi}} \int_a^b e^{-(1/2)t^2} \, dt$$

can be computed approximately from the tables as $\Phi(b) - \Phi(a)$. The calculation (as stated earlier) can only be approximate, because the integral cannot be evaluated via antiderivatives and refined numerical techniques of integration must be used.

The graph of $\phi(x)$ is similar to the one in Figure 1, except that $\mu = 0$ and $\sigma = 1$, and it is symmetric about the y-axis, since $\phi(-x) = \phi(x)$. The graph is given in Figure 2. The graph of the probability distribution function Φ is given in Figure 3.

Because $\phi(x)$ is a symmetric function, we can prove the following theorem.

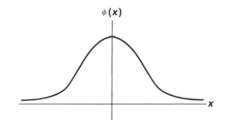

Figure 2

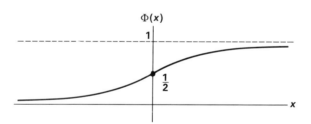

Figure 3

Theorem 2

$\Phi(-x) = 1 - \Phi(x)$.

■ **PROOF:** By definition $\Phi(-x) = \int_{-\infty}^{-x} \phi(u) \, du$ and $\Phi(x) = \int_{-\infty}^{x} \phi(u) \, du$. Let us assume that $x > 0$. We shall show first that

(4) $$\int_{-\infty}^{-x} \phi(u) \, du = \int_{x}^{\infty} \phi(u) \, du.$$

Now

$$\int_{-\infty}^{-x} \phi(u) \, du = \lim_{-R \to \infty} \int_{-R}^{-x} \phi(u) \, du.$$

If we substitute $-t$ for u, then

$$\int_{-R}^{-x} \phi(u)\, du = (-1) \int_{R}^{x} \phi(-t)\, dt.$$

But $\phi(-t) = \phi(t)$, and because the minus sign changes the limits of integration, we obtain

$$\int_{-R}^{-x} \phi(t)\, dt = \int_{x}^{R} \phi(t)\, dt.$$

Therefore, when we take the limit, $\int_{-\infty}^{-x} \phi(t)\, dt = \int_{x}^{\infty} \phi(t)\, dt.$ Now

$$\Phi(-x) + \Phi(x) = \int_{-\infty}^{-x} \phi(t)\, dt + \int_{-\infty}^{x} \phi(t)\, dt$$

$$= \int_{x}^{\infty} \phi(t)\, dt + \int_{-\infty}^{x} \phi(t)\, dt = \int_{-\infty}^{\infty} \phi(t)\, dt = 1,$$

and the assertion of the theorem follows. ∎

Example 1

If Z is a normal random variable with $\mu = 0$ and $\sigma = 1$, find the values of the following probabilities using Table 5 of the Appendix:

(a) $P(-1 < Z < 1) = \Phi(+1) - \Phi(-1)$
$\qquad\qquad\qquad\quad = 0.8413 - 0.1587$
$\qquad\qquad\qquad\quad = 0.6826.$

The probability is represented by the area of the shaded region in Figure 4.

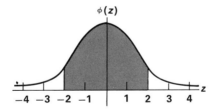

Figure 4

(b) $P(Z > 2) = 1 - P(Z \leq 2)$
$\qquad\qquad\quad = 1 - \Phi(2)$
$\qquad\qquad\quad = 0.0228.$

The probability is represented by the area of the shaded region in Figure 5. ∎

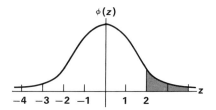

Figure 5

If X is a normal random variable with expectation μ and standard deviation σ, then we can express the distribution function $F(\cdot)$ for X in terms of Φ. Let $Z = (X - \mu)/\sigma$, so that

$$X = \sigma Z + \mu.$$

Therefore,

$$P(X \le b) = P(\sigma Z + \mu \le b)$$

$$= P\left(Z \le \frac{b - \mu}{\sigma}\right).$$

Consequently, we may write

(5) $$F(b) = \Phi\left(\frac{b - \mu}{\sigma}\right).$$

and this is valid for any real number b. Thus

$$P(a < X < b) = F(b) - F(a)$$

$$= \Phi\left(\frac{b - \mu}{\sigma}\right) - \Phi\left(\frac{a - \mu}{\sigma}\right).$$

As soon as the values of $\Phi(\cdot)$ are known, the probabilities for any normal random variable can be calculated.

Theorem 3

$F(\mu) = \frac{1}{2}$.

PROOF: From (5) with $b = \mu$ we have

$$F(\mu) = \Phi((\mu - \mu)/\sigma) = \Phi(0).$$

However, from Theorem 2, $\Phi(0) = 1 - \Phi(0)$; thus $\Phi(0) = \frac{1}{2}$. Therefore, $F(\mu) = \frac{1}{2}$. ∎

Example 2

If X is a normal random variable with expected value $\mu = 12$ and standard deviation $\sigma = 2$, express the following probabilities in terms of the standardized probability distribution functions and find the values from Table 5 in the Appendix.

(a) $P(10 \leq X \leq 12)$.

Let $Z = (X - 12)/2$; then $X = 2Z + 12$. Hence

$$P(10 \leq X \leq 12) = P(10 \leq 2Z + 12 \leq 12)$$
$$= P(-2 \leq 2Z \leq 0)$$
$$= P(-1 \leq Z \leq 0)$$
$$= \Phi(0) - \Phi(-1)$$
$$= 0.5 - 0.1587$$
$$= 0.3413.$$

Alternatively, we could use equation (5) as follows:

$$P(10 \leq X \leq 12) = F(12) - F(10)$$

$$= \Phi\left(\frac{12 - 12}{2}\right) - \Phi\left(\frac{10 - 12}{2}\right)$$

$$= \Phi(0) - \Phi(-1).$$

(b) Find a number a such that

$$P(|X - 12| < a) = 0.90.$$

Rewriting the statement as

$$P(12 - a < X < 12 + a) = 0.90,$$

we obtain by (5) the statement

$$\Phi\left(\frac{12 + a - 12}{2}\right) - \Phi\left(\frac{12 - a - 12}{2}\right) = 0.90;$$

that is,

$$\Phi\left(\frac{a}{2}\right) - \Phi\left(\frac{-a}{2}\right) = 0.90.$$

But, by Theorem 2, $\Phi(-a/2) = 1 - \Phi(a/2)$; therefore,

$$2\Phi\left(\frac{a}{2}\right) - 1 = 0.90,$$

$$\Phi\left(\frac{a}{2}\right) = 0.95.$$

In Table 5 of the Appendix we find that $\Phi(1.64) = 0.9495$ and $\Phi(1.65) = 0.9505$, so interpolating between these numbers, $\Phi(1.645) = 0.95$. Hence if $a = 3.29$, then

$$P(|X - 12| < 3.29) = 0.90. \qquad \blacksquare$$

Example 3

Suppose the grade-point average of all students at a college is a normal random variable X with $\mu = 2.6$ and $\sigma = 0.45$. A student is placed on academic probation if his grade point is below 2.0. What is the probability that a student selected at random is on probation?

The probability to be calculated is just $P(X < 2.0)$, which in terms of the distribution function F is $F(2)$. Using equation (5) this is equal to $\Phi[(2 - 2.6)/0.45] = \Phi(-1.33)$. From Table 5 of the Appendix, $\Phi(-1.33) = 0.0918$. Thus the probability that a randomly selected student is on probation is approximately 0.09. $\quad\blacksquare$

Example 4

In the production of clothesline rope, the breaking strength of the rope will vary. Let us suppose that the breaking strength, measured in pounds, is a normal random variable X with $\mu = 100$ and $\sigma = 4$. Rope is sold in coils of 100 feet, and if X exceeds 95, a profit of \$0.25 is realized. If $X \leq 95$, the rope is considered defective; however, it may be used for a different purpose for which the profit per coil amounts to \$0.10. Find the expected profit per coil.

Let Y be the finite random variable with the two values 10 and 25. We must determine the probability point function for Y.

$$p_Y(10) = P(X \leq 95) = F_X(95) = \Phi\left(\frac{95 - 100}{4}\right).$$

Hence

$$p_Y(10) = \Phi(-\tfrac{5}{4}) = 0.1056.$$

Thus

$$p_Y(25) = 0.8944.$$

Since $E(Y) = 10p_Y(10) + 25p_Y(25)$, we obtain

$$E(Y) = 1.056 + 22.38 = 23.436. \qquad \blacksquare$$

Example 5

A standardized psychological test is given to a class of 35 "normal" children. It is reasonable to assume that the score a person makes on the test can be represented by a normal random variable X with $\mu = 82$ and $\sigma = 6$. Write expressions for the following:

(a) the probability that all scores are less than 94.
(b) the probability that the lowest score will be less than 64.
(c) the expected number of scores between 76 and 88.

The 35 scores can be considered to be 35 independent repeated trials of random experiment. Thus, for (a), if $p = P(X < 94)$, the answer is p^{35}. But

$$P(X < 94) = F_X(94) = \Phi\left(\frac{94 - 82}{6}\right)$$

$$= \Phi(2)$$

$$= 0.9772,$$

so $P(\text{all scores less than } 94) = (0.9772)^{35} = 0.446$. To find the probability that the lowest score is less than 64, turn the problem around and find the probability that the smallest score is greater than 64; that is, $P(\text{at least one score} < 64) = 1 - P(\text{no score} < 64)$. If $p = P(X > 64)$, then

$$p = 1 - F_X(64)$$

$$= 1 - \Phi\left(\frac{64 - 82}{6}\right)$$

$$= 1 - \Phi(-3)$$

$$= \Phi(3)$$

$$= 0.9987.$$

Therefore,

$$P(\text{no score} < 64) = (0.9987)^{35}$$

$$= 0.96,$$

and $P(\text{at least one score} < 64) = 0.04$.

For part (c), $P(76 < X < 88) = F_X(88) - F_X(76)$, and using the standardized normal random variable we may write

$$F_X(88) - F_X(76) = \Phi(1) - \Phi(-1)$$

$$= 0.68.$$

Let Z be a binomial random variable with $p = 0.68$ and $n = 35$; that is, Z denotes the number of successes in 35 trials. The expected value of Z, $E(Z) = 35(0.68) = 23.8$. Rounding this to 24 gives us the expected number. ∎

Problems

1. If X is a normal random variable with $\mu = -1$ and $\sigma = 3$, express the probability $P(-1 < X < 2)$ in terms of the standardized probability functions and find its value from a table.

2. If X is a normal random variable with $\mu = -1$ and $\sigma = 3$, find a number b such that $P(|X + 1| < b) = 0.95$.

3. If X is a normal random variable with $\mu - 5$ and $\sigma - \frac{1}{2}$, express the probability $P(|X - 5| > 1)$ in terms of the standardized probability functions and find its value from a table.

4. If X is a normal random variable with $\mu = 3$ and $\sigma = 2$, find a number b so that $P(X > b) = 2P(X \leq b)$.

5. Let us assume that the breaking strength of cotton fabric is a normal random variable X with expected value 150 and variance 16. A bolt of material is considered defective if $X < 146$. What is the probability that a bolt chosen at random is not defective?

6. The diameter of a telephone cable is considered to be a normal random variable with $\mu = 0.8$ and variance 0.0004. A cable is considered defective if the diameter differs from its expected value by more than 0.025. What is the probability of obtaining a defective cable?

7. The temperature in a chemical reaction (measured in degrees centigrade) is a normal random variable with expected value 15 and variance 9. What is the probability the temperature will be between 12 and 17?

8 The college-board verbal score for the students at a particular institution can be considered a normal random variable with expectation 625 and standard deviation 75. What is the probability that a score will fall below 500?

9. The annual snowfall is known to be a random variable with expectation 46 inches and a standard deviation 4 inches. For what annual snowfall in inches is the probability 0.10 that it will be exceeded?

10. Refer to Problem 9. Over a 50-year period, in how many years would we expect the annual snowfall to be between 40 and 50 inches?

11. If X and Y are independent normal random variables, with the expected value $\mu = 8$ and the standard deviation $\sigma = 2$ for X and the expected value $\mu = 8$ and the standard deviation $\sigma = 3$ for Y, find

   (a) the probability that the smaller of X and Y is less than 10.
   (b) the probability that the larger of X and Y is less than 10.

12. The true volume of a bottle of a soft drink can be considered to be a normal random variable with $\mu = 16.02$ ounces and $\sigma = 0.01$ ounces. What is the probability a bottle contains less than 16 ounces?

13. The heights of all college men can be considered a normal random variable X with expectation 69.0 inches and standard deviation 2.5 inches. For what interval of heights around the mean would the probability be 0.5 that X lies in the interval?

14. The hourly wage of workers in an industry is considered to be a normal random variable X with an expectation of $3.80 and a standard deviation of $0.50. In a large sample of workers what percentage receive wages between $3.00 and $4.00 per hour?

15. The radius of a circle is a normal random variable with $\mu = 1$ and variance 0.04. Find the probability density function for the area of the circle.

16. If we can assume that the length of life in hours of a flashlight battery is a normal random variable X with $\mu = 120$ and $\sigma = 36$, what is the expected number of batteries in a sample of size 160 that would

    (a) have a length of life between 85 and 135 hours?
    (b) last longer than 130 hours?

17. If the test grade in a particular class can be considered to be a normal random variable with $E(X) = 79$ and $V(X) = 25$, what is the expected number of students in a class of size 200 that did not receive a grade between 75 and 85?

18. If X is a normal random variable with expectation 0 and variance 1, find the probability density function of $Y = |X|$.

19. A specific type of manufactured nail has a length that can be considered to be a normal random variable X with expectation 3.2 and standard deviation 0.1. From a large supply of these nails a new lot is determined by discarding all nails with a length greater than 3. Let Y be the random variable representing the length of the nails in the new lot. Find the probability distribution function $F_Y(\cdot)$ (in integral form).

20. The length of life of two different brands of light bulbs are considered to be normal random variables X_1 and X_2 with $\mu_1 = 40$, $\sigma_1 = 6$ and $\mu_2 = 42$, $\sigma_2 = 3$. If the light bulb is to be used for a 45-hour period, which brand is preferred?

21. The inside diameter of a washer can be considered to be a normal random variable X with expectation 10 millimeters and variance 1 millimeter. If the washer diameter lies between 9 and 11 millimeters, the manufacturer realizes a profit of c_1 dollars, while if $X < 9$, there is a loss of c_2 dollars, and if $X > 11$, there is a loss of c_3 dollars. Find the expected profit per washer.

22. Refer to Problem 21.   Suppose the manufacturing process is such that the expectation μ for the inside diameter can be adjusted within reasonable limits.   Determine the value of μ that will maximize the expected profit.   (*Hint*: Find the expected profit as a function of μ and then find the maximum value of the function.)

23. Six independent observations of a normal random X with $\mu = 1$ and $\sigma = 1$ are made.   What is the probability that exactly two of the results are negative?

24. The weight of a student at a school can be considered to be a normal random variable X with expectation 150 and standard deviation 10. If it is known that the weight of a particular student is less than 150, what is the probability that the student weighs more than 145?

25. A random sample of size 5 is taken from a normal random variable X with expectation 12 and variance 4.   What is the probability that

    (a) the maximum of the sample exceeds 15?
    (b) the minimum of the sample is less than 10?

26. If X is a normal random variable with expectation μ and standard deviation σ, find an expression for $P(\mu - k\sigma < X < \mu + k\sigma)$ in terms of the standardized distribution function, where k is a positive real constant.   Find the value of the probability for $k = 1$, 2, 3.

27. Prove: If X is a normal random variable with expectation μ and standard deviation σ and $Y = aX + b$, a and b real constants with $a > 0$, then Y is a normal random variable with mean $a\mu < b$ and standard deviation $a\sigma$.

28. Prove: If X is a normal random variable with expectation μ and standard deviation σ and $Y = (x - \mu)/\sigma$, then Y is a normal random variable with mean 0 and standard deviation 1.   (*Hint*: Find the probability density function for Y.)

10.3 Exponential Probability Function

The exponential probability function arises in applications of probability theory to random phenomena that in one way or another are the first occurrence of an event in an interval of time.   Specific examples are the length of life of a light bulb, or the time it takes for an electronic component to fail, or the time for the first accident to occur.

Definition 1: Exponential probability function

Let X be a continuous random variable with a probability density function specified by

$$f(x) = \begin{cases} \alpha e^{-\alpha x}, & \text{for } x > 0, \\ 0, & \text{otherwise,} \end{cases}$$

where α is a positive constant. We say that X is an exponential random variable with parameter α.

The function f is an acceptable probability density function, for the improper integral

$$\int_{-\infty}^{\infty} f(x)\, dx = \int_0^{\infty} \alpha e^{-\alpha x}\, dx$$

has the value 1. This is evident since $-e^{-\alpha x}$ is an antiderivative of $\alpha e^{-\alpha x}$. Hence

$$\int_0^{\infty} \alpha e^{-\alpha x}\, dx = \lim_{R \to \infty} \int_0^{R} \alpha e^{-\alpha x}\, dx = \lim_{R \to \infty} -e^{-\alpha x} \Big|_0^{R}$$

$$= \lim_{R \to \infty} (1 - e^{-\alpha x}) = 1.$$

The probability distribution function $F(\cdot)$ can also be readily determined:

$$F(x) = \int_0^{x} \alpha e^{-\alpha t}\, dt, \qquad x > 0,$$

so

$$F(x) = -e^{-\alpha t} \Big|_0^{x} = 1 - e^{-\alpha x}.$$

Consequently,

$$F(x) = \begin{cases} 0, & x < 0, \\ 1 - e^{-\alpha x}, & x \geq 0. \end{cases}$$

The graph of the probability density function is given in Figure 1 and the graph of the probability distribution function is given in Figure 2.

The exponential probability distribution is closely related to the Poisson probability distribution, even though the former is continuous and the latter discrete.

Let us consider a Poisson random variable Y with a parameter α. Since α is the mean rate of occurrence of Y over a unit interval of time (or length), then over the interval $[0, x]$ the parameter would have the value αx, and the corresponding probability point function is

$$p(n) = \frac{(\alpha x)^n e^{-\alpha x}}{n!}, \qquad n = 0, 1, 2, \ldots.$$

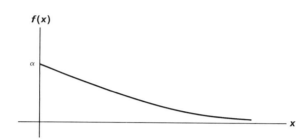

Figure 1

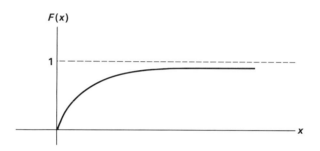

Figure 2

Now consider the special case $n = 0$, that is, the probability of no occurrences in the interval from 0 to x, $p(0) = e^{-\alpha x}$. Interpreting this last expression as a function of x, we can say that the probability that the first occurrence of the original Poisson random variable, after x units of time (or length), will be $1 - e^{-\alpha x}$. But as a function of x, this is the probability distribution function for an exponential random variable X:

$$F(x) = P(X \le x) = 1 - e^{-\alpha x}.$$

Thus the random variable X denotes the time that elapses from the beginning of the observation to the first occurrence of the event.

The expectation and the variance for the exponential probability function follow directly from the definitions.

Theorem 1

If X is an exponential random variable with parameter α, then

$$E(X) = \frac{1}{\alpha} \quad \text{and} \quad V(X) = \frac{1}{\alpha^2}.$$

As in the case with the normal random variable, the proof of Theorem 1 is an exercise in the use of integration by parts and l'Hospital's rule, and so omitted. For the details see [17, p. 626].

Example 1

Suppose the length of life of a light bulb can be considered to be an exponential random variable X with a parameter $\alpha = \frac{1}{100}$. Find the length of time t such that the probability is 0.90 that the light bulb will operate for a time greater than t.

The probability density function for X is

$$f(x) = \begin{cases} \frac{1}{100} e^{-x/100}, & x > 0, \\ 0, & \text{otherwise,} \end{cases}$$

and the probability distribution function is

$$F(x) = \begin{cases} 0, & x < 0, \\ 1 - e^{-x/100}, & x > 0. \end{cases}$$

Since we must determine t so that

$$P(X > t) = 0.90,$$

express the probability in terms of the probability distribution function:

$$P(X > t) = 1 - F(t).$$

Therefore, t must satisfy

$$1 - (1 - e^{-t/100}) = 0.90;$$

that is,

$$e^{-t/100} = 0.90.$$

From a table of value of exponential functions or using natural logarithms, we find that

$$\frac{t}{100} = 0.10;$$

hence $t = 10$ units of time. ∎

Example 2

The length of telephone conversations has been found to be an exponential random variable with an expected value of 3 minutes. What is the probability of a call lasting more than 10 minutes?

Let the random variable X denote the length of the telephone conversation. Since the expected value is 3, $1/\alpha = 3$; that is, $\alpha = \frac{1}{3}$.

The probability density function is

$$f(x) = \begin{cases} \frac{1}{3} e^{-x/3}, & x > 0, \\ 0, & \text{otherwise,} \end{cases}$$

and the probability distribution function is

$$F(x) = \begin{cases} 0, & x < 0, \\ 1 - e^{-x/3}, & x \geq 0. \end{cases}$$

The probability that X exceeds 10 is

$$P(X > 10) = 1 - F(10)$$
$$= 1 - (1 - e^{-10/3})$$
$$= e^{-10/3}$$
$$= 0.035 \text{ (approximately).}$$ ∎

Example 3

The number of customers arriving at a checkout counter in a super-market can be taken to be a Poisson random variable with a parameter $\alpha = 30$ per hour (see Example 4 in Section 8.3). A customer has just been checked out. What is the probability that it is at least 6 minutes before the next customer arrives? What is the probability that it is at most 4 minutes?

The mean rate of occurrence for this random event is 30 per hour, so if we set $\alpha = 30$ in the probability distribution function, $F(x) = 1 - e^{-30x}$ (hours is the unit of measurement for x). The first question is answered by evaluating $P(X > \frac{6}{60})$, that is, the probability that the time to the first occurrence is greater than 6 minutes. Since the complement of the event $\{X > \frac{6}{60}\}$ is $\{X \leq \frac{6}{60}\}$, we can write

$$P(X > \tfrac{6}{60}) = 1 - F(\tfrac{6}{60}).$$

But

$$F(0.1) = 1 - e^{-30(0.1)},$$

so

$$P(X > \tfrac{6}{60}) = e^{-3},$$

and from a table of exponential values

$$P(X > \tfrac{6}{60}) = 0.05.$$

For the second question we need to evaluate $P(X \leq \frac{4}{60})$, which is $F(\frac{1}{15})$. Hence

$$P(X \leq \tfrac{4}{60}) = 1 - e^{-30(1/15)}$$
$$= 1 - e^{-2}$$
$$= 1 - 0.082$$
$$= 0.918.$$ ∎

The exponential probability functions have a property that is particularly applicable in the study of "time-to-failure" models. Suppose a component, such as an electron tube, for which the exponential probability distribution function appropriately describes the length of life, has operated satisfactorily for b units of time. What is the probability that the component will operate for d more units of time?

In terms of conditional probability for the events $\{X > b\}$ and $\{X > d + b\}$, this can be written

$$P(X > d + b \mid X > b) = ?$$

where X denotes the length of life of the component. However, assuming that $P(X > b) > 0$, we can use the definition of conditional probability to write

$$P(X > d + b \mid X > b) = \frac{P(X > d + b)}{P(X > b)}$$

$$= \frac{e^{-\alpha(d + b)}}{e^{-\alpha b}}$$

$$= e^{-\alpha d}$$

$$= P(X > d).$$

Thus the conditional probability of operating for $(d + b)$ units of time after operating for b units of time is the same as the original probability of operating d units of time. This can be interpreted to mean that if the length of life of a component obeys an exponential probability function, the component is not subject to wear or fatigue. Conversely, it can be shown that if a random variable that is always nonnegative satisfies the relationship $P(X > d + b \mid X > b) = P(X > d)$ for b and d, it must be an exponential random variable (see [16]).

Problems

1. If X is an exponential random variable with $\alpha = \frac{1}{2}$, find the expected value, the standard deviation, and $P(1 < X < 3)$.

2. If X is an exponential random variable with $\alpha = 2$, find the expected value, the standard deviation, and $P(|X - \frac{1}{2}| < \frac{1}{2})$.

3. The length of time between auto accidents at a particular intersection can be considered to be an exponential random variable with expected value 10 hours. What is the probability that there will be no accidents for 24 hours?

4. The length of the time period between sales of a certain item (measured in days) can be considered to be an exponential random variable with expected value $\frac{4}{5}$. What is the probability that the time between successive sales is greater than 2 days? Less than 4 days?

5. The time in minutes between the arrival of automobiles at the toll-gate on a turnpike is an exponential random variable with parameter 6. What is the probability that the time between successive arrivals is between 1 and 2 minutes?

6. The length of life in months of a component in a computer can be considered to be an exponential random variable X with parameter $\alpha = 2$. If the component fails, the computer is down, so a new component is installed every month. What is the probability that the computer will be down because of this component?

7. If the length of life of a computer component is an exponential random variable with parameter 0.01, and six such components operate independently in the computer, what is the probability that at least one half of them are operating 200 hours after installation?

8. If X is an exponential random variable with parameter α, find the probability that X exceeds the expected value of X.

9. Let Z_t denote a Poisson random variable for the number of times lightning strikes Essex County in t days, and assume that the parameter is βt. Let X be the length of time between successive lightning strikes. Show that X is an exponential random variable with parameter β.

10. Refer to Problem 9. If $\beta = 10$, what is the probability that the time between successive strikes will exceed 5 days? Be less than 10 days?

11. If X is an exponential random variable with parameter α, show that

$$P(t \leq X \leq t + h) = \alpha h e^{-\alpha t_1},$$

where $t < t_1 < t + h$.

12. If X is an exponential random variable with parameter α, find the probability density function for $Y = X^2$.

13. If X is the length of life of a device, the function $R(x) = P(X > x)$ is called the *reliability function*. For X an exponential random variable with parameter α, find $R(x)$ and show that $f(x)/R(x) = \alpha$, where $f(x)$ is probability density function for X.

14. Suppose that the cost to produce an item with a length of life governed by an exponential probability density function is a function of the mean time to failure, $4/\alpha^2$. Furthermore, suppose there is a profit of b dollars for each unit of time that the time operates without failure. Find the maximum of the expected profit with respect to α.

15. An electronic device can be manufactured by two different processes with a cost of $3 per unit for the first process and a cost of $6 per

unit for the second process.   The mean length of life for the first process is 100 hours and the mean length of life for the second process is 150 hours.   However, if the device fails before 200 hours, a penalty of $10 per unit must be paid by the manufacturer.   Which process will have the lower expected cost per unit?

16. The length of life in months of a component in a computer can be considered to be an exponential random variable X with parameter $\alpha = 2$.   As a part of the preventive maintenance program the component is replaced after m months.   If the life $X < m$, the cost is $5(m - X)$; otherwise the cost is $2(X - m)$.   Find the value of m such that the expected value $E(Y)$ is minimum.

10.4 Law of Large Numbers and Central Limit Theorem

In this, the final section of the book, we shall discuss two important theorems of vital importance both to the theory and to the applications.   The first of these, the law of large numbers, is a generalized version of Theorem 2 in Section 7.7 for continuous random variables.   The second theorem generalizes a result first obtained in 1733 and extended by Laplace in 1812, the approximation of the binomial probability function by the normal probability function.   As a prelude to both theorems, we extend the Chebyshev inequality of Section 7.6 to continuous random variables.

Theorem 1

Let X be a continuous random variable with an expected value $E(X)$ and a variance $V(X)$.   Let c be a positive number.   Then the probability that a value of X occurs that differs from $E(X)$ by more than c is less than or equal to $V(X)/c^2$ that is,

$$P(|X - E(X)| > c) \leq \frac{V(X)}{c^2}.$$

PROOF: As in Section 7.6, we start with the expression for the variance, where $f(\cdot)$ is the probability density function for the continuous random variable X:

$$V(X) = \int_{-\infty}^{\infty} [x - E(X)]^2 f(x)\ dx.$$

Since we are primarily interested in the event $\{|X - E(X)|\} > c$, let us rewrite the integral for $V(X)$ with this event in mind.

$$V(X) = \int_{-\infty}^{E(X) - c} [x - E(X)]^2 f(x) \, dx + \int_{E(X) - c}^{E(X) + c} [x - E(X)]^2 f(x) \, dx$$

$$+ \int_{E(X) + c}^{\infty} [x - E(X)]^2 f(x) \, dx.$$

The middle integral has a positive integrand, so that dropping it from the equation yields the inequality,

$$V(X) \geq \int_{-\infty}^{E(X) - c} [x - E(X)]^2 f(x) \, dx + \int_{E(X) - c}^{\infty} [x - E(X)]^2 f(x) \, dx.$$

In the interval $(-\infty, E(X) - c)$, $x < (E(X) - c)$, so $(x - E(X)) < -c$; while in the interval $(E(X) + c, \infty)$, $(E(X) + c) < x$, so $c < (x - E(X))$. Together these two inequalities yield $|x - E(X)| > c$. Therefore,

$$V(X) \geq \int_{-\infty}^{E(X) - c} c^2 f(x) \, dx + \int_{E(X) + c}^{\infty} c^2 f(x) \, dx$$

$$\geq c^2 \left[\int_{-\infty}^{E(X) - c} f(x) \, dx + \int_{E(X) + c}^{\infty} f(x) \, dx \right].$$

The first integral is $P(X < (E(X) - c))$; hence $P((X - E(X)) < -c)$, while the second is $P(X > (E(X) + c))$; hence $P((X - E(X)) > c)$. Together they give $P(|X - E(X)| > c)$. Therefore,

$$V(X) \geq c^2 P(|X - E(X)| > c),$$

and the desired conclusion follows:

$$P(|X - E(X)| > c) \leq \frac{V(X)}{c^2}. \qquad \blacksquare$$

Chebyshev's inequality may be used for continuous random variables much as it was used for finite random variables. However, in order to apply it to sample random variables that are continuous, Theorem 2 of Section 7.3 and Theorem 2 of Section 7.4 need to be extended to continuous random variables. A reasonable justification of these results at this time necessitates several digressions of considerable magnitude, such as multiple integrals. We shall therefore only state the extensions.

Theorem 2

If $X_1, X_2, \ldots, X_n$ are n random variables defined on the sample space with expectations $E(X_k)$ $k = 1, 2, \ldots, n$, then

$$E\left(\sum_{k=1}^{r} X_k \right) = \sum_{k=1}^{n} E(X_k).$$

Theorem 3

If $X_1, X_2, \ldots, X_n$ are n mutually independent random variables defined on the same sample space with variances $V(X_k)$, $k = 1, 2, \ldots, n$, then

$$V\left(\sum_{k=1}^{n} X_k\right) = \sum_{i=1}^{n} V(X_i).$$

These theorems state that the expectation of the sum of a finite set of continuous random variables is the sum of the expectations and the variance of the sum of a finite set of mutually independent continuous random variables is the sum of the variances. We recognize that there are several things here that need precise formulation, but we choose to rely on the reader's previous experience with the terms for discrete random variables to provide a plausible interpretation.

If the n random variables in Theorems 2 and 3 are sample random variables, then in addition to the usual assumption of independence, all the X_i have the same expectation $E(X)$ and variance $V(X)$. Let $\bar{X}$ denote the sample mean for n sample random variables, exactly as stated in Definition 1 of Section 7.6; thus

$$\bar{X} = \frac{1}{n} \sum_{k=1}^{n} X_i.$$

By Theorem 1 of Section 7.6 we have

$$E(\bar{X}) = E(X),$$

$$V(\bar{X}) = \frac{1}{n} V(X).$$

Theorem 2 of Section 7.6 is essentially the mathematical formulation of the common expression "the larger the sample, the better the results." At this time we can restate the theorem for continuous random variables using the limit concept. Rereading the paragraphs preceding Theorem 2 of Section 7.6 might be desirable.

Theorem 4

Law of Large Numbers for the Sample Mean Let X be a random variable with mean $E(X)$ and variance $V(X)$. If $\bar{X}$ is the sample mean for a set of sample random variables of size n for X, then for any $\varepsilon > 0$,

$$\lim_{n \to \infty} P(|\bar{X} - E(X)| > \varepsilon) = 0.$$

PROOF: Since $\bar{X}$ is the sample mean for a set of sample random variables, $E(\bar{X}) = E(X)$ and $V(\bar{X}) = (1/n)V(X)$. Applying Chebyshev's inequality to the random variable $\bar{X}$ we obtain

$$P(|\bar{X} - E(X)| > \varepsilon) \leq \frac{V(X)}{n\varepsilon^2}.$$

Since $V(X)/\varepsilon^2$ is a constant, we know that

$$\lim_{n \to \infty} \frac{V(X)}{\varepsilon^2} \frac{1}{n} = 0,$$

and since the probability is nonnegative, the result follows easily by the squeeze principle for convergent sequences.   ∎

Chebyshev's inequality can also be written in the form

$$P(|\bar{X} - E(X)| \leq \varepsilon) = 1 - P(|\bar{X} - E(X)| > \varepsilon),$$

in which case the conclusion of Theorem 4 must be that

$$\lim_{n \to \infty} P(|\bar{X} - E(X)| \leq \varepsilon) = 1.$$

Observe that the sequence depends on n because $\bar{X}$ does. The convergence implied in this limit statement is not quite the same as the convergence used in the definition of convergence for sequences of real numbers. The statement does not say that the sequence $\{\bar{X}\}$ converges to $E(X)$ as the sample size increases. What it does say is that the sequence $\{P(|\bar{X} - E(X)| < \varepsilon)\}$ converges to 1 as $n \to \infty$. That is, for n sufficiently large, the probability that the value of the sample mean will be within ε of $E(X)$ is close to 1. This does not guarantee that $\bar{X}$ will actually be close to $E(X)$, but it does guarantee that the probability that $\bar{X}$ will be close to $E(X)$ is near 1.

A very useful application of the law of large numbers for the sample mean is to use the sample mean for obtaining estimates of parameters in probability functions.

Example 1

Suppose that a certain type of light bulb has a length of life in hours that can be described by an exponential random variable X with unknown parameter α. Since $E(X) = 1/\alpha$, an estimate for $E(X)$ gives an estimate for α. To make the procedure precise, take $\varepsilon = 1.0$ in Theorem 4 and determine the sample size n necessary to give a probability of 0.90 that $|\bar{X} - 1/\alpha| < 1.0$.

Since X is an exponential random variable, $E(X) = 1/\alpha$ and $V(X) = 1/\alpha^2$. Chebyshev's inequality for $\bar{X}$ reads

$$P(|\bar{X} - 1/\alpha| < 1.0) \geq 1 - \frac{(1/\alpha^2)}{n(1.0)^2}.$$

We want to determine n so that

$$1 - \frac{1}{n\alpha^2} \geq 0.90.$$

Hence we find that

$$n\alpha^2 \geq 10,$$

$$n \geq \frac{10}{x^2}.$$

We are stuck here unless we can make a reasonable guess regarding α. For example, if we can assume that $1/\alpha \geq 50$, then for a sample size n such that

$$n \geq 10(50)^2 = 25{,}000,$$

the probability that we are within 1 unit of $1/\alpha$ is at least 0.90. Thus our determination is finished as soon as we have a lower bound for $1/\alpha$. ∎

As we saw in Example 3, Section 5.2, if S_n denotes the number of successes in n Bernoulli trials with p the probability of success in each trial, then S_n is a binomial random variable with a probability point function

$$p(k) = \binom{n}{k} p^k q^{n-k}, \qquad k = 0, 1, \ldots, n.$$

More often, however, we are interested in determining the probability that S_n lies between two values; for example,

$$P(a < S_n < b) = \sum_{k > a}^{k < b} \binom{n}{k} p^k q^{n-k},$$

where k is taken over all integral values between a and b. Attempts to approximate this kind of probability when n is large is what led to the development of the normal probability density and distribution functions.

Theorem 5

DeMoivre–Laplace If S_n is a binomial random variable with parameter p, then for real numbers a and b,

$$P(a < S_n < b) \doteq \Phi\left(\frac{b - np + \frac{1}{2}}{\sqrt{npq}}\right) - \Phi\left(\frac{a - np - \frac{1}{2}}{\sqrt{npq}}\right),$$

where Φ is the normal probability distribution function with mean 0 and variance 1 and where $\doteq$ indicates "approximately equal."

This theorem says that, for large n, the probability that S_n is between a and b is approximately equal to the integral of the normal probability function between the limits $[(a - np + \frac{1}{2})/npq, (b - np + \frac{1}{2})/npq]$. The value of the integral is, of course, the difference between the two values of the normal probability distribution function.

Example 2

Suppose a fair die is tossed 6000 times. The probability that the number of tosses that result in a 6 is between 980 and 1020 is the sum

$$\sum_{k=980}^{1020} \binom{6000}{k} \left(\frac{1}{6}\right)^k \left(\frac{5}{6}\right)^{6000-k}.$$

An approximation to this sum is provided by Theorem 5, where $a = 980$, $b = 1020$, $p = \frac{1}{6}$, and $n = 6000$, so

$$\frac{b - np + \frac{1}{2}}{\sqrt{npq}} = 0.71,$$

$$\frac{a - np - \frac{1}{2}}{\sqrt{npq}} = -0.71.$$

Thus

$$P(980 < S_n < 1020) \doteq \Phi(0.71) - \Phi(-0.71)$$
$$= 0.52. \quad \blacksquare$$

Theorem 5 can be proved using only elementary techniques. However, the proof is lengthy and we shall omit it. (For an elementary rigorous proof the reader should consult [15], pp. 234–242.)

Because the normal probability density function is a continuous function and the binomial probability point function is a discrete function, some adjustment of this approximation is required. The factors of $\frac{1}{2}$ on the limits of the integral are there for that purpose and are known as a "continuity" correction. The statement of Theorem 5 without the correction factors gives a slightly less accurate approximation but a form that is easier to handle.

Let S_n^* be the standardized binomial random variable for S_n; that is,

$$S_n^* = \frac{S_n - np}{\sqrt{npq}},$$

and let us find an approximation for the probability $P(\alpha < S_n^* < \beta)$. First express the inequality in terms of S_n; that is,

$$\alpha < \frac{S_n - np}{\sqrt{npq}} < \beta,$$

or

$$\alpha\sqrt{npq} + np < S_n < \beta\sqrt{npq} + np,$$

so

$$P(\alpha < S_n^* < \beta) = P(\alpha\sqrt{npq} + np < S_n < \beta\sqrt{npq} + np).$$

Now use Theorem 5 on the right probability with

$$a = \alpha\sqrt{npq} + np,$$
$$b = \beta\sqrt{npq} + np.$$

Thus

$$P(\alpha\sqrt{npq} + np < S_n < \beta\sqrt{npq} + np) \doteq \Phi(\beta) - \Phi(\alpha),$$

since

$$\frac{a - np}{\sqrt{npq}} = \frac{\alpha\sqrt{npq} + np - np}{\sqrt{npq}} = \alpha,$$

$$\frac{b - np}{\sqrt{npq}} = \frac{\beta\sqrt{npq} + np - np}{\sqrt{npq}} = \beta.$$

Hence $P(\alpha < S_n^* < \beta) \doteq \Phi(\beta) - \Phi(\alpha)$.

From this statement we can declare that the standardized binomial random variable is approximately a normal random variable. This approximation is very good if $n \geq 100$, and it is still quite useful if n is as small as 25 or 30.

Example 3

A balanced coin is flipped 400 times. Determine the number x such that the probability that the number of heads is between $200 - x$ and $200 + x$ is approximately 0.85.

Let S_n be the binomial random variable that denotes the number of successes in 400 Bernoulli trials with $p = \frac{1}{2}$. Then we have $P(A) = 0.85$ for the event $A = \{200 - x < S_{400} < 200 + x\}$. However, $np = 200$ and $\sqrt{npq} = 10$, so the event A is equivalent to the event

$$\left(\frac{-x}{10} < \frac{S_{400} - 200}{10} < \frac{x}{10}\right).$$

But the latter event is approximately equal to

$$\Phi\left(\frac{x}{10}\right) - \Phi\left(\frac{-x}{10}\right) = 2\Phi\left(\frac{x}{10}\right) - 1.$$

We therefore have

$$2\Phi\left(\frac{x}{10}\right) - 1 = 0.85,$$

or

$$\Phi\left(\frac{x}{10}\right) = 0.925.$$

In a table of normal probability distribution function values, $\Phi(1.4) = 0.9251$, so $x/10 = 1.4$, and $x = 14$ is a reasonable result. ∎

The approximation in Theorem 5 can also be made for a sample mean of a general random variable.   In the law of large numbers we assert that as n increases, the probability that $\bar{X}$ is close to $E(X)$ approaches 1.   But a stronger statement is possible.   One can assert that the sample mean $\bar{X}$ is approximately a normal random variable.   This statement is a special case of a powerful and general theorem of probability known as the central limit theorem.

Theorem 6

Central Limit Theorem If $\bar{X}_n$ is the sample mean of a sample of size n for a random variable X with mean μ and variance σ^2, then for any real numbers a and b with $a < b$,

$$\lim_{n \to \infty} P\left(a < \frac{\bar{X}_n - \mu}{\sigma/\sqrt{n}} < b\right) = \frac{1}{2\pi} \int_a^b e^{-t^2/2}\, dt.$$

Once again we shall omit the proof, for in this case the mathematics needed is well beyond the scope of this text.   The DeMoivre–Laplace theorem is a special case of Theorem 6, which in turn is a special case of a more general form.   In practical and theoretical statistical determinations, Theorem 6 provides effective techniques by which estimates of parameters are made.

Example 4

Suppose that the length of life in hours of a computer component can be represented by an exponential random variable X with the parameter $\alpha = \frac{1}{1000}$.   A random sample of size 100 of these components is selected from the output of the production line.   Find an approximation to the probability that the expected value of the sample will be between 975 and 1025.

Since X is an exponential random variable with $\mu = 1000$ and $\sigma = 1000$, the event $\{975 < \bar{X} < 1025\}$ is equivalent to the event

$$\left\{\frac{-25}{100} < \frac{\bar{X} - 1000}{100} < \frac{25}{100}\right\}.$$

Therefore, using Theorem 6 with $a = -0.25$, $b = 0.25$,

$$P(975 < \bar{X} < 1025) = \Phi(0.25) - \Phi(-0.25)$$

$$\doteq 0.1974.   \quad ∎$$

Example 5

The time that a certain freshman arrives at his eight o'clock class seems to be uniform over the interval 7:58 A.M. to 8:04 A.M. A sample of size 100 of the arrival times is taken. Find the approximate probability that the sample mean will lie between 7:59 and 8.01.

Let X be the uniform random variable over the interval $[0, 6]$, where 0 corresponds to the time 7:58 A.M. Then $E(X) = 3$ and $V(X) = 3$, so the event

$$\{1 < \bar{X} < 3\}$$

is equivalent to the event

$$\left\{ \frac{1-3}{\sqrt{3/10}} < \frac{\bar{X} - 3}{\sqrt{3/10}} < \frac{3-3}{\sqrt{3/10}} \right\}.$$

Using Theorem 6 with $a = -11.5$, $b = 0$,

$$P(1 < \bar{X} < 3) = \Phi(0) - \Phi(-11.5).$$

Now $\Phi(-11.5)$ is positive, but so small that it is negligible, and $\Phi(0) = 0.5$. Thus

$$P(1 < \bar{X} < 3) = 0.5. \quad \blacksquare$$

Example 6

(See [1], p. 158.) Wire nails are produced automatically by a machine that cuts a continuous wire into regulated lengths. An adjustment has been made to the machine and so a sample of 64 nails is selected and measured. The average length for the sample is 2.05 inches, and there is reason to accept a variance of 0.04 inch from previous experience. A natural question to ask at this point is the following: What is the probability that the true expected value of the length of the nails is within 0.05 inch of this sample mean 2.05 inches?

Unfortunately, the question is incorrect and unanswerable. If X denotes the length of a nail produced by this machine, then $E(X) = \mu$ is a constant, $V(X) = \sigma^2$ is a constant, and $\bar{X}$ is a random variable associated with X. You cannot ask for the probability about a constant taking on values, because it has one and only one value. You can ask for the probability that a random variable lies in an interval, and the question can be answered before the sample is taken.

If $\bar{X}$ is the sample mean for a sample of size 64 of a random variable X of unknown expectation μ and known standard deviation $\sigma = 0.2$, what is the probability that $\bar{X}$ does not deviate from μ by more than 0.05? That is, determine

$$P(-0.05 < \bar{X} - \mu < 0.05).$$

The event $\{-0.05 < \bar{X} - \mu < 0.05\}$ is equivalent to the event

$$\left(\frac{-0.05}{0.2/\sqrt{64}} < \frac{\bar{X} - \mu}{0.2/\sqrt{64}} < \frac{0.05}{0.2/\sqrt{64}} \right).$$

Now using the central limit theorem for an approximation we can write

$$
\begin{aligned}
P(-0.05 < \bar{X} - \mu < 0.05) &\doteq \Phi(2.0) - \Phi(-2.0) \\
&\doteq 2\Phi(2.0) - 1 \\
&\doteq 0.9544.
\end{aligned}
$$

Consequently, the probability that the observed value 2.05 differs from μ by less than 0.05 is approximately 0.9544.   ∎

Problems

1. Compute the exact value of $P(|X - E(X)| > \sqrt{V(X)})$ when X is a uniform variable over the interval $[0, 1]$ and compare the result to the upper bound obtained from Chebyshev's inequality.

2. Compute the exact value of $P(|X - E(X)| > 2\sqrt{V(X)})$ when X is a normal random variable with mean $E(X)$ and variance $V(X)$ and then compare this to the upper bound obtained from Chebyshev's inequality.

3. One hundred numbers are selected at random from the interval $[2, 4]$ and the average is formed. What is the expected value and the variance of the random variable that denotes the outcome of this experiment?

4. A sample of size n is taken from a large lot of nails where the length in inches of the nail can be considered to be a normal random variable X of unknown expected value μ but the variance is known to be less than $\frac{1}{4}$ inch. Use Chebyshev's inequality to determine the sample size necessary to give a probability of 0.90 that $\bar{X}$ will be within $\frac{1}{8}$ inch of the expected value μ.

5. A sample of size 50 is taken from a large collection of people whose weights can be considered to be a normal random variable X of unknown expected value μ. Let us assume that the variance is less than 10 pounds. Calculate a lower bound on the probability that the sample mean $\bar{X}$ does not differ from μ by more than 2 pounds using Chebyshev's inequality.

6. Thirty people independently measure the same object. Let us assume that these calculations can be represented by a normal random variable X with unknown expected value μ and a variance less than $\frac{1}{4}$ inch. Determine a lower bound on the probability that the average measurement differs from the expected value μ by less than 0.15 inch. (Use Chebyshev's inequality.)

7. Apply Chebyshev's inequality to the sample mean $\bar{X}$ for a sample of size n of the random variable X with expected value $E(X)$ (or μ) and variance $V(X)$ (or σ^2) and find a lower bound for

$$P(|\bar{X} - E(X)| < c).$$

8. Use the DeMoivre–Laplace theorem to find an approximate value for the probability that a sample of size 100 taken from a large lot of washers, where it is assumed that about 5 percent are defective, will have fewer than 4 defectives.

9. A computer has 100 components each of which has a probability of 0.95 that it will operate properly for a given time period. Each component operates independently of the others, and as long as at least 80 function properly, the computer will function properly. What is the probability the computer works? (Use the DeMoivre–Laplace theorem.)

10. A fair die is tossed 600 times. Determine the number x such that the probability that the number of 3's is between $100 - x$ and $100 + x$ is approximately 0.75.

11. Suppose that a certain radio tube has a probability of 0.2 of operating more than 300 hours. If 500 such tubes are tested, find an approximate value for the probability that more than 150 tubes function more than 300 hours.

12. A sample size of n is taken from a large lot of padlocks of which 3 percent are defective. What is the probability that at most 5 percent are defective if $n = 6$? $n = 100$?

13. An urn contains 99 red balls and 1 black ball. A large number of draws are made with replacement, and the ratio of number of black balls drawn to the total draws is recorded. How large should n be to guarantee that the probability is 0.90 that the ratio differs from 0.01 by less than 0.05? (Use the DeMoivre–Laplace theorem.)

14. If a sample of 100 numbers is selected at random from the interval $[-1, 1]$, find an approximate value for the probability that the sample mean will be positive.

15. If a sample of size 100 is taken of exponential random variable X with parameter 4 denoting the length of time in days between sales of a certain item, find an approximate value for the probability that the sample mean will be less than 1.

16. A sample of size 10 is taken of a normal random variable with expectation 12 and variance 16. What is the probability the sample mean is within 1 of 12?

17. A sample of size 90 is taken from the collection of high-grade bonds whose values may be considered to be approximately a normal random variable X. The expectation $E(X)$ is unknown but the

standard deviation can be taken to be $10. The actual value of the sample mean is $200. What is the probability that the observed value is within $1 of $E(X)$?

18. A particular variety of bird has a weight that may be considered to be a normal random variable X with unknown expectation $E(X)$, but the standard deviation is not more than 1 gram. Thirty birds are captured and the actual value of the sample mean is 89 grams. What is the probability that the observed value of the sample mean differs from $E(X)$ by more than 6 grams?

19. In digital-computer addition each number is rounded to the nearest integer. Suppose that each rounding error is a sample random variable for the uniform random variable X over the interval $[-0.5, 0.5]$. What is the probability that in 1500 additions the magnitude of the total error exceeds 15?

20. Refer to Problem 19. Determine the number of additions so that the probability is 0.90 that the magnitude of the total error is less than 10.

21. The number of inoperative milk trucks each day at a particular dairy may be considered to be a Poisson random variable X with parameter 2. What is the probability that in 30 days independently chosen the total number of inoperative trucks is between 50 and 60?

22. Write out the statement of the central limit theorem for the random variable S that is the sum of n sample random variables of the random variable X with mean $E(X)$ and variance $V(X)$.

23. The weight in pounds of the luggage of an airline passenger may be assumed to be a normal random variable X with $\mu = 38$ and $\sigma = 2$. What is the probability that the total weight of luggage for 80 passengers will be less than 3200 pounds?

Bibliography

The following list of books is not intended to be exhaustive or selective. The reader is encouraged to browse through other books with similar titles.

Probability and Statistics

1. H. D. Brunk, *An Introduction to Mathematical Statistics*, Ginn/Blaisdell, Waltham, Mass., 1960.
2. P. A. Chapman and R. A. Schaufele, *Elementary Probability Models and Statistical Inference*, Ginn/Blaisdell, Waltham, Mass., 1970.
3. H. Cramer, *The Elements of Probability Theory and Some of Its Applications*, John Wiley & Sons, Inc., New York, 1955.
4. W. J. Dixon and F. J. Massey, Jr., *Introduction to Statistical Analysis*, McGraw-Hill Book Company Inc., New York, 3rd ed., 1969.
5. W. Feller, *An Introduction to Probability Theory and Its Applications*, Vol. 1, John Wiley & Sons, Inc., New York, 2nd ed., 1957.
6. R. A. Gangolli and D. Ylvisaker, *Discrete Probability*, Harcourt Brace Jovanovich Inc., New York, 1967.
7. S. Goldberg, *Probability: An Introduction*, Prentice-Hall, Inc., Englewood Cliffs, N.J., 1960.
8. J. B. Johnston, G. Price, and F. S. Van Vleck, *Sets, Functions, and Probability*, Addison-Wesley Publishing Company, Inc., Reading, Mass., 1968.
9. M. Kac, *Lectures on Probability*, Mathematical Association of America: Cooperative Seminar Notes (not yet published).
10. A. N. Kolmogorov, *Foundations of the Theory of Probability Translation*, ed. by N. Morrison, Chelsea Publishing Co., Inc., New York, 1956.
11. P. L. Meyer, *Introductory Probability and Statistical Applications*, Addison-Wesley Publishing Company, Inc., Reading, Mass., 1965.
12. F. C. Mills, *Statistical Methods*, Holt, Rinehart & Winston, New York, 3rd ed., 1955.
13. J. R. McCord, III, and R. M. Moroney, Jr., *Introduction to Probability Theory*, The Macmillan Company, New York, 1964.
14. F. Mosteller, R. E. K. Rourke, and G. Thomas, Jr., *Probability and Statistics*, Addison-Wesley Publishing Company, Inc., Reading, Mass., 1961.

15. J. Neyman, *First Course in Probability and Statistics*, Holt, Rinehart & Winston, Inc., New York, 1950.
16. E. Parzen, *Modern Probability Theory and Its Applications*, John Wiley & Sons, Inc., New York, 1960.

Probability and Calculus
17. B. R. Gelbaum and J. G. March, *Mathematics for the Social and Behavioral Sciences: Probability, Calculus and Statistics*, W. B. Saunders Company, Philadelphia, 1969.
18. E. R. Mullins, Jr., and D. Rosen, *Probability and Calculus*, Bogden & Quigley, Inc., Tarrytown-on-Hudson, N.Y., 1971.

Finite Mathematics
19. J. G. Kemeny, J. L. Snell, and G. L. Thompson, *Introduction to Finite Mathematics*, Prentice-Hall, Inc., Englewood Cliffs, N.J., 2nd ed., 1966.
20. H. J. Ryser, *Combinatorial Mathematics*, The Mathematical Association of America, Washington, D.C., 1963.

Calculus
21. T. M. Apostol, *Calculus*, Ginn/Blaisdell, Waltham, Mass., 1961.
22. J. G. Cedar and D. L. Outcalt, *A Short Course in Calculus*, Worth Publishers, Inc., New York, 1968.
23. A. E. Taylor, *Advanced Calculus*, Ginn/Blaisdell, Waltham, Mass., 1955.

appendix a

Real Numbers and Inequalities

We have assumed in this book that the reader has a working knowledge of the real number system and the laws of arithmetic. It is unlikely that a reader would be totally ignorant of the whole numbers $(1, 2, \ldots)$ or fractions $(\frac{1}{2}, \frac{1}{3}, \frac{2}{3}, \ldots)$ and the rules for these numbers. Hence we shall briefly review these ideas by listing some of the axioms which numbers obey. A rigorous and logical presentation of the real-number system would be lengthy, and the interested reader is encouraged to consult specialized texts.

Some of the important subcollections of real numbers are as follows:

1. The set of whole numbers, also called the counting numbers, is the set $1, 2, 3, \ldots$ and is denoted by Z^+.
2. The set of integers is the set of whole numbers, their negatives, and 0, that is $0, 1, -1, 2, -2, 2, -3, \ldots$, and is denoted by Z.
3. The set of rational numbers is the set of all fractions p/q, where p and q are elements of Z and $q \neq 0$, and is denoted by Q.
4. The set of irrational numbers is the set of real numbers that are not rational numbers, and is denoted by I.

There is no simple way to define irrational numbers. Although one can say that an infinite nonrepeating decimal belongs to I, such a definition is useless for identifying elements of I, because it is clearly impossible to write down all the digits in an infinite decimal expansion. We shall not attempt a definition of irrational numbers here, although we shall assume that the reader has met them and used them. By definition, therefore, the rational and irrational numbers make up the set of real numbers that we denote by the symbol R.

The *real-number system* consists of the real numbers and two arithmetic operations called addition (represented by $+$) and multiplication (denoted by $\cdot$) such that the following properties are satisfied

1. For every pair of real numbers a and b, $a + b$ and $a \cdot b$ are real numbers.
2. Commutative law. If $a, b \in R$, then $a + b = b + a$; $a \cdot b = b \cdot a$.
3. Associative law. If $a, b, c \in R$, then $(a + b) + c = a + (b + c)$; $(a \cdot b) \cdot c = a \cdot (b \cdot c)$.
4. Identity elements. There exist real numbers called zero and one with $0 \neq 1$ such that $a + 0 = a$; $a \cdot 1 = a$, for all $a \in R$.

5. Inverse elements. For every $a \in R$ there exists an element $b \in R$ such that $a + b = 0$. The element b is called the additive inverse of a and is written as $-a$.

For every $a \in R$, $a \neq 0$, there exists an element $c \in R$ such that $a \cdot c = 1$. The element c is called the multiplicative inverse of a and is written as $1/a$.

6. Distributive law. For all $a, b, c \in R$, $a \cdot (b + c) = a \cdot b + a \cdot c$, and we say that multiplication distributes over addition.

Definition 1: Field

A set of elements that satisfy the laws (1)–(6) is called a field.

Our assumption, therefore, is that the real numbers form a field. Other properties of real numbers that can be deduced from properties (1)–(6) are these:

(a) $b \cdot 0 = 0$.
(b) $(-a)(-b) = a \cdot b$.
(c) $-0 = 0$.
(d) $-(-a) = a$.
(e) $1 \cdot a = a$.

Definition 2: Ordered field

A field F is said to be ordered if and only if there exists a subset of elements of F, which we denote by F_p, called "positive" elements, with the following properties:

(a) If $a, b \in F_p$, then $a + b \in F_p$.
(b) If $a, b \in F_p$, then $a \cdot b \in F_p$.
(c) For all $a \in F$, only one of the conditions $a \in F_p$, $-a \in F_p$, or $a = 0$ holds.

The positive elements of a field are by definition those elements that satisfy properties (a)–(c) of Definition 2. Our formal assumption about the real numbers is contained in the following axiom:

Axiom

The field of real numbers is an ordered field.

The set Z^+ for example, must be in the set of positive real numbers, because the sum and product of numbers in Z^+ are also in Z^+. The integer (-1) is not in the set of positive numbers, because $(-1)(-1) = 1$. However, statement (c) of Definition 2 asserts that either 1 or -1 is positive, not both. Hence if -1 were positive, it would follow that 1 is positive, which contradicts statement (c). Observe that we have used the phrase "1 is positive" as a familiar form of "1 is in the set of positive numbers."

Definition 3: Greater than and less than

Let $a, b \in R$, and let R_p denote the set of positive numbers of R.

(a) We say that a is greater than b if and only if $a - b \in R_p$; that is, $a - b$ is positive and we write $a > b$.

(b) We say that b is less than a if and only if $a > b$, that is, $a - b \in R_p$, and write $b < a$.

If we put a line under the symbol $>$, then $a \geq b$ means $a > b$ or $a = b$.

The pertinent facts about inequalities are contained in the next theorem.

Theorem 1

Let $a, b, c, d \in R$, and let R_p denote the set of positive elements. If $a > b$, then

(a) $a + c > b + c$.

(b) $a \cdot c > b \cdot c$, if $c \in R_p$.

(c) $a \cdot c < b \cdot c$, if $-c \in R_p$.

(d) If $c > d$, then $a + c > b + d$.

(e) If $b \in R_p$, then $a^2 > b^2$.

(f) If $b \in R_p$, then $a^n > b^n$, for all integers $n > 0$.

PROOF: We shall prove (b) and (e) and leave the others as exercises. [Part (f) is harder than the others and is also valid when n is a positive real number.]

(b) Since $a > b$, we know by definition of $>$ that $a - b \in R_p$. We are given that $c \in R_p$; therefore by (b) of Definition 2 we have that $(a - b) \cdot c \in R_p$. The distributive law for R yields $a \cdot c - b \cdot c \in R_p$, so by definition of $>$, it follows that $a \cdot c > b \cdot c$.

(e) The proof of this statement requires a different approach. We first try to acquire some hindsight by assuming that the conclusion is true and deducing from it a statement that we know is true. We then try to reverse our steps, starting with the known true statement until we reach the assumed conclusion. Suppose, therefore, that we assume that $a^2 > b^2$ is true. The original hypotheses still hold, namely, $a > b$ and $b \in R_p$. By definition of $>$ we can assert that $a^2 - b^2 \in R_p$. Consequently, $(a - b)(a + b) \in R_p$; that is, the product of $a - b$ and $a + b$ is positive. Now we know how to prove the theorem.

Since a and $b \in R_p$, we know that $a + b \in R_p$. By hypothesis, $a - b \in R_p$. Therefore, by (b) of Definition 2, we must have $(a - b)$ $(a + b) \in R_p$. Hence $a^2 - b^2 \in R_p$, and $a^2 > b^2$. ∎

Each of the statements of the theorem has a verbalization that is easier to memorize than the symbolic formula. For example, (a) says "Adding a number to both sides of an inequality does not change the *sense* of the inequality." For (c) we might say, "Multiplying both sides of an inequality by a negative number reverses the sense of the inequality,"

Henceforth, if $a \in R_p$, we shall say that a is positive, and if $-a \in R_p$, we shall say that a is negative. Symbolically, we shall write $a > 0$ or $a < 0$.

Definition 4: Between

Given three real numbers a, b, and c, we say that b is between a and c if and only if either

(a) $a < c$, $a < b$, and $b < c$. (We then write $a < b < c$.)
(b) $a > c$, $a > b$, and $b > c$. (We then write $a > b > c$.)

An important theorem about real numbers asserts that between every two real numbers there is an irrational number and a rational number. We do not prove this theorem.

Example 1

For what real numbers x is $x(x + 1) \leq 0$?

The product of two numbers is negative or zero if (a) $x \leq 0$ and $x + 1 \geq 0$; or (b) $x \geq 0$ and $x + 1 \leq 0$. Condition (b) cannot occur because $x \leq -1$ and $x \geq 0$ is impossible. Condition (a) asserts that $x \leq 0$ and $x \geq -1$. Hence for x such that $-1 \leq x \leq 0$, the inequality is solved.

The same analysis can be conveniently represented in tabular form, see Table 1, where the left-hand column indicates the range of values for which the factors listed in the top row assume the sign that is written in the table.

Table 1

	x	$x+1$	$x(x+1)$
$x \leq -1$	$-$	$-$	$+$
$-1 \leq x \leq 0$	$-$	$+$	$-$
$0 \leq x$	$+$	$+$	$+$

Sometimes we do not have to distinguish between a number and its negative and only need to know the positive value (that is, the numerical value). This is true, for example, when we talk about the distance between two points and say that point A is two units to the left or right of B. Thus if α is a real number, we define the numerical value of α.

Definition 5: Absolute value

The absolute value of a real number α is written $|\alpha|$ and defined by

$$|\alpha| = \begin{cases} \alpha, & \alpha \text{ is positive or zero,} \\ -\alpha, & \alpha \text{ is negative.} \end{cases}$$

Clearly, $|-2| = -(-2) = 2$, and $|2| = 2$.

Theorem 2

If b is a positive real number, then $|\alpha| < b$ means that $\alpha < b$ and $\alpha > -b$; that is, $-b < \alpha < b$.

PROOF: Suppose that α is positive. Then $\alpha = |\alpha| < b$; hence $\alpha < b$. Suppose that α is negative; then $-\alpha = |\alpha| < b$; hence $\alpha > -b$. Therefore, $|\alpha| < b$ is the set of real numbers such that $\alpha < b$ and $\alpha > -b$. ∎

Example 2
(a) Solve the inequality $|x-2| < 7$. By Theorem 2, the inequality means $(x-2) < 7$ and $(x-2) > -7$. Hence we deduce that for $x < 9$ and $x > -5$, the inequality is satisfied; that is, $-5 < x < 9$.
(b) Solve the inequality $|3x+4| \geq 2$. If we can solve the inequality $|3x+4| < 2$, then the original inequality will be satisfied by all other real numbers. By Theorem 2 we solve the alternative

inequality: either (1) $(3x + 4) < 2$ or (2) $(3x + 4) > -2$. Condition 1 results in the inequality $x < -\frac{2}{3}$, while condition 2 gives us $x > -2$. Therefore, all real numbers that do not satisfy the inequality $-2 < x < -\frac{2}{3}$ will satisfy the original inequality. These are the numbers $x \geq -\frac{2}{3}$ and $x \leq -2$. ∎

Problems

In Problems 1–5 find all real numbers x for which the statement is true.

1. (a) $x + 3 < 7x - 2$ (b) $x(x - 3) > 0$

2. (a) $3 \leq 2x + 5 \leq 7$ (b) $x - 1 < 2x - 7$

3. (a) $x(x - 1) \leq 0$ (b) $(x - 3) \cdot (x + 2) > 0$

4. (a) $x^2 - 1 < 0$ (b) $-12 + 7x - x^2 > 0$

5. (a) $5(x - 2)^2 \geq 0$ (b) $2 < x + 4 < 8$

6. Show that if $a > 0$, $a + (1/a) > 2$. (This says that a positive number plus its reciprocal is greater than 2.) $a \neq 1$.

7. If $b \in R_p$ and $a > b$, show that $a + b \in R_p$.

8. If $a \in R_p$ and $b \notin R_p$, then $a \cdot b \notin R_p$.

9. Show that the square of any real number is positive.

10. Prove Theorem 1(a).

11. Prove Theorem 1(c).

12. Prove Theorem 1(d).

13. Prove if α and β are real numbers, then

$$|\alpha\beta| = |\alpha| \cdot |\beta|,$$
$$|\alpha + \beta| \leq |\alpha| + |\beta|.$$

14. Give a suitable definition of $|x| > b$.

In Problems 15–17 find all values of x that satisfy the inequality.

15. (a) $|x - 2| < 7$ (b) $|x| < 3$

16. (a) $|x - 3| > 2$ (b) $|x - 1| \leq 6$

17. (a) $|3x + 5| < 4$ (b) $|2x - 3| \geq 7$

18. Devise an appropriate geometric interpretation of the symbol $|x - 2| < \frac{1}{9}$, that is, in terms of points on the number line.

19. Show that if $a < b$, then $a < (a + b)/2 < b$.

20. If x is a positive rational number such that $x < \sqrt{2}$, show that there exists a rational y such that $x < y < \sqrt{2}$.   [*Hint*: Let $y = 4x/(x^2 + 2)$ and verify that (a) y is a rational number, (b) $x < y$, and (c) $y^2 < 2$.]

21. Prove: $a - b < a - c$ if and only if $b > c$.

22. Show that the sum and product of two rational numbers are rational numbers.

appendix b

Mathematical Induction

The economist observes that high taxes cut down on consumer spending and from this asserts, "In times of inflation increase taxes to cut down on consumer spending." A geneticist observes that seeds obtained by crossing plants with two different traits (for example, tall and short) produced short, tall, and medium-sized offspring in a constant proportion. From this information he asserts probability laws of transmitting parental characteristics to offspring. This is the process of inducing a conclusion from observation and particular instances.

A proof by mathematical induction, however, is a substantially different idea from the notion of an inductive argument that the scientist, physical or social, might use, although the process is similar. In those disciplines the researcher performs an experiment and obtains a result. If the same result occurs after several repetitions of the experiment, the scientist might be willing to make a conjecture that the result always occurs, and after many successes with this experiment he might even be willing to claim that the result will always occur. Unfortunately, there is always the uncertainty of knowing how many repetitions are required in order to validate the assertion.

In mathematics, students frequently try this experimental procedure of testing a mathematical formula for some values of the variable, and on the basis of a few verifications (or experiments) asserts that the formula is true for every value of the variable. It will soon be obvious how dangerous such a practice can be.

Suppose that we wanted to add the first 100 positive integers, denoting the sum by S_{100}. That is, $S_{100} = 1 + 2 + \cdots + 100$. We would write the sum of the first n positive integers as $S_n = 1 + 2 + \cdots + n$. Although for each n the sum could be obtained by tediously carrying out the addition, there is a slick formula that gives the sum immediately (we shall prove this later):

(1)
$$1 + 2 + \cdots + n = \frac{n(n+1)}{2}.$$

The two sides of (1) together make up a statement about the set Z^+. The set of integers for which the statement is true is called the *truth set* for the statement. If a formula or statement is not true for any value of n, then we say that the truth set is the empty set. Mathematical induction is concerned with ascertaining the truth set for a formula.

Let us look at the formula

(2) $$1 + 2 + \cdots + n = \frac{n(n+1)}{2} + (n-1)(n-2)(n-3).$$

This statement happens to be true for the values $n = 1, 2, 3$. For example, if $n = 2$, $1 + 2 = 2(2+1)/2 + (2-1)(2-2)(2-3) = 3 + 0 = 3$. For the value $n = 4$, we see that

$$1 + 2 + 3 + 4 = 4 \cdot \tfrac{5}{2} + 3 \cdot 2 \cdot 1 = 10 + 6,$$

which is nonsense because $1 + 2 + 3 + 4 = 10$.

With very little effort we can now create a formula that is valid for 1 billion consecutive values of n but is false for the billion and first value. We merely have to tack on enough factors of the form $(n- \ \ )$ to the right side of (2).

Look at this assertion:

(3) $n^2 - n + 41$ is a prime number.

Recall that a prime is a positive integer that can be represented as a product of only two integers: 1 and itself. For the first few values of n we see that $1^2 - 1 + 41 = 41 = \text{prime}; 2^2 - 2 + 43 = \text{prime}; 3^2 - 3 + 41 = 47 = \text{prime}$. Indeed, for the next values of n up to $n = 40$, the statement is correct; that is, $n^2 - n + 41$ is a prime. But for $n = 41$ we see that $41^2 - 41 + 41 = 41^2$ is a composite number. As a result we say that in general the assertion given in (3) is false. The moral of the discussion then is that a mathematical truth cannot be established by showing that it happens to be true for some values of the variable.

As a matter of notation we shall denote by $P(n)$ the assertion that the formula makes when the value n is substituted in the statement. Thus, in (1), $P(1)$ means $1 = 1(1+1)/2$, $P(2)$ means $1 + 2 = 2(2+1)/2$, and $P(k)$ means $1 + 2 + \cdots + k = k(k+1)/2$.

It is easy to verify that both $P(1)$ and $P(2)$ are true statements. In fact, we can obtain the statement for $P(2)$ by adding 2 to the statement for $P(1)$: $1 + 2 = 1(1+1)/2 + 2 = 2(\frac{1}{2} + 1) = 2(2+1)/2$. In a similar way we can obtain $P(3)$ by adding 3 to both sides of the statement $P(2)$. It seems reasonable that we could get to the truth of $P(n)$ in a finite number of steps. To obtain the truth of $P(k+1)$ let us add $k+1$ to both sides of $P(k)$. The left side gives us $1 + 2 + \cdots + k + (k+1)$; on the right we find that $k(k+1)/2 + (k+1) = (k+1)(k+2)/2$, which is the result we would hope to obtain, that is, given $P(k)$ we obtained from it $P(k+1)$. Thus if $P(k)$ is true, we have proved for this problem that $P(k+1)$ is also true. Since we also were able to verify that $P(1)$ is also true, our little proof says that $P(2)$ is true; then $P(3)$ and so on up to the line of integers. These two actions—(a) verification of $P(1)$, and (b) proof of the propositon that if $P(k)$ is true, then $P(k+1)$ is true—are the ingredients of a proof by mathematical induction. The plausibility of the proof lies in the realization that since the assertion has been verified for the value $n = 1$, the implication $P(k)$ implies $P(k + 1)$ says that the statement is true for $n = 2$. Since it is true for $n = 2$, the same implication gives us the truth for $n = 3$. Continuing in this way we are willing to assert that the statement is true for all positive integers; that is, the truth set is Z^+ (the set of positive integers). This plausibility stems directly from the following:

Axiom

If a set S of integers contains the positive integer m_0, and if, moreover, S contains the integer $k + 1$ whenever it contains the integer $k \geq m_0$, then S contains all the integers $\geq m_0$.

The way in which we use this axiom to prove the statement in (1) was to show that $(1 = m_0)$ was in the truth set of $P(n)$, and that if k was in the truth set, so was $k + 1$. The axiom then says that the truth set contains all integers ≥ 1. Interestingly enough, this axiom can be proved as a theorem if we use the appropriate axiom system for the real numbers.

In a formal way we define

Definition 1: Proof by mathematical induction

A proposition is said to be proved by mathematical induction if the truth set of $P(n)$

(a) contains 1 [$P(1)$ is true], and
(b) when it contains the integer k, it also contains the integer $k + 1$ [$P(k)$ implies $P(k + 1)$, $k \geq 1$].

A more general version of proof by mathematical induction is obtained if in Definition 1(a) we replace 1 by n_0, the smallest integer for which $P(n)$ is true.

Both parts of the proof are necessary. We have already shown that just verifying a proposition, for a few specific values, can be disastrous. It is not surprising, therefore, that the second part of the induction proof is not enough to provide a valid proof by itself. Let's look at a proposition that we know is impossible and false, but where we will be able to prove: "If it is true for the case k, then it is true for the case $k + 1$." Suppose that we assert that

$$(4) \qquad\qquad 1 + 2 + \cdots + n = \frac{n(n + 1)}{2} + 4,$$

which is clearly false for $n = 1, 2, 3, \ldots$. If the kth case is true, then $1 + 2 + \cdots + k = k(k + 1)/2 + 4$. If we add $k + 1$ to both sides of this expression, we obtain

$$1 + 2 + \cdots + k + (k + 1) = \frac{k(k + 1)}{2} + (k + 1) + 4 = (k + 1)\frac{(k + 2)}{2} + 4,$$

which is precisely $P(k + 1)$. Hence if $P(k)$ is true, we can deduce that $P(k + 1)$ is true, but we cannot find a first value for which the assertion is true. If we could, then something would be wrong with the proof of (1).

Mathematical induction has been compared to standing a set of dominoes or toy soldiers in a line and giving the first one a push. If the kth soldier falls, so does the $(k + 1)$th.

Example 1

Prove $P(n): (1 + x)^n \geq 1 + nx$, where x is real and $x > -1$.

PROOF: The verification when $n = 1$ is obvious since $(1 + x)^1 = 1 + 1 \cdot x$. Hence the truth set of $P(n)$ contains 1. Suppose that the truth set contains k; does it contain $k + 1$? We must show that $(1 + x)^{k+1} \geq 1 + (k + 1)x$. Since by assumption $P(k)$ is $(1 + x)^k \geq 1 + kx$, we can obtain the left side of $P(k + 1)$ by multiplying both sides of $P(k)$ by $(1 + x)$ (notice that $1 + x$ is positive because we imposed the restriction that $x > -1$). Thus

$$(1 + x)^{k+1} \geq (1 + x)^k(1 + x) \geq (1 + kx)(1 + x) = 1 + (k + 1) \cdot x + kx^2.$$

Now k is positive and so is x^2, so that if we forget to include kx^2 on the right, the resulting number will be smaller; that is,

$$1 + (k + 1)x + kx^2 \geq 1 + (k + 1)x.$$

We have therefore deduced that $P(k)$ implies $P(k + 1)$. The principle of mathematical induction now assures us that the original inequality is valid for all integers ≥ 1. ∎

Induction is not a panacea for proving theorems, although it can be used to prove relations which are true for particular values of n and which we would like to show is true for all positive values of n.

As an example of how the inductive technique can suggest a formula, suppose we wanted to find the sum of the first 100 odd integers. We observe for the first few cases that

$$\begin{aligned}
1 &= 1 &&= 1^2, \\
1 + 3 &= 4 &&= 2^2, \\
1 + 3 + 5 &= 9 &&= 3^2, \\
1 + 3 + 5 + 7 &= 16 = 4^2,
\end{aligned}$$

and we observe that $1, 4, 9$, and 16 are all squares, as indicated on the extreme right. Since the sum of the first three odd integers is the square of 3 while the sum of the first four odd integers is the square of 4, the statement that the sum of the first 100 odd integers is the square of 100 has much to recommend it.

Indeed, for an arbitrary odd integer $(2n - 1)$ we should try as a statement $P(n)$

$$1 + 3 + 5 + \cdots + (2n - 1) = n^2.$$

We have already verified this relation when $n = 1$; that is, $P(1): 1 = 1^2$. Assume that $P(k): 1 + 3 + \cdots + (2k - 1) = k^2$ is true and we try to deduce that $P(k + 1): 1 + 3 + \cdots + (2k + 1) = (k + 1)^2$ is true. By adding $2k + 1$ to both sides of $P(k)$ we obtain

$$1 + 3 + \cdots + (2k - 1) + (2k + 1) = k^2 + 2k + 1.$$

It should be clear that this is just $P(k + 1)$. Our conjecture turned out to be a good one.

There are also other uses for mathematical induction, especially in making definitions. For example, what is meant by x^{k+1}, where k is a positive integer? At some stage in our education someone wrote $x^{k+1} = x \cdot x \cdot x \cdot x \cdots x$, where there are $(k + 1)$ x's, which is most unappealing, for there is always the problem of knowing which x we multiply by first (unless a generalized associative law has been proved). Inductively, however, we can easily define $x^2 = x \cdot x$. Then we define $x^{k+1} = x^k \cdot x$. Hence if x^k is defined, so is x^{k+1}. But x^2 is defined, that is, the set of numbers for which x^{k+1} is defined contains 1, and, moreover, x^{k+1} is defined when x^k is, so when x^n is defined for k it is defined for $k + 1$. Therefore, by the induction axiom it is defined for all positive integers ≥ 1.

Example 2

Define $\sum_{i=1}^{n} x_i$ for all positive integers n. For $n = 1$, put $\sum_{i=1}^{1} x_i = x_1$. For $n = k + 1$, we put

$$\sum_{i=1}^{k+1} x_i = \sum_{i=1}^{k} x_i + x_{k+1}.$$

Hence by definition $\sum_{i=1}^{2} x_i = \sum_{i=1}^{1} x_i + x_2$ is well defined and is equal to $x_1 + x_2$. Continuing in this way we obtain inductively $\sum_{i=1}^{n} x_i = x_1 + x_2 + \cdots + x_n$. This little development might seem heavy handed for defining a fairly straightforward symbol, but it does illustrate the essential character of an inductive definition. The phrase "continuing in this way" is really a camouflage for an inductive statement. ∎

Problems

In Problems 1–4 without using the $\sum$ notation, write out the expression.

1. (a) $\displaystyle\sum_{k=2}^{7} x_k$ (b) $\displaystyle\sum_{k=1}^{5} k$

2. (a) $\displaystyle\sum_{i=2}^{4} x^i$ (b) $\displaystyle\sum_{j=0}^{5} (2+3j)$

3. (a) $\displaystyle\sum_{t=1}^{3} t^t$ (b) $\displaystyle\sum_{r=1}^{4} \frac{1}{r^2}$

4. $\displaystyle\sum_{r \text{ is a divisor of } 12} r$

5. What is meant by $\displaystyle\sum_{n=1}^{15} 1$, $\displaystyle\sum_{n=7}^{10} 1$, and $\displaystyle\sum_{n=1}^{k} 1$?

6. Compute (a) $\displaystyle\sum_{j=1}^{4} \frac{1}{j}$;   (b) $\displaystyle\sum_{n=0}^{3} \frac{(-1)^n}{n+1}$

7. Rewrite the following using the $\sum$ notation.

   (a) $2+4+6+8$ (b) $6^2+7^2+8^2+9^2+10^2$
   (c) $1+3+5+\cdots+2n-1$ (d) $1+2^3+3^3+\cdots+n^3$

8. Compute

   (a) $\displaystyle\sum_{r=1}^{8} (-1)^r$ (b) $\displaystyle\sum_{r=1}^{n} (-1)^r$

In Problems 9–22 establish the truth of the stated proposition using mathematical induction.

9. Prove $\sum_{k=1}^{n} k^2 = n(n+1)(2n+1)/6$.

10. Prove: For every positive integer n, 2 is a factor of n^2+n.

11. Prove: For every positive integer n, $\cos n\pi = (-1)^n$.

12. Prove: $\sum_{k=1}^{n} (4k-2) = 2n^2$.

13. Prove: For every positive integer n, $n(n+1)(n+2)$ is divisible by 3.

14. Prove: $\sum_{k=1}^{n} cx^k = c \sum_{k=1}^{n} x^k$, where c is a constant.

15. Prove: For every positive integer n, $5^n - 1$ is divisible by 4.

16. Prove: $2^n < n(n-1)(n-2)\cdots 3\cdot 2\cdot 1$ for $n \geq 4$.

17. Prove: For n points in a plane, no three of which are collinear, there are $n(n-1)/2$ lines determined by pairs of these points.

18. Prove: $\sum_{i=1}^{n} (a_i - a_{i-1}) = a_n - a_0$. This is called a telescoping sum.

19. Prove: $(x - y)$ divides $x^n - y^n$ for n a positive integer and $x \neq y$.
    [*Hint*: $x^n - y^n = x^{n-1}(x - y) + yx^{n-1} - y^n$.]

20. Prove: $\sum_{k=0}^{n} x^k = (1 - x^{n+1})/(1 - x)$.

21. Prove: $\sum_{k=1}^{n} k^3 = (\sum_{k=1}^{n} k)^2$.

22. Prove: If $a_1, a_2, \ldots, a_n$ are all real numbers greater than -1 and have the same sign, then

    $$(1 + a_1)(1 + a_2) \cdots (1 + a_n) \geq 1 + a_1 + \cdots + a_n.$$

    What conclusion can you deduce when the a_i's are equal?

23. Give an inductive definition of the symbol $\sum_{k=1}^{n} r^k$.

24. Experiment with the product $(1 - \frac{1}{4})(1 - \frac{1}{9}) \cdots (1 - 1/n^2)$ and obtain a simplification that can be used as a conjectured value of the product. Prove your conjecture by induction.

25. Problem 9 can be proved without using an induction argument by observing that $k^3 - (k-1)^3 = 3k^2 - 3k + 1$. Sum both sides for $k = 1$ to n. The left side becomes

    $$\sum_{k=1}^{n} (k^3 - (k-1)^3) = n^3$$

    and the right side becomes

    $$3 \sum_{k=1}^{n} k^2 - 3 \sum_{k=1}^{n} k + \sum_{k=1}^{n} 1 = 3 \sum_{k=1}^{n} k^2 - 3 \frac{n(n+1)}{2} + n.$$

    Now finish the problem.

26. Try the procedure of Problem 25 to obtain

    $$\sum_{k=1}^{n} k^3 = \frac{n^4}{4} + \frac{n^3}{2} + \frac{n^2}{4}.$$

appendix c

Two-Dimensional
Analytic Geometry

C.1 Coordinate Line and Plane

The idea of a $1:1$ (one to one) correspondence neatly describes the relationship between points on a line and the real numbers.

> ### Definition 1: One-to-one correspondence
>
> Two sets A and B are said to be in one-to-one correspondence (written as $1:1$) if and only if for every element of A there is associated a unique element of B, and for every element of B there is only one element of A that is associated with it.

Example 1

(a) The sets $\{1, 2, 3\}$ and $\{a, b, c\}$ can be put into a $1:1$ correspondence:

$$1 \leftrightarrow a,$$
$$2 \leftrightarrow b,$$
$$3 \leftrightarrow c.$$

(b) The sets $A = \{1, 2, 3\}$ and $B = \{a, b\}$ cannot be put into a $1:1$ correspondence, because A has more elements than B.

(c) The sets $A = \{1, 2, 3, \ldots\}$ and $B = \{2, 4, 6, \ldots\}$ can be put into a $1:1$ correspondence by associating $n \leftrightarrow 2n$. ∎

Theorem 1

There exists a $1:1$ correspondence between the real numbers and the points on a line.

We omit the proof of this theorem because it is difficult to do rigorously at this level. However, we indicate how the correspondence can be accomplished.

On a line (drawn with a straightedge) select a reference point and mark it 0. Then select another arbitrary point and call it 1. Conventionally, the point called 1 is taken to the right of the point called 0 if the line is not vertical, and above 0 if the line is vertical. The length of the line segment between the two points called 0 and 1 is called the *unit* length. The point that corresponds to the number -1 is now obtained as that point on the line that has the same unit distance from 0 but in the other direction (see Figure 1). The

Figure 1

points of the line that will correspond to the integers are obtained by marking off the unit length successively as in Figure 2.

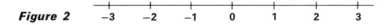

Figure 2

Definition 2: Coordinate of a point

The coordinate of a point P on a line is the real number x that corresponds to it by Theorem 1.

The coordinate of the point will be used as another name for the point. Thus the point A whose coordinate is a will be called the point a.

Once the integers are located on the line, the rational coordinates are easily obtained. We illustrate the method. To find the point whose coordinate is $\frac{1}{2}$, construct lines as in Figure 3, forming the triangle AOB. The line from C parallel to AB cuts OA at the point $\frac{1}{2}$, which can be verified by theorems from plane geometry. The points $\frac{1}{3}$ and $\frac{2}{3}$ are obtained similarly,

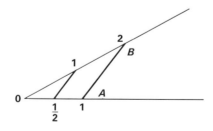

Figure 3

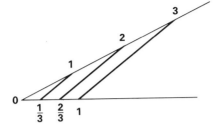

Figure 4

as in Figure 4. After the rational points in the interval (0, 1) are obtained all rational points are obtained by translating the unit interval.

Irrational numbers cannot in general be constructed, although some can. For example, $\sqrt{2}$ is constructed as the hypotenuse of a $45°$ triangle. Alternatively, $\sqrt{2}$ can be indicated approximately by taking the sequence of rational approximations, 1.4, 1.414, ... It is the correspondence between points and irrational numbers that is difficult.

Definition 3: The real-number line

The line on which each point has been made to correspond to a real number is called the real-number line.

Definition 4: Distance between points

The distance between two points P_1 and P_2 on a line is the absolute value of the difference between their coordinates x_1 and x_2. We write the distance as $d = |x_1 - x_2| = |P_1 P_2|$.

Example 2

The distance between $a = 2$ and $b = 5$ is $d = |5 - 2| = 3$, and between $a = |7.3 - (-1)| = 8.3$. ∎

Theorem 2

The coordinate of the midpoint between the points with coordinates a and b is given by $(a + b)/2$.

PROOF: Suppose that $a < b$ (that is, the real number $a <$ real number b). The distance between a and b is $d = |b - a| = b - a$. The midpoint is at $a + d/2 = a + (b - a)/2 = (a + b)/2$. If $a > b$, then $d = |b - a| = -(b - a) = a - b$. The midpoint is therefore at $b + d/2 = b + (a - b)/2 = (a + b)/2$. ∎

Example 3

The midpoint between the points 3 and 8 is $(3 + 8)/2 = \frac{11}{2}$, between the points 0 and 1 is $\frac{1}{2}$, and between the points 4 and -3 is $(4 - 3)/2 = \frac{1}{2}$. ∎

Points in the plane can also be identified by real numbers. Divide the plane into quarters by two lines that intersect at right angles. Conventionally, one of the lines is taken horizontal, and the other is then vertical. The horizontal line is called the x-axis, and the vertical line is called the y-axis. Set up a coordinate system on each line. The point of intersection is called the origin and is given the symbol $(0, 0)$. An arbitrary point in the plane is identified by the number pair (x, y) in the following way. Given any point P find the point A on the x-axis that is the foot of the perpendicular from P to the x-axis. Let x be the coordinate of A. Similarly, let y be the coordinate of the point B on the y-axis that is the foot of the perpendicular from P to the y-axis. If we are given a pair (x, y), we can reverse the procedure and locate the corresponding point P as the intersection of the two lines perpendicular to the x-axis and y-axis, respectively. Thus the first member of the couple denotes the distance of the point along the x-axis, and the second member of the couple denotes the distance along the y-axis. The points $(1, 3)$ and $(-4, 2)$ are identified in Figure 5. The first member of the number pair is called the x-coordinate of the point and the second member the y-coordinate. Observe that the couples $(1, 2)$ and $(2, 1)$ are different points. Hence the number pairs are called *ordered couples*, or *ordered pairs*.

The notion of ordered pairs allows us to associate every point in the plane with a pair of real numbers and conversely. Again we shall omit the proof of the important

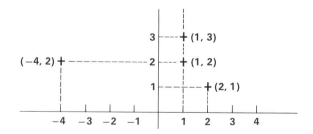

Figure 5

Theorem 3

There exists a one-to-one correspondence between points in the plane and the ordered pairs of real numbers.

We shall identify a point in the plane by its corresponding ordered couple. Hence the point $(2, 1)$ refers to the point whose x-coordinate is 2 and y-coordinate is 1. Because of Theorem 3, such geometric notations as distance between point, parallel lines, perpendicular lines, etc., can be expressed analytically. We discuss some of these ideas next.

Theorem 4

If (x_1, y_1) and (x_2, y_2) are two points in the plane, then the distance between them is given by

$$D = ((x_1 - x_2)^2 + (y_1 - y_2)^2)^{1/2}.$$

PROOF: If $x_1 \neq x_2$ and $y_1 \neq y_2$, the point (x_2, y_1) forms with the two given points a right triangle. The lengths of the two legs are clearly $|y_2 - y_1|$ and $|x_2 - x_1|$ (Figure 6). Therefore, by the theorem of

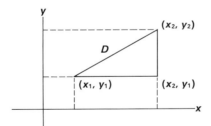

Figure 6

Pythagoras, we obtain the desired result. If $x_1 = x_2$, then $D = ((y_x - y_1)^2)^{1/2} = |y_2 - y_1|$. If $y_1 = y_2$, then $D = |x_2 - x_1|$.   ∎

Example 4

The square of the distance between the points $(-1, 2)$ and $(2, -3)$ is
$$D^2 = (-1 - 2)^2 + (2 - (-3))^2 = 34. \quad ∎$$

Using similar triangles, we find that the coordinates of the midpoint of the line segment joining the two points (x_1, y_1) and (x_2, y_2) (Figure 7) are given by

$$x = \frac{x_1 + x_2}{2}, \qquad y = \frac{y_1 + y_2}{2}.$$

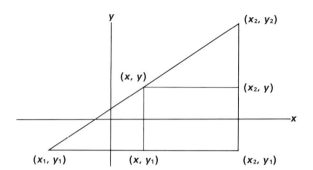

Figure 7

Problems

1. Let A, B, C, D, and E be the points on a coordinate line with respective coordinate 0, 1, 3, $-\frac{1}{2}$, and $-\frac{4}{3}$. Find each of the following: $|AB|$, $|BC|$, $|EC| + |CA|$, $|DA| - |BD|$, and $|AB| + |BC|$.

2. Find the distance between the points $\frac{11}{10}$ and $\frac{3}{7}$. Find the midpoint between these points.

3. Show how to construct geometrically the coordinate of a point that is at a distance of 3 from the point 0.

4. Divide geometrically the interval $(0, 1)$ into seven equal parts; that is, find the coordinates $\frac{1}{7}$, $\frac{2}{7}$, $\frac{3}{7}$, ...

5. Show that if r is an arbitrary number between 0 and 1, the point $x = a + [r(b - a)]$ is the coordinate of a point that is a fraction r of the distance from a to b.

6. Find the coordinate of the point that is one third of the distance from 2.5 to 7.

7. If the coordinates of A and B are 3 and -6.3, respectively, find the coordinates of the points of trisection of AB.

8. If A, B, and C are any three points on a coordinate line, under what conditions is it true that $|AB| + |BC| = |AC|$?

9. The coordinates of A and B are 7 and -3, respectively. Find the coordinate of the point P if $|AP|/|PB| = \sqrt{2}$.

10. If P, Q, and R are the midpoints of the segments BC, CA, and AB, where A, B, and C are any three points on a coordinate line, prove that the midpoint of CR coincides with the midpoint of PQ.

11. Find the midpoint of the line segment determined by $(2, 7)$ and $(-3, -4)$.

12. Find the distance between the points $(1, -1)$ and $(-3, 1)$.

13. Given the points $(1, 0)$, $(-2, 3)$, and $(-1, 5)$, find the length of the line segment from the third point to the midpoint of the line segment determined by the first two points.

14. If the coordinates of A, B, and C are $(0, -1)$, $(4, 0)$, and $(3, 4)$, respectively, show that triangle ABC is a right triangle.

In Problems 15–18 determine analytically whether or not all the given points lie on a line.

15. $A(1, 0)$, $B(0, 1)$, $C(2, -1)$ 16. $A(-2, 1)$, $B(0, 5)$, $C(-1, 2)$

17. $A(-2, -1)$, $B(-1, 2)$, $C(1, 5)$, $D(2, 7)$

18. $A(-2, 3)$, $B(0, 2)$, $C(2, 0)$

19. If C is the point of intersection of a horizontal line through $A(-1, 2)$ and a vertical line through $B(2, -1)$, find (a) the coordinates of C and (b) the length of segment AB.

20. The distance between the point $A(5, -2)$ and the point B, whose y-coordinate is 1, is 4 units.   Locate the point B.

21. If the coordinates of A, B, and C are $(1, -2)$, $(-4, 2)$, and $(1, 6)$, respectively, show that the triangle is isosceles.

22. Refer to Problem 21.   Find the area of the triangle.

23. If the midpoint of line segment AB is $C(-4, -3)$ and the point A has coordinates $(8, -5)$, find the coordinates of point C.

24. Show that the quadrilateral $A(3, 2)$, $B(0, 5)$, $C(-3, 2)$, and $D(0, -1)$ is a square.

25. Find the length of the median of the triangle with vertices $A(4, 1)$, $B(-5, 2)$, and $C(3, -7)$ on the vertex A.

26. Show that the points $(-2, 5)$, $(2, 1)$, and $(x, 3 - x)$ lie on the same line for every value of x.

27. Find the coordinates of the point on the x-axis that is equidistant from $A(-4, 6)$ and $B(14, -2)$.

28. Show that the point $A(4, 1)$ is the center of a circle that passes through $(0, -2)$, $(7, -3)$, and $(8, -2)$.

29. The x-coordinate of a point is twice the y-coordinate and the point is equidistant from $A(-3, 1)$ and $B(8, -2)$.   Find its coordinates.

30. Construct a line parallel to the x-axis and 3 units below it, and let $P(x, y)$ be an arbitrary point.   Write an expression for the distance from P to the foot of the perpendicular from P to the line.

C.2 The Line

In this section we turn our attention to finding an analytic expression for a straight line. The equation of a line in a plane can assume many forms, depending on what are given to determine the line.

Definition 1: Slope

The slope of a line that connects two points (x_1, y_1) and (x_2, y_2) is given by the expression

$$(1) \qquad\qquad m = \frac{y_2 - y_1}{x_2 - x_1}, \qquad x_2 \neq x_1.$$

If $x_2 = x_1$, the two points are on a line parallel to the y-axis. If $y_2 = y_1$, then the two points are on a line parallel to the x-axis.

The slope is also verbalized as the rise per run. Figure 1 pictures this notion.

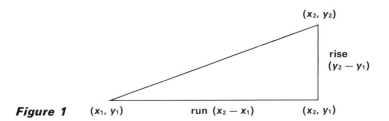

Figure 1 (x_1, y_1) run $(x_2 - x_1)$ (x_2, y_1)

Definition 2: Algebraic definition of a line

The set of points (x, y) are said to lie on a line if for any fixed point (x_1, y_1) the slope of the line between (x_1, y_1) and the set (x, y) is a constant. That is, $(y - y_1)/(x - x_1)$ is a constant for every point (x, y) in the set.

In order to find the equation of a line we need to know either two points on the line, or one point on the line and the slope. If two points on the line are known, we can obtain the slope from (1), so that obtaining a formula using one point and the slope contains the formula that uses two points as a special case.

The method for finding the equation of a line is apparent from the definition: If (a, c) is the given point on this line and m is the slope, then any point (x, y) on the line together with the given point must satisfy the equation

$$m = \frac{y - c}{x - a}.$$

Hence

$$y = mx - ma + c.$$

Since $c - ma$ is made up of constants, let us call it b.   Then the equation has the form

(2) $$y = mx + b.$$

We have just proved

Theorem 1

The points on a line that is not parallel to the y-axis satisfy an equation of the form $y = mx + b$, where m is the slope of the line, and the point $(0, b)$ is the intersection of the line with the y-axis.

The point $(0, b)$ is called the *y-intercept*.   The *x-intercept* is the point where the line intersects the x-axis, and its coordinates must be $(-b/m, 0)$.

Example 1
Find the equation of the line that contains the points $(3, 1)$ and $(-2, -3)$.
We find that the slope $m = (-3 - 1)/(-2 - 3) = \frac{4}{5}$.   Therefore the line has the form $y = \frac{4}{5}x + b$.   Since $b = 1 - \frac{4}{5} \cdot 3 = -\frac{11}{5}$, the equation of the line is

$$y = \tfrac{4}{5}x - \tfrac{11}{5}.$$

Alternatively, the equation can be written as $5y - x + 11 = 0$.   ∎

The most general linear equation has the form

(3) $$Ax + By + C = 0,$$ not both A and B equal to 0.

If $B \neq 0$, we can write $y = (-A/B)x - (C/B)$; hence the points on a line satisfy an equation in the shape of (3) provided that the line is not parallel to the y-axis.   In Theorem 2 we show that the points that satisfy an equation such as (3) must lie on a line.

Theorem 2

The set of points that satisfy an equation in the shape $Ax + By + C = 0$, not both A and B zero, lie on a line.

PROOF: If $A = 0$, then the points $(x, -C/B)$ satisfy the equation; hence the set $\{x, -C/B\}$ lies on a line that is parallel to the x-axis.

If $B = 0$, then the points $(-C/A, y)$ (observe that if $B = 0$, $A \neq 0$) satisfy the equation and these points are on a line parallel to the x-axis, because they all have the same coordinate.

If both A and B are not zero, let (x_1, y_1) and (x_2, y_2) be any two points that satisfy (3). Then

$$Ax_1 + By_1 = -C,$$
$$Ax_2 + By_2 = -C.$$

Subtracting the second equation from the first, we obtain $A(x_1 - x_2) + B(y_1 - y_2) = 0$, so that $(y_2 - y_1)/(x_2 - x_1) = -A/B$. From this we can conclude that the slope defined by any two points that satisfy (3) is a constant. Therefore, these points, by Definition 1, where we keep one of the points fixed, lie on a line. ∎

Example 2

Sketch the graph of the equation $x + \frac{1}{2}y = \frac{1}{2}$. This equation has the shape of (3); therefore it is a line. Solving for $y: y = -2x + 1$, we find that the slope is -2, y-intercept is 1. The x-intercept is $(\frac{1}{2}, 0)$. The two intercepts determine the graph as in Figure 2. ∎

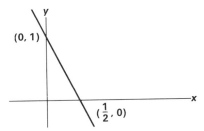

Figure 2

Example 3

Find the equation of the line that contains the points $(1, 2)$ and $(3, 4)$. We first determine the slope: $m = (4 - 2)/(3 - 1) = 1$. Using $(1, 2)$ as the fixed point, we write $(y - 2)/(x - 1) = 1$; hence the equation is $y = x + 1$. ∎

Parallel lines are characterized as lines having the same slope.

Definition 3: Parallel lines

Two lines that are not parallel to the y-axis are parallel to each other if and only if they have the same slope.

Theorem 3

Two lines $A_1x + B_1y + C_1 = 0$ and $A_2x + B_2y + C_2 = 0$ are parallel if and only if for $B_1 \neq 0$, $B_2 \neq 0$, $A_1/B_1 = A_2/B_2$; that is, $A_1B_2 = A_2B_1$.

PROOF: If the two lines are parallel, then by definition the slopes are equal and the first of the equalities holds. If the equality holds, then the slopes are equal, and the lines are parallel. ∎

Definition 4: Perpendicular lines

Two intersecting lines, neither of whose slopes is zero, are perpendicular if and only if the product of their slopes is -1; that is, if m_1 and m_2 are the two slopes, then $m_1m_2 = -1$.

Theorem 4

Two lines $A_1x + B_1y + C_1 = 0$, $A_2x + B_2y + C_2 = 0$ are perpendicular if and only if for $B_1 \neq 0$, and $B_2 \neq 0$, the relation $A_1A_2 + B_1B_2 = 0$ holds.

PROOF: The conditions $B_1 \neq 0$ and $B_2 \neq 0$, together with $A_1A_2 + B_1B_2 = 0$, imply that neither of the lines are parallel to the coordinate axes; hence we can rewrite the equations of the lines as $y = m_1x + b_1$ and $y = m_2x + b_2$, where $m_1 = -A_1/B_1$ and $m_2 = -A_2/B_2$.

If the two lines are perpendicular, then, by definition, $m_1m_2 = -1$, and it is seen that the desired relation holds. Conversely, if the relation $A_1A_2 + B_1B_2 = 0$ holds, it is a straightforward calculation to show that the product of the slopes is -1. ∎

Example 4

Find the equations of the lines through $(-2, 5)$ that are (a) parallel and (b) perpendicular to the line $y = -3x + 1$.

The slope of the given line is -3, so a line parallel to it will have the form $y = -3x + b$. Since $(-2, 5)$ must satisfy the equation, $5 = -6 + b$ and $b = 11$. Therefore, $y = -3x + 11$ is the line parallel.

The line perpendicular to the given line must have a slope that is the negative reciprocal of -3, that is, $\frac{1}{3}$. Thus $y = \frac{1}{3}x + b$ is the general form, and we find from $5 = -\frac{2}{3} + b$ that $b = \frac{17}{3}$. Hence $y = \frac{1}{3}x + \frac{17}{3}$, or $x - 3y = -17$ is the required line.   ∎

An alternative way of defining slope uses the fact that every line that is not parallel to the x-axis makes an angle with the x-axis. The angle measured from the x-axis to the line in the counterclockwise direction is called the *angle of inclination* of the line. The values of the angle will range from 0 to 180°. Each line determines a unique value of the angle of inclination, and parallel lines will certainly have the same angle of inclination.

If we denote the angle of inclination by α, we define the tangent of α as the slope m given in (1) and write $m = \tan \alpha$. We shall discuss this relation in Appendix D, the trigonometry section.

An interesting question is how to tell whether a point is above or below a given line. The next theorem provides an answer.

Theorem 5

If $B > 0$, then the point (a, b) is above the line $Ax + By = C$ if $Aa + Bb - C > 0$ and below the line if $Aa + Bb - C < 0$.

PROOF: We prove only part of the theorem. Suppose (a, b) is above the line. Drop a line perpendicular to the x-axis that cuts the given line in the point (a, y). If $B > 0$ and $b > y$, then $Bb > By$. Add Aa to both sides and subtract C from both sides. Then $Aa + Bb - C > Aa + By - C$. The right side of the inequality is zero, because (a, y) is on the line (Figure 3). The other part of the proof is similar.   ∎

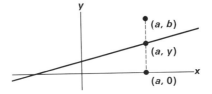

Figure 3

Problems

In Problems 1–10 find the equation of the line satisfying the given conditions.

1. Slope $= \frac{1}{2}$ and $(1, 5)$ is on the line.

2. Slope $=0$ and $(-3, -2)$ is on the line.

3. $(-1, 1)$ and $(3, 2)$ are on the line.

4. Slope $=-3$ and the x-intercept is $(3, 0)$.

5. Slope $=2$ and the y-intercept is $(0, -1)$.

6. The x-intercept is $(4, 0)$ and the y-intercept is $(0, 3)$.

7. Tan $\alpha = \frac{3}{2}$ and the point $(1, 1)$ is on the line.

8. The y-intercept is $(-7, 0)$ and the point $(3, 6)$ is on the line.

9. Slope $=-4$ and the x-intercept is $(0, 0)$.

10. Slope $=0$ and the y-intercept is $(0, -5)$.

11. If the line has the equation $2x + 3y + 5 = 0$, find the slope and the y-intercept.

12. If the line has the equation $\frac{3}{2}x - 4y + 2 = 0$, find the slope and the x-intercept.

13. If the line has the equation $2x + 7y + 1 = 0$, find the x- and y-intercepts.

14. Find the equation of the line passing through $(1, 2)$ and parallel to the line $x + 5y - 3 = 0$.

15. Find the equation of the line passing through $(-2, -3)$ and perpendicular to the line $3x - 7y + 4 = 0$.

16. Find the point of intersection of the two lines $x - y + 2 = 0$ and $2x + 5y + 7 = 0$.

17. Show that the two lines are either parallel or perpendicular or neither.

   (a) $2x + 5y + 1 = 0$ (b) $x + 3y - 1 = 0$
          $x + \frac{5}{2}y + 7 = 0$ $-3x + 2y + 1 = 0$

18. Find the equation of the line passing through $(4, -2)$ and parallel to the line through the points $(3, 2)$ and $(5, 7)$.

19. Find the equation of the line passing through $(-1, -3)$ and perpendicular to the line through the points $(2, -1)$ and $(5, 7)$.

20. Find the equation of the perpendicular bisector of the line segment joining $(5, 2)$ and $(-1, 3)$.

21. Show that the triangle $A(3, 2)$, $B(1, 1)$, and $C(-1, 5)$ is a right triangle.

22. Determine whether or not the three points all lie on the same straight line: $A(2, -1)$, $B(5, 3)$, $C(-7, 4)$.

23. Draw a figure to describe the region of points (if any) that satisfy the inequalities

$$x + 3y - 6 > 0 \quad \text{and} \quad 2x - y + 5 > 0.$$

24. Find the distance from the point (2, 5) to the line $y = -x + 3$.

25. Show that the equation of the line perpendicular to $Ax + By + C = 0$ and through the point (a, b) is $Bx - Ay = Ba - Ab$.

26. Let $A(x_1, y_1)$, $B(x_2, y_2)$, $C(x_3, y_3)$, and $D(x_4, y_4)$ be any four noncollinear points in the coordinate plane. Show that the quadrilateral formed by the midpoints of the successive sides is a parallelogram.

27. Let $A(x_1, y_1)$, $B(x_2, y_2)$, and $C(x_3, y_3)$ be three noncollinear points in the plane. Show that the point $((x_1 + x_2 + x_3)/3, (y_1 + y_2 + y_3)/3)$ is the point of intersection of the medians of the triangle ABC.

C.3   Graphing Functions

We review here some of the elementary theory of graphing functions. Given a function $y = f(x)$, our object will be to sketch easily the relation between the variables x and y, and in so doing to discover the behavior of the curve that represents the function. A point-by-point plot is generally impossible, and we try to keep the number of points that must be plotted to as few as possible. For functions of the form $f(x) = mx + b$, we only need to find two points to sketch the graph. More complicated functions cannot be plotted so facilely. As an aid to graphing we investigate four characteristics.

1. The *extent*. By the extent of the function we mean the set of values that the domain and range can assume.

Example 1

If $f(x) = (1 - x^2)^{1/2}$, x can take on the values $|x| \leq 1$, while the values of $f(x)$ are between zero and 1 inclusive.   ∎

Example 2

If $f(x) = x/(1 - x^2)$, x can take on all values except ± 1, and $f(x)$ takes on all values between $-\infty$ and $+\infty$. At $x = \pm 1, f(x)$ is undefined because the denominator is zero. For x near $+1$ or -1, the values of $f(x)$ are arbitrarily large positively and negatively.   ∎

2. The *intercepts*. The intercepts are those points where the graph intersects the two coordinate axes. The *x*-intercept is found by putting $y = 0$ and solving for *x*, while the *y*-intercept is obtained by putting $x = 0$ and solving for *y*.

Example 3

If *f* is the function in Example 1, the *x*-intercept are $x = \pm 1$, while the *y*-intercept occurs at $y = 1$.

If *f* is the function in Example 2, the *y*-intercept is seen to be at $(0, 0)$ which is also the *x*-intercept. ▮

3. *Symmetry*. When we talk about symmetry in painting, we usually mean a picture that has a mirror likeness on either side of a line or plane, with respect to a point. Sometimes the imagination needs to be stretched in order to see artistic symmetry. In mathematics we have formal statements.

Definition 1: Symmetry of a function with respect to the y-axis

A function $y = f(x)$ is symmetric with respect to the *y*-axis if and only if $f(x) = f(-x)$.

Definition 2: Symmetry of a function with respect to the origin

A function $y = f(x)$ is symmetric with respect to the origin if and only if $f(-x) = -f(x)$.

Example 4

The function $y = x/(x^2 + 1)$ is symmetric to the origin. The function $y = x^2/(x^2 + 1)$ is symmetric with respect to the *y*-axis. The graph of the two functions are given in Figures 1 and 2. Writing the equation of

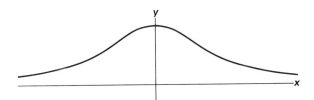

Figure 1

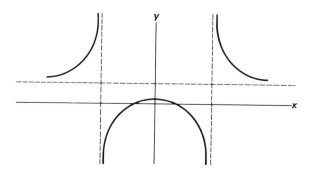

Figure 2

the graph of a function as $f(x, y) = 0$, we define the symmetry of a graph with respect to the axes and origin.

Definition 3: Symmetry of a graph with respect to the origin

The graph of the equation $f(x, y) = 0$ is symmetric with respect to $(0, 0)$ if $f(x, y) = f(-x, -y) = 0$; that is, if both (x, y) and $(-x, -y)$ satisfy the equation.

Example 5

If $f(x, y) = 2x^2 + 3y^2 + 7x + 5y = 0$, then

$$f(-x, -y) = 2(-x)^2 + 3(-y)^2 + 7(-x)(-y) + 5$$

$$= 2x^2 + 3y^2 + 7xy + 5 = 0.$$

Hence it is symmetric with respect to the origin.

Example 6

If $y = f(x) = x^3/(1 + x^2)$, then with x and y replaced by $-x$ and $-y$, we have

$$-y = \frac{(-x)^3}{1 + (-x)^2}; \qquad \text{hence } y = \frac{(x)^3}{1 + x^2};$$

that is, $f(-x, -y) = f(x, y)$, and so every point (x, y) on the graph is symmetric with respect to the origin to the point $(-x, -y)$. ∎

Example 7

The equation $f(x, y) = x^2 + y^2 - x + 2 = 0$ is not symmetric with respect to $(0, 0)$, because $(-x)^2 + (-y)^2 - (-x) + 2 = 0$ is different from the original equation. ▮

The graph of a function has an additional symmetry.

Definition 4: Symmetry of a graph with respect to the x-axis

The graph of the equation $f(x, y) = 0$ is symmetric with respect to the x-axis if the points (x, y) and $(x, -y)$ both satisfy the equation.

Example 8

$x^2 + y^2 + 2x = 6$ is symmetric with respect to the x-axis because the equation is unchanged when $(x, -y)$ replaces (x, y),

$$x^2 + (-y)^2 + 2x = 6.$$

However, the equation $x^2 + y^2 + 2y = 6$ is not symmetric with respect to the x-axis because $x^2 + (-y)^2 + 2(-y) = 6$ is not the same as the original equation. ▮

Definition 5: Symmetry of a graph with respect to the y-axis

The graph of the equation $f(x, y) = 0$ is symmetric with respect to the y-axis if the points $(-x, y)$ and (x, y) both satisfy the equation.

The idea of symmetry with respect to the coordinate axes and the origin can be a very big help in sketching graphs. If symmetry does exist we only need to sketch part of the graph and take the appropriate reflection in the line or point.

4. *Asymptotes.* The graph of $f(x) = 1/x$ has among other properties one that says that as x gets close to 0, the graph gets arbitrarily close to the y-axis. The graph of $y = 1/(x - 1)$ gets arbitrarily close to the line $x = 1$ as x gets close to 1. Similarly, as x increases without bound, the graph of $y = 1/(x - 1)$ gets arbitrarily close to the line $y = 0$. These are examples of a graph being asymptotic to a line.

Definition 6: Linear asymptotes

The line $y = mx + b$ is a linear asymptote to the graph of $y = f(x)$ if and only if $|f(x) - (mx + b)|$ decreases to zero as x increases to infinity. The line $x = 1$ is *vertical asymptote* if and only if $f(x)$ increases to infinity as x gets arbitrarily close to $x = a$. The line $y = a$ is a *horizontal asymptote* if and only if $|f(x) - a|$ decreases to 0 as x increases to infinity.

Example 9

The function $f(x) = (2x^2 + 3x + 2)/(x + 1)$ has as linear asymptote the line $y = 2x + 1$, because $f(x)$ can be written as $f(x) = (2x + 1) + 1/(x + 1)$.

The function $f(x) = (x + 1)/x$ has as linear asymptote the line $y = 1$, for f can be written as $f(x) = 1 + 1/x$.

The function $y = 1/x$ has $x = 0$ as a vertical asymptote, for y increases to infinity as x gets arbitrarily close to 0. ∎

We remark here that the study of asymptotes is most efficiently undertaken with the aid of the calculus.

Example 10

Sketch the graph of $y = 8/(x^2 + 4)$.

Intercepts: Put $y = 0$; no solution for x.
              Put $x = 0$; solution $y = 2$.

Extent: There are no values of x for which y is not defined but y is always positive. The largest value of y will occur when the denominator has its smallest value. This occurs when $x = 0$ and so 2 is the largest value of y.

Symmetry: Since x only appears to an even power, the graph is symmetric with respect to the y-axis.

Asymptotes: There are no vertical asymptotes, but $y = 0$ is a horizontal asymptote (see Figure 3). ∎

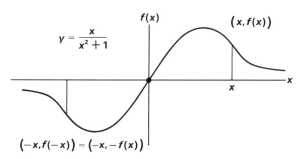

$$y = \frac{x}{x^2 + 1}$$

$f(x)$

$(x, f(x))$

$(-x, f(-x)) = (-x, -f(x))$

Figure 3

Example 11

Sketch the graph of $y = (x^2 - 1)/(x^2 - 4)$.

Intercepts: Put $y = 0$; then $x = +1, -1$.
                  Put $x = 0$; then $y = \frac{1}{4}$.

Extent: x can take on all values except $x = \pm 2$. y is positive when $x^2 - 1 > 0$ and $x^2 - 4 > 0$, or when $x^2 - 1 < 0$ and $x^2 - 4 < 0$. In the first case $x^2 > 4$, which means that $x < -2$ or $x > 2$. In the second case $x^2 < 1$, which occurs when $-1 < x < 1$.

Symmetry: Since x appears only in even powers, the graph is symmetric with respect to the y-axis.

Asymptotes: Vertical asymptotes occur at $x = 2$ and $x = -2$. The line $y = 1$ is a horizontal asymptote. (see Figure 4).   **❚**

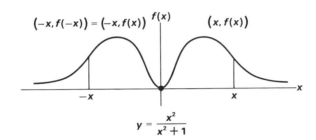

Figure 4

Problems

In Problems 1–24 discuss the intercepts, extent, symmetry, and asymptotes and sketch the graph.

1. $x^2 + y^2 = 4$ 2. $y = 5x^3$

3. $y = (x - 1)^2$ 4. $y = x(1 - x^2)^2$

5. $x^2 - y^2 = 1 - x$ 6. $y = x^2 + \dfrac{1}{x}$

7. $y = \dfrac{x + 1}{x - 1}$ 8. $x + y = 3x^2y$

9. $4x^2 + 9y = 0$ 10. $y = x^2 - x^4$

11. $y = \dfrac{y}{x - 1}$ 12. $y = \dfrac{x}{x^2 + 1}$

13. $y^2 - x^2 = 0$ 14. $y = \dfrac{1}{x(x^2 - 1)}$

15. $y = x^4 + 3$

16. $y = \dfrac{x^2}{x^2 - 4}$

17. $y = \dfrac{x^2 + 2}{x + 2}$

18. $y = \dfrac{x^2 + 1}{x^2}$

19. $y = 4x - x^3$

20. $y = 4 - x^3$

21. $y = \dfrac{x}{x^2 - 3x + 2}$

22. $y = \dfrac{x + 2}{x^2 - 1}$

23. $2xy + 1 = x^2(1 - y)$

24. $x^2(y - x) = 1 - y$

Table 1 · Binomial Probability Function, $\binom{n}{x}p^x(1-p)^{n-x}$[a]

n	x	.05	.10	.15	.20	.25	.30	.35	.40	.45	.50
1	0	.9500	.9000	.8500	.8000	.7500	.7000	.6500	.6000	.5500	.5000
	1	.0500	.1000	.1500	.2000	.2500	.3000	.3500	.4000	.4500	.5000
2	0	.9025	.8100	.7225	.6400	.5625	.4900	.4225	.3600	.3025	.2500
	1	.0950	.1800	.2550	.3200	.3750	.4200	.4550	.4800	.4950	.5000
	2	.0025	.0100	.0225	.0400	.0625	.0900	.1225	.1600	.2025	.2500
3	0	.8574	.7290	.6141	.5120	.4219	.3420	.2746	.2160	.1664	.1250
	1	.1354	.2430	.3251	.3840	.4219	.4410	.4436	.4320	.4084	.3750
	2	.0071	.0270	.0574	.0960	.1406	.1890	.2389	.2880	.3341	.3750
	3	.0001	.0010	.0034	.0080	.0156	.0270	.0429	.0640	.0911	.1250
4	0	.8145	.6561	.5220	.4096	.3164	.2401	.1785	.1296	.0915	.0625
	1	.1715	.2916	.3685	.4096	.4219	.4116	.3845	.3456	.2995	.2500
	2	.0135	.0486	.0975	.1536	.2109	.2646	.3105	.3456	.3675	.3750
	3	.0005	.0036	.0115	.0256	.0469	.0756	.1115	.1536	.2005	.2500
	4	.0000	.0001	.0005	.0016	.0039	.0081	.0150	.0256	.0410	.0625
5	0	.7738	.5905	.4437	.3277	.2373	.1681	.1160	.0778	.0503	.0312
	1	.2036	.3280	.3915	.4096	.3955	.3602	.3124	.2592	.2059	.1562
	2	.0214	.0729	.1382	.2048	.2637	.3087	.3364	.3456	.3369	.3125
	3	.0011	.0081	.0244	.0512	.0879	.1323	.1811	.2304	.2757	.3125
	4	.0000	.0004	.0022	.0064	.0146	.0284	.0488	.0768	.1128	.1562
	5	.0000	.0000	.0001	.0003	.0010	.0024	.0053	.0102	.0185	.0312
6	0	.7351	.5314	.3771	.2621	.1780	.1176	.0754	.0467	.0277	.0156
	1	.2321	.3543	.3993	.3932	.3560	.3025	.2437	.1866	.1359	.0938
	2	.0305	.0984	.1762	.2458	.2966	.3241	.3280	.3110	.2780	.2344
	3	.0021	.0146	.0415	.0819	.1318	.1852	.2355	.2765	.3032	.3125
	4	.0001	.0012	.0055	.0154	.0330	.0595	.0951	.1382	.1861	.2344
	5	.0000	.0001	.0004	.0015	.0044	.0102	.0205	.0369	.0609	.0938
	6	.0000	.0000	.0000	.0001	.0002	.0007	.0011	.0041	.0083	.0156
7	0	.6983	.4783	.3206	.2097	.1335	.0824	.0490	.0280	.0152	.0078
	1	.2573	.3720	.3960	.3670	.3115	.2471	.1848	.1306	.0872	.0547
	2	.0406	.1240	.2097	.2753	.3115	.3177	.2985	.2613	.2140	.1641
	3	.0036	.0230	.0617	.1147	.1730	.2269	.2679	.2903	.2918	.2734
	4	.0002	.0026	.0109	.0287	.0577	.0972	.1442	.1935	.2388	.2734
	5	.0000	.0002	.0012	.0043	.0115	.0250	.0466	.0774	.1172	.1641
	6	.0000	.0000	.0001	.0004	.0013	.0036	.0084	.0172	.0320	.0547
	7	.0000	.0000	.0000	.0000	.0001	.0002	.0006	.0016	.0037	.0078
8	0	.6634	.4305	.2725	.1678	.1001	.0576	.0319	.0168	.0084	.0039
	1	.2793	.3826	.3847	.3355	.2670	.1977	.1373	.0896	.0548	.0312
	2	.0515	.1488	.2376	.2936	.3115	.2965	.2587	.2090	.1569	.1094
	3	.0054	.0331	.0839	.1468	.2076	.2541	.2786	.2787	.2568	.2188
	4	.0004	.0046	.0185	.0459	.0865	.1361	.1875	.2322	.2627	.2734
	5	.0000	.0004	.0026	.0092	.0231	.0467	.0808	.1239	.1719	.2188
	6	.0000	.0000	.0002	.0011	.0038	.0100	.0217	.0413	.0703	.1094
	7	.0000	.0000	.0000	.0001	.0004	.0012	.0033	.0079	.0164	.0312
	8	.0000	.0000	.0000	.0000	.0000	.0001	.0002	.0007	.0017	.0039

[a] Entries in the table are values of $\binom{n}{x}p^x(1-p)^{n-x}$ for the indicated values of n, x, and p. When $p > 0.5$, the value of $\binom{n}{x}p^x(1-p)^{x-n}$ for a given n, x, and p is obtained by finding the tabular entry for the given n, with $n-x$ in place of the given x, and $1-p$ in place of the given p.

For extensive tables of $\binom{n}{x}p^x(1-p)^{n-x}$, see *Tables of the Binomial Probability Distribution*, National Bureau of Standards, Applied Mathematics Series 6, Washington, D.C., 1950. Used with permission from *Handbook of Probability and Statistics with Tables* by R. S. Burington and D. C. May. Copyright 1953, by McGraw-Hill, Inc., McGraw-Hill Book Company, New York.

Table 1 389

						p					
n	x	.05	.10	.15	.20	.25	.30	.35	.40	.45	.50
9	0	.6302	.3874	.2316	.1342	.0751	.0404	.0207	.0101	.0046	.0030
	1	.2985	.3874	.3679	.3020	.2253	.1556	.1004	.0605	.0339	.0176
	2	.0629	.1722	.2597	.3020	.3003	.2668	.2162	.1612	.1110	.0703
	3	.0077	.0446	.1069	.1762	.2336	.2668	.2716	.2508	.2119	.1641
	4	.0006	.0074	.0283	.0661	.1168	.1715	.2194	.2508	.2600	.2461
	5	.0000	.0008	.0050	.0165	.0389	.0735	.1181	.1672	.2128	.2461
	6	.0000	.0001	.0006	.0028	.0087	.0210	.0424	.0743	.1160	.1641
	7	.0000	.0000	.0000	.0003	.0012	.0039	.0098	.0212	.0407	.0703
	8	.0000	.0000	.0000	.0000	.0001	.0004	.0013	.0035	.0083	.0176
	9	.0000	.0000	.0000	.0000	.0000	.0000	.0001	.0003	.0008	.0020
10	0	.5987	.3487	.1969	.1074	.0563	.0282	.0135	.0060	.0025	.0010
	1	.3151	.3874	.3474	.2684	.1877	.1211	.0752	.0403	.0207	.0098
	2	.0746	.1937	.2759	.3020	.2816	.2335	.1757	.1209	.0763	.0400
	3	.0105	.0574	.1298	.2013	.2503	.2668	.2522	.2150	.1665	.1172
	4	.0010	.0112	.0401	.0881	.1460	.2001	.2377	.2508	.2384	.2051
	5	.0001	.0015	.0085	.0264	.0584	.1029	.1536	.2007	.2340	.2461
	6	.0000	.0001	.0012	.0055	.0162	.0368	.0689	.1115	.1596	.2051
	7	.0000	.0000	.0001	.0008	.0031	.0090	.0212	.0425	.0746	.1172
	8	.0000	.0000	.0000	.0001	.0004	.0014	.0043	.0106	.0229	.0439
	9	.0000	.0000	.0000	.0000	.0000	.0001	.0005	.0016	.0042	.0093
	10	.0000	.0000	.0000	.0000	.0000	.0000	.0000	.0001	.0003	.0010
11	0	.5688	.3138	.1673	.0859	.0422	.0198	.0088	.0036	.0014	.0005
	1	.3293	.3835	.3248	.2362	.1549	.0932	.0518	.0266	.0125	.0054
	2	.0867	.2131	.2866	.2953	.2581	.1998	.1395	.0887	.0513	.0269
	3	.0137	.0710	.1517	.2215	.2581	.2568	.2254	.1774	.1259	.0806
	4	.0014	.0158	.0536	.1107	.1721	.2201	.2428	.2365	.2060	.1611
	5	.0001	.0025	.0132	.0388	.0803	.1321	.1830	.2207	.2360	.2256
	6	.0000	.0003	.0023	.0097	.0268	.0566	.0985	.1471	.1931	.2256
	7	.0000	.0000	.0003	.0017	.0064	.0173	.0379	.0701	.1128	.1611
	8	.0000	.0000	.0000	.0002	.0011	.0037	.0102	.0234	.0462	.0806
	9	.0000	.0000	.0000	.0000	.0001	.0005	.0018	.0052	.0126	.0269
	10	.0000	.0000	.0000	.0000	.0000	.0000	.0002	.0007	.0021	.0054
	11	.0000	.0000	.0000	.0000	.0000	.0000	.0000	.0000	.0002	.0005
12	0	.5404	.2824	.1422	.0687	.0317	.0138	.0057	.0022	.0008	.0002
	1	.3413	.3766	.3012	.2062	.1267	.0712	.0368	.0174	.0075	.0029
	2	.0988	.2301	.2924	.2835	.2323	.1678	.1088	.0639	.0339	.0161
	3	.0173	.0852	.1720	.2326	.2581	.2397	.1954	.1419	.0923	.0537
	4	.0021	.0213	.0683	.1329	.1936	.2311	.2367	.2128	.1700	.1208
	5	.0002	.0038	.0193	.0532	.1032	.1585	.2039	.2270	.2225	.1934
	6	.0000	.0005	.0040	.0155	.0401	.0792	.1281	.1766	.2124	.2256
	7	.0000	.0000	.0006	.0033	.0115	.0291	.0591	.1009	.1489	.1934
	8	.0000	.0000	.0001	.0005	.0024	.0078	.0199	.0420	.0762	.1208
	9	.0000	.0000	.0000	.0001	.0004	.0015	.0048	.0125	.0277	.0537
	10	.0000	.0000	.0000	.0000	.0000	.0002	.0008	.0025	.0068	.0161
	11	.0000	.0000	.0000	.0000	.0000	.0000	.0001	.0003	.0010	.0029
	12	.0000	.0000	.0000	.0000	.0000	.0000	.0000	.0000	.0001	.0002
13	0	.5133	.2542	.1209	.0550	.0238	.0097	.0037	.0013	.0004	.0001
	1	.3512	.3672	.2774	.1787	.1029	.0540	.0259	.0113	.0045	.0016
	2	.1109	.2448	.2937	.2680	.2059	.1388	.0836	.0453	.0220	.0095
	3	.0214	.0997	.1900	.2457	.2517	.2181	.1651	.1107	.0660	.0349
	4	.0028	.0277	.0838	.1535	.2097	.2337	.2222	.1845	.1350	.0873
	5	.0003	.0055	.0266	.0691	.1258	.1803	.2154	.2214	.1989	.1571
	6	.0000	.0008	.0063	.0230	.0559	.1030	.1546	.1968	.2169	.2095
	7	.0000	.0001	.0011	.0058	.0186	.0442	.0833	.1312	.1775	.2095
	8	.0000	.0000	.0001	.0011	.0047	.0142	.0336	.0656	.1089	.1571
	9	.0000	.0000	.0000	.0001	.0009	.0034	.0101	.0243	.0495	.0873

Table 1

						p					
n	x	.05	.10	.15	.20	.25	.30	.35	.40	.45	.50
	10	.0000	.0000	.0000	.0000	.0001	.0006	.0022	.0065	.0162	.0349
	11	.0000	.0000	.0000	.0000	.0000	.0001	.0003	.0012	.0036	.0095
	12	.0000	.0000	.0000	.0000	.0000	.0000	.0000	.0001	.0005	.0016
	13	.0000	.0000	.0000	.0000	.0000	.0000	.0000	.0000	.0000	.0001
14	0	.4877	.2288	.1028	.0440	.0178	.0068	.0024	.0008	.0002	.0001
	1	.3593	.3559	.2539	.1539	.0832	.0407	.0181	.0073	.0027	.0009
	2	.1229	.2570	.2912	.2501	.1802	.1134	.0634	.0317	.0141	.0056
	3	.0259	.1142	.2056	.2501	.2402	.1943	.1366	.0845	.0462	.0222
	4	.0037	.0349	.0998	.1720	.2202	.2290	.2022	.1549	.1040	.0611
	5	.0004	.0078	.0352	.0860	.1468	.1963	.2178	.2066	.1701	.1222
	6	.0000	.0013	.0093	.0322	.0734	.1262	.1759	.2066	.2088	.1833
	7	.0000	.0002	.0019	.0092	.0280	.0618	.1082	.1574	.1952	.2095
	8	.0000	.0000	.0003	.0020	.0082	.0232	.0510	.0918	.1398	.1833
	9	.0000	.0000	.0000	.0003	.0018	.0066	.0183	.0408	.0762	.1222
	10	.0000	.0000	.0000	.0000	.0003	.0014	.0049	.0136	.0312	.0611
	11	.0000	.0000	.0000	.0000	.0000	.0002	.0010	.0033	.0093	.0222
	12	.0000	.0000	.0000	.0000	.0000	.0000	.0001	.0005	.0019	.0056
	13	.0000	.0000	.0000	.0000	.0000	.0000	.0000	.0001	.0002	.0009
	14	.0000	.0000	.0000	.0000	.0000	.0000	.0000	.0000	.0000	.0001
15	0	.4633	.2059	.0874	.0352	.0134	.0047	.0016	.0005	.0001	.0000
	1	.3658	.3432	.2312	.1319	.0668	.0305	.0126	.0047	.0016	.0005
	2	.1348	.2669	.2856	.2309	.1559	.0916	.0476	.0219	.0090	.0032
	3	.0307	.1285	.2184	.2501	.2252	.1700	.1110	.0634	.0318	.0139
	4	.0049	.0428	.1156	.1876	.2252	.2186	.1792	.1268	.0780	.0417
	5	.0006	.0105	.0449	.1032	.1651	.2061	.2123	.1859	.1404	.0916
	6	.0000	.0019	.0132	.0430	.0917	.1472	.1906	.2066	.1914	.1527
	7	.0000	.0003	.0030	.0138	.0393	.0811	.1319	.1771	.2013	.1964
	8	.0000	.0000	.0005	.0035	.0131	.0348	.0710	.1181	.1647	.1964
	9	.0000	.0000	.0001	.0007	.0034	.0116	.0298	.0612	.1048	.1527
	10	.0000	.0000	.0000	.0001	.0007	.0030	.0096	.0245	.0515	.0916
	11	.0000	.0000	.0000	.0000	.0001	.0006	.0024	.0074	.0191	.0417
	12	.0000	.0000	.0000	.0000	.0000	.0001	.0004	.0016	.0052	.0139
	13	.0000	.0000	.0000	.0000	.0000	.0000	.0001	.0003	.0010	.0032
	14	.0000	.0000	.0000	.0000	.0000	.0000	.0000	.0000	.0001	.0005
	15	.0000	.0000	.0000	.0000	.0000	.0000	.0000	.0000	.0000	.0000
16	0	.4401	.1853	.0743	.0281	.0100	.0033	.0010	.0003	.0001	.0000
	1	.3706	.3294	.2097	.1126	.0535	.0228	.0087	.0030	.0009	.0002
	2	.1463	.2745	.2775	.2111	.1336	.0732	.0353	.0150	.0056	.0018
	3	.0359	.1423	.2285	.2463	.2079	.1465	.0888	.0468	.0215	.0085
	4	.0061	.0514	.1311	.2001	.2252	.2040	.1553	.1014	.0572	.0278
	5	.0008	.0137	.0555	.1201	.1802	.2099	.2008	.1623	.1123	.0667
	6	.0001	.0028	.0180	.0550	.1101	.1649	.1982	.1983	.1684	.1222
	7	.0000	.0004	.0045	.0197	.0524	.1010	.1524	.1889	.1969	.1746
	8	.0000	.0001	.0009	.0055	.0197	.0487	.0923	.1417	.1812	.1964
	9	.0000	.0000	.0001	.0012	.0058	.0185	.0442	.0840	.1318	.1746
	10	.0000	.0000	.0000	.0002	.0014	.0056	.0167	.0392	.0755	.1222
	11	.0000	.0000	.0000	.0000	.0002	.0013	.0049	.0142	.0337	.0667
	12	.0000	.0000	.0000	.0000	.0000	.0002	.0011	.0040	.0115	.0278
	13	.0000	.0000	.0000	.0000	.0000	.0000	.0002	.0008	.0029	.0085
	14	.0000	.0000	.0000	.0000	.0000	.0000	.0000	.0001	.0005	.0018
	15	.0000	.0000	.0000	.0000	.0000	.0000	.0000	.0000	.0001	.0002
	16	.0000	.0000	.0000	.0000	.0000	.0000	.0000	.0000	.0000	.0000
17	0	.4181	.1668	.0631	.0225	.0075	.0023	.0007	.0002	.0000	.0000
	1	.3741	.3150	.1893	.0957	.0426	.0169	.0060	.0019	.0005	.0001
	2	.1575	.2800	.2673	.1914	.1136	.0581	.0260	.0102	.0035	.0010
	3	.0415	.1556	.2359	.2393	.1893	.1245	.0701	.0341	.0144	.0052
	4	.0076	.0605	.1457	.2093	.2209	.1868	.1320	.0796	.0441	.0182

Table 1   391

| | | | | | | p | | | | | |
n	x	.05	.10	.15	.20	.25	.30	.35	.40	.45	.50
	5	.0010	.0175	.0668	.1361	.1914	.2081	.1849	.1379	.0875	.0472
	6	.0001	.0039	.0236	.0680	.1276	.1784	.1991	.1839	.1432	.0944
	7	.0000	.0007	.0065	.0267	.0668	.1201	.1685	.1927	.1841	.1484
	8	.0000	.0001	.0014	.0084	.0279	.0644	.1134	.1606	.1883	.1855
	9	.0000	.0000	.0003	.0021	.0093	.0276	.0611	.1070	.1540	.1855
	10	.0000	.0000	.0000	.0004	.0025	.0095	.0263	.0571	.1008	.1484
	11	.0000	.0000	.0000	.0001	.0005	.0026	.0090	.0242	.0525	.0944
	12	.0000	.0000	.0000	.0000	.0001	.0006	.0024	.0081	.0215	.0472
	13	.0000	.0000	.0000	.0000	.0000	.0001	.0005	.0021	.0068	.0182
	14	.0000	.0000	.0000	.0000	.0000	.0000	.0001	.0004	.0016	.0052
	15	.0000	.0000	.0000	.0000	.0000	.0000	.0000	.0001	.0003	.0010
	16	.0000	.0000	.0000	.0000	.0000	.0000	.0000	.0000	.0000	.0001
	17	.0000	.0000	.0000	.0000	.0000	.0000	.0000	.0000	.0000	.0000
18	0	.3972	.1501	.0536	.0180	.0056	.0016	.0004	.0001	.0000	.0000
	1	.3763	.3002	.1704	.0811	.0338	.0126	.0042	.0012	.0003	.0001
	2	.1683	.2835	.2556	.1723	.0958	.0458	.0190	.0069	.0022	.0006
	3	.0473	.1680	.2406	.2297	.1704	.1046	.0547	.0246	.0095	.0031
	4	.0093	.0700	.1592	.2153	.2130	.1681	.1104	.0614	.0291	.0117
	5	.0014	.0218	.0787	.1507	.1988	.2017	.1664	.1146	.0666	.0327
	6	.0002	.0052	.0301	.0816	.1436	.1873	.1941	.1655	.1181	.0708
	7	.0000	.0010	.0091	.0350	.0820	.1376	.1792	.1892	.1657	.1214
	8	.0000	.0002	.0022	.0120	.0376	.0811	.1327	.1734	.1864	.1669
	9	.0000	.0000	.0004	.0033	.0139	.0386	.0794	.1284	.1694	.1855
	10	.0000	.0000	.0001	.0008	.0042	.0149	.0385	.0771	.1248	.1669
	11	.0000	.0000	.0000	.0001	.0010	.0046	.0151	.0374	.0742	.1214
	12	.0000	.0000	.0000	.0000	.0002	.0012	.0047	.0145	.0354	.0708
	13	.0000	.0000	.0000	.0000	.0000	.0002	.0012	.0045	.0134	.0327
	14	.0000	.0000	.0000	.0000	.0000	.0000	.0002	.0011	.0039	.0117
	15	.0000	.0000	.0000	.0000	.0000	.0000	.0000	.0002	.0009	.0031
	16	.0000	.0000	.0000	.0000	.0000	.0000	.0000	.0000	.0001	.0006
	17	.0000	.0000	.0000	.0000	.0000	.0000	.0000	.0000	.0000	.0001
	18	.0000	.0000	.0000	.0000	.0000	.0000	.0000	.0000	.0000	.0000
19	0	.3774	.1351	.0456	.0144	.0042	.0011	.0003	.0001	.0000	.0000
	1	.3774	.2852	.1529	.0685	.0268	.0093	.0029	.0008	.0002	.0000
	2	.1787	.2852	.2428	.1540	.0803	.0358	.0138	.0046	.0013	.0003
	3	.0533	.1796	.2428	.2182	.1517	.0869	.0422	.0175	.0062	.0018
	4	.0112	.0798	.1714	.2182	.2023	.1491	.0909	.0467	.0203	.0074
	5	.0018	.0266	.0907	.1636	.2023	.1916	.1468	.0933	.0497	.0222
	6	.0002	.0069	.0374	.0955	.1574	.1916	.1844	.1451	.0949	.0518
	7	.0000	.0014	.0122	.0443	.0974	.1525	.1844	.1797	.1443	.0961
	8	.0000	.0002	.0032	.0166	.0487	.0981	.1489	.1797	.1771	.1442
	9	.0000	.0000	.0007	.0051	.0198	.0514	.0980	.1464	.1771	.1762
	10	.0000	.0000	.0001	.0013	.0066	.0220	.0528	.0976	.1449	.1762
	11	.0000	.0000	.0000	.0003	.0018	.0077	.0233	.0532	.0970	.1442
	12	.0000	.0000	.0000	.0000	.0004	.0022	.0083	.0237	.0529	.0961
	13	.0000	.0000	.0000	.0000	.0001	.0005	.0024	.0085	.0233	.0518
	14	.0000	.0000	.0000	.0000	.0000	.0001	.0006	.0024	.0082	.0222
	15	.0000	.0000	.0000	.0000	.0000	.0000	.0001	.0005	.0022	.0074
	16	.0000	.0000	.0000	.0000	.0000	.0000	.0000	.0001	.0005	.0018
	17	.0000	.0000	.0000	.0000	.0000	.0000	.0000	.0000	.0001	.0003
	18	.0000	.0000	.0000	.0000	.0000	.0000	.0000	.0000	.0000	.0000
	19	.0000	.0000	.0000	.0000	.0000	.0000	.0000	.0000	.0000	.0000
20	0	.3585	.1216	.0388	.0115	.0032	.0008	.0002	.0000	.0000	.0000
	1	.3774	.2702	.1368	.0576	.0211	.0068	.0020	.0005	.0001	.0000
	2	.1887	.2852	.2293	.1369	.0669	.0278	.0100	.0031	.0008	.0002
	3	.0596	.1901	.2428	.2054	.1339	.0716	.0323	.0123	.0040	.0011
	4	.0133	.0898	.1821	.2182	.1897	.1304	.0738	.0350	.0139	.0046

Table 1

						p					
n	x	.05	.10	.15	.20	.25	.30	.35	.40	.45	.50
	5	.0022	.0319	.1028	.1746	.2023	.1789	.1272	.0746	.0365	.0148
	6	.0003	.0089	.0454	.1091	.1686	.1916	.1713	.1244	.0746	.0370
	7	.0000	.0020	.0160	.0545	.1124	.1643	.1844	.1659	.1221	.0739
	8	.0000	.0004	.0046	.0222	.0609	.1144	.1614	.1797	.1623	.1201
	9	.0000	.0001	.0011	.0074	.0271	.0654	.1158	.1597	.1771	.1602
	10	.0000	.0000	.0002	.0020	.0099	.0308	.0686	.1171	.1593	.1762
	11	.0000	.0000	.0000	.0005	.0030	.0120	.0336	.0710	.1185	.1602
	12	.0000	.0000	.0000	.0001	.0008	.0039	.0136	.0355	.0727	.1201
	13	.0000	.0000	.0000	.0000	.0002	.0010	.0045	.0146	.0366	.0739
	14	.0000	.0000	.0000	.0000	.0000	.0002	.0012	.0049	.0150	.0370
	15	.0000	.0000	.0000	.0000	.0000	.0000	.0003	.0013	.0049	.0148
	16	.0000	.0000	.0000	.0000	.0000	.0000	.0000	.0003	.0013	.0046
	17	.0000	.0000	.0000	.0000	.0000	.0000	.0000	.0000	.0002	.0011
	18	.0000	.0000	.0000	.0000	.0000	.0000	.0000	.0000	.0000	.0002
	19	.0000	.0000	.0000	.0000	.0000	.0000	.0000	.0000	.0000	.0000
	20	.0000	.0000	.0000	.0000	.0000	.0000	.0000	.0000	.0000	.0000

Table 2   393

Table 2 · Summed Binomial Probability Function $\sum_{x=x'}^{x=n}\binom{n}{x}p^{x}(1-p)^{n-x}$ [a]

		p									
n	x'	.05	.10	.15	.20	.25	.30	.35	.40	.45	.50
2	1	.0975	.1900	.2775	.3600	.4375	.5100	.5775	.6400	.6975	.7500
	2	.0025	.0100	.0225	.0400	.0625	.0900	.1225	.1600	.2025	.2500
3	1	.1426	.2710	.3859	.4880	.5781	.6570	.7254	.7840	.8336	.8750
	2	.0072	.0280	.0608	.1040	.1562	.2160	.2818	.3520	.4252	.5000
	3	.0001	.0010	.0034	.0080	.0156	.0270	.0429	.0640	.0911	.1250
4	1	.1855	.3439	.4780	.5904	.6836	.7599	.8215	.8704	.9085	.9375
	2	.0140	.0523	.1095	.1808	.2617	.3483	.4370	.5248	.6090	.6875
	3	.0005	.0037	.0120	.0272	.0508	.0837	.1265	.1792	.2415	.3125
	4	.0000	.0001	.0005	.0016	.0039	.0081	.0150	.0256	.0410	.0625
5	1	.2262	.4095	.5563	.6723	.7627	.8319	.8840	.9222	.9497	.9688
	2	.0226	.0815	.1648	.2627	.3672	.4718	.5716	.6630	.7438	.8125
	3	.0012	.0086	.0266	.0579	.1035	.1631	.2352	.3174	.4069	.5000
	4	.0000	.0005	.0022	.0067	.0156	.0308	.0540	.0870	.1312	.1875
	5	.0000	.0000	.0001	.0003	.0010	.0024	.0053	.0102	.0185	.0312
6	1	.2649	.4686	.6229	.7379	.8220	.8824	.9246	.9533	.9723	.9884
	2	.0328	.1143	.2235	.3446	.4661	.5798	.6809	.7667	.8364	.8906
	3	.0022	.0158	.0473	.0989	.1694	.2557	.3529	.4557	.5585	.6562
	4	.0001	.0013	.0059	.0170	.0376	.0705	.1174	.1792	.2553	.3438
	5	.0000	.0001	.0004	.0016	.0046	.0109	.0223	.0410	.0692	.1094
	6	.0000	.0000	.0000	.0001	.0002	.0007	.0018	.0041	.0083	.0156
7	1	.3017	.5217	.6794	.7903	.8665	.9176	.9510	.9720	.9848	.9922
	2	.0444	.1497	.2834	.4233	.5551	.6706	.7662	.8414	.8976	.9375
	3	.0038	.0257	.0738	.1480	.2436	.3529	.4677	.5801	.6836	.7734
	4	.0002	.0027	.0121	.0333	.0706	.1260	.1998	.2898	.3917	.5000
	5	.0000	.0002	.0012	.0047	.0129	.0288	.0556	.0963	.1529	.2266
	6	.0000	.0000	.0001	.0004	.0013	.0038	.0090	.0188	.0357	.0625
	7	.0000	.0000	.0000	.0000	.0001	.0002	.0006	.0016	.0037	.0078
8	1	.3366	.5695	.7275	.8322	.8999	.9424	.9681	.9832	.9916	.9961
	2	.0572	.1896	.3428	.4967	.6329	.7447	.8309	.8936	.9368	.9648
	3	.0058	.0381	.1052	.2031	.3215	.4482	.5722	.6846	.7799	.8555
	4	.0004	.0050	.0214	.0563	.1138	.1941	.2936	.4059	.5230	.6367
	5	.0000	.0004	.0029	.0104	.0273	.0580	.1061	.1737	.2604	.3633
	6	.0000	.0000	.0002	.0012	.0042	.0113	.0253	.0498	.0885	.1445
	7	.0000	.0000	.0000	.0001	.0004	.0013	.0036	.0085	.0181	.0352
	8	.0000	.0000	.0000	.0000	.0000	.0001	.0002	.0007	.0017	.0039
9	1	.3698	.6126	.7684	.8658	.9249	.9596	.9793	.9899	.9954	.9980
	2	.0712	.2252	.4005	.5638	.6997	.8040	.8789	.9295	.9615	.9805
	3	.0084	.0530	.1409	.2618	.3993	.5372	.6627	.7682	.8505	.9102
	4	.0006	.0083	.0339	.0856	.1657	.2703	.3911	.5174	.6386	.7461
	5	.0000	.0009	.0056	.0196	.0489	.0988	.1717	.2666	.3786	.5000
	6	.0000	.0001	.0006	.0031	.0100	.0253	.0536	.0994	.1658	.2539
	7	.0000	.0000	.0000	.0003	.0013	.0043	.0112	.0250	.0498	.0898
	8	.0000	.0000	.0000	.0000	.0001	.0004	.0014	.0038	.0091	.0195
	9	.0000	.0000	.0000	.0000	.0000	.0000	.0001	.0003	.0008	.0020

[a] Entries in the table are values of $\sum_{x=x'}^{x=n}\binom{n}{x}p^{x}(1-p)^{n-x}$ for the indicated values of n, x', and p. When $p > 0.5$, the value of $\sum_{x=x'}^{x=n}\binom{n}{x}p^{x}(1-p)^{n-x}$ for a given n, x', and p is equal to 1 minus the tabular entry for the given n, with $n-x'+1$ in place of the given value of x', and $1-p$ in place of the given value of p.

For extensive tables of $\sum_{x=x'}^{x=n}\binom{n}{x}p^{x}(1-p)^{n-x}$ see *Tables of the Binomial Probability Distribution*, National Bureau of Standards, Applied Mathematics Series 6, Washington, D.C., 1950. Used with permission from *Handbook of Probability and Statistics with Tables* by R. S. Burington and D. C. May. Copyright 1953 by McGraw-Hill, Inc., McGraw-Hill Book Company, New York.

						p					
n	x'	.05	.10	.15	.20	.25	.30	.35	.40	.45	.50
10	1	.4013	.6513	.8031	.8926	.9437	.9718	.9865	.9940	.9975	.9990
	2	.0861	.2639	.4557	.6242	.7560	.8507	.9140	.9536	.9767	.9893
	3	.0115	.0702	.1798	.3222	.4744	.6172	.7384	.8327	.9004	.9453
	4	.0010	.0128	.0500	.1209	.2241	.3504	.4862	.6177	.7340	.8281
	5	.0001	.0016	.0099	.0328	.0781	.1503	.2485	.3669	.4956	.6230
	6	.0000	.0001	.0014	.0064	.0197	.0473	.0949	.1662	.2616	.3770
	7	.0000	.0000	.0001	.0009	.0035	.0106	.0260	.0548	.1020	.1719
	8	.0000	.0000	.0000	.0001	.0004	.0016	.0048	.0123	.0274	.0547
	9	.0000	.0000	.0000	.0000	.0000	.0001	.0005	.0017	.0045	.0107
	10	.0000	.0000	.0000	.0000	.0000	.0000	.0000	.0001	.0003	.0010
11	1	.4312	.6862	.8327	.9141	.9578	.9802	.9912	.9964	.9986	.9995
	2	.1019	.3026	.5078	.6779	.8029	.8870	.9394	.9698	.9861	.9941
	3	.0152	.0896	.2212	.3826	.5448	.6873	.7999	.8811	.9348	.9673
	4	.0016	.0185	.0694	.1611	.2867	.4304	.5744	.7037	.8089	.8867
	5	.0001	.0028	.0159	.0504	.1146	.2103	.3317	.4672	.6029	.7256
	6	.0000	.0003	.0027	.0117	.0343	.0782	.1487	.2465	.3669	.5000
	7	.0000	.0000	.0003	.0020	.0076	.0216	.0501	.0994	.1738	.2744
	8	.0000	.0000	.0000	.0002	.0012	.0043	.0122	.0293	.0610	.1133
	9	.0000	.0000	.0000	.0000	.0001	.0006	.0020	.0059	.0148	.0327
	10	.0000	.0000	.0000	.0000	.0000	.0000	.0002	.0007	.0022	.0059
	11	.0000	.0000	.0000	.0000	.0000	.0000	.0000	.0000	.0002	.0005
12	1	.4596	.7176	.8578	.9313	.9683	.9862	.9943	.9978	.9992	.9998
	2	.1184	.3410	.5565	.7251	.8416	.9150	.9576	.9804	.9917	.9968
	3	.0196	.1109	.2642	.4417	.6093	.7472	.8487	.9166	.9579	.9807
	4	.0022	.0256	.0922	.2054	.3512	.5075	.6533	.7747	.8655	.9270
	5	.0002	.0043	.0239	.0726	.1576	.2763	.4167	.5618	.6956	.8062
	6	.0000	.0005	.0046	.0194	.0544	.1178	.2127	.3348	.4731	.6128
	7	.0000	.0001	.0007	.0039	.0143	.0386	.0846	.1582	.2607	.3872
	8	.0000	.0000	.0001	.0006	.0028	.0095	.0255	.0573	.1117	.1938
	9	.0000	.0000	.0000	.0001	.0004	.0017	.0056	.0153	.0356	.0730
	10	.0000	.0000	.0000	.0000	.0000	.0002	.0008	.0028	.0079	.0193
	11	.0000	.0000	.0000	.0000	.0000	.0000	.0001	.0003	.0011	.0032
	12	.0000	.0000	.0000	.0000	.0000	.0000	.0000	.0000	.0001	.0002
13	1	.4867	.7458	.8791	.9450	.9762	.9903	.9963	.9987	.9996	.9999
	2	.1354	.3787	.6017	.7664	.8733	.9363	.9704	.9874	.9951	.9983
	3	.0245	.1339	.2704	.4983	.6674	.7975	.8868	.9421	.9731	.9888
	4	.0031	.0342	.0967	.2527	.4157	.5794	.7217	.8314	.9071	.9539
	5	.0003	.0065	.0260	.0991	.2060	.3457	.4995	.6470	.7721	.8666
	6	.0000	.0009	.0053	.0300	.0802	.1654	.2841	.4256	.5732	.7095
	7	.0000	.0001	.0013	.0070	.0243	.0624	.1295	.2288	.3563	.5000
	8	.0000	.0000	.0002	.0012	.0056	.0182	.0462	.0977	.1788	.2905
	9	.0000	.0000	.0000	.0002	.0010	.0040	.0126	.0321	.0698	.1334
	10	.0000	.0000	.0000	.0000	.0001	.0007	.0025	.0078	.0203	.0461
	11	.0000	.0000	.0000	.0000	.0000	.0001	.0003	.0013	.0041	.0112
	12	.0000	.0000	.0000	.0000	.0000	.0000	.0000	.0001	.0005	.0017
	13	.0000	.0000	.0000	.0000	.0000	.0000	.0000	.0000	.0000	.0001
14	1	.5123	.7712	.8972	.9560	.9822	.9932	.9976	.9992	.9998	.9999
	2	.1530	.4154	.6433	.8021	.8990	.9525	.9795	.9919	.9971	.9991
	3	.0301	.1584	.3521	.5519	.7189	.8392	.9161	.9602	.9830	.9935
	4	.0042	.0441	.1465	.3018	.4787	.6448	.7795	.8757	.9368	.9713
	5	.0004	.0092	.0467	.1298	.2585	.4158	.5773	.7207	.8328	.9102
	6	.0000	.0015	.0115	.0439	.1117	.2195	.3595	.5141	.6627	.7880
	7	.0000	.0002	.0022	.0116	.0383	.0933	.1836	.3075	.4539	.6047
	8	.0000	.0000	.0003	.0024	.0103	.0315	.0753	.1501	.2586	.3953
	9	.0000	.0000	.0000	.0004	.0022	.0083	.0243	.0583	.1189	.2120
	10	.0000	.0000	.0000	.0000	.0003	.0017	.0060	.0175	.0426	.0898

Table 2 395

n	x'					p					
		.05	.10	.15	.20	.25	.30	.35	.40	.45	.50
	11	.0000	.0000	.0000	.0000	.0000	.0002	.0011	.0039	.0114	.0287
	12	.0000	.0000	.0000	.0000	.0000	.0000	.0001	.0006	.0022	.0065
	13	.0000	.0000	.0000	.0000	.0000	.0000	.0000	.0001	.0003	.0009
	14	.0000	.0000	.0000	.0000	.0000	.0000	.0000	.0000	.0000	.0001
15	1	.5367	.7941	.9126	.9648	.9866	.9953	.9984	.9995	.9999	1.0000
	2	.1710	.4510	.6814	.8329	.9198	.9647	.9858	.9948	.9983	.9995
	3	.0362	.1841	.3958	.6020	.7639	.8732	.9383	.9729	.9893	.9963
	4	.0055	.0556	.1773	.3518	.5387	.7031	.8273	.9095	.9576	.9824
	5	.0006	.0127	.0617	.1642	.3135	.4845	.6481	.7827	.8796	.9408
	6	.0001	.0022	.0168	.0611	.1484	.2784	.4357	.5968	.7392	.8491
	7	.0000	.0003	.0036	.0181	.0566	.1311	.2452	.3902	.5478	.6964
	8	.0000	.0000	.0006	.0042	.0173	.0500	.1132	.2131	.3465	.5000
	9	.0000	.0000	.0001	.0008	.0042	.0152	.0422	.0950	.1818	.3036
	10	.0000	.0000	.0000	.0001	.0008	.0037	.0124	.0338	.0769	.1509
	11	.0000	.0000	.0000	.0000	.0001	.0007	.0028	.0093	.0255	.0592
	12	.0000	.0000	.0000	.0000	.0000	.0001	.0005	.0019	.0063	.0176
	13	.0000	.0000	.0000	.0000	.0000	.0000	.0001	.0003	.0011	.0037
	14	.0000	.0000	.0000	.0000	.0000	.0000	.0000	.0000	.0001	.0005
	15	.0000	.0000	.0000	.0000	.0000	.0000	.0000	.0000	.0000	.0000
16	1	.5599	.8147	.9257	.9719	.9900	.9967	.9990	.9997	.9999	1.0000
	2	.1892	.4853	.7161	.8593	.9365	.9739	.9902	.9967	.9990	.9997
	3	.0429	.2108	.4386	.6482	.8029	.9006	.9549	.9817	.9934	.9979
	4	.0070	.0684	.2101	.4019	.5950	.7541	.8661	.9349	.9719	.9894
	5	.0009	.0170	.0791	.2018	.3698	.5501	.7108	.8334	.9147	.9616
	6	.0001	.0033	.0235	.0817	.1897	.3402	.5100	.6712	.8024	.8949
	7	.0000	.0005	.0056	.0267	.0796	.1753	.3119	.4728	.6340	.7228
	8	.0000	.0001	.0011	.0070	.0271	.0744	.1594	.2839	.4371	.5982
	9	.0000	.0000	.0002	.0015	.0075	.0257	.0671	.1423	.2559	.4018
	10	.0000	.0000	.0000	.0002	.0016	.0071	.0229	.0583	.1241	.2272
	11	.0000	.0000	.0000	.0000	.0003	.0016	.0062	.0191	.0486	.1051
	12	.0000	.0000	.0000	.0000	.0000	.0003	.0013	.0049	.0149	.0384
	13	.0000	.0000	.0000	.0000	.0000	.0000	.0002	.0009	.0035	.0106
	14	.0000	.0000	.0000	.0000	.0000	.0000	.0000	.0001	.0006	.0021
	15	.0000	.0000	.0000	.0000	.0000	.0000	.0000	.0000	.0001	.0003
	16	.0000	.0000	.0000	.0000	.0000	.0000	.0000	.0000	.0000	.0000
17	1	.5819	.8332	.9369	.9775	.9925	.9977	.9993	.9998	1.0000	1.0000
	2	.2078	.5182	.7475	.8818	.9499	.9807	.9933	.9979	.9994	.9999
	3	.0503	.2382	.4802	.6904	.8363	.9226	.9673	.9877	.9959	.9988
	4	.0088	.0826	.2444	.4511	.6470	.7981	.8972	.9536	.9816	.9936
	5	.0012	.0221	.0987	.2418	.4261	.6113	.7652	.8740	.9404	.9755
	6	.0001	.0047	.0319	.1057	.2347	.4032	.5803	.7361	.8529	.9283
	7	.0000	.0008	.0083	.0377	.1071	.2248	.3812	.5522	.7098	.8338
	8	.0000	.0001	.0017	.0109	.0402	.1046	.2128	.3595	.5257	.6855
	9	.0000	.0000	.0003	.0026	.0124	.0403	.0994	.1989	.3374	.5000
	10	.0000	.0000	.0000	.0005	.0031	.0127	.0383	.0919	.1834	.3145
	11	.0000	.0000	.0000	.0001	.0006	.0032	.0120	.0348	.0826	.1662
	12	.0000	.0000	.0000	.0000	.0001	.0007	.0030	.0106	.0301	.0717
	13	.0000	.0000	.0000	.0000	.0000	.0001	.0006	.0025	.0086	.0245
	14	.0000	.0000	.0000	.0000	.0000	.0000	.0001	.0005	.0019	.0064
	15	.0000	.0000	.0000	.0000	.0000	.0000	.0000	.0001	.0003	.0012
	16	.0000	.0000	.0000	.0000	.0000	.0000	.0000	.0000	.0000	.0001
	17	.0000	.0000	.0000	.0000	.0000	.0000	.0000	.0000	.0000	.0000
18	1	.6028	.8499	.9464	.9820	.9944	.9984	.9996	.9999	1.0000	1.0000
	2	.2265	.5497	.7759	.9009	.9605	.9858	.9954	.9987	.9997	.9999
	3	.0581	.2662	.5203	.7287	.8647	.9400	.9764	.9918	.9975	.9993
	4	.0109	.0982	.2798	.4990	.6943	.8354	.9217	.9672	.9880	.9962
	5	.0015	.0282	.1206	.2836	.4813	.6673	.8114	.9058	.9589	.9846

n	x'	.05	.10	.15	.20	.25	.30	.35	.40	.45	.50
						p					
	6	.0002	.0064	.0419	.1329	.2825	.4656	.6450	.7912	.8923	.9519
	7	.0000	.0012	.0118	.0513	.1390	.2783	.4509	.6257	.7742	.8811
	8	.0000	.0002	.0027	.0163	.0569	.1407	.2717	.4366	.6085	.7597
	9	.0000	.0000	.0005	.0043	.0193	.0596	.1391	.2632	.4222	.5927
	10	.0000	.0000	.0001	.0009	.0054	.0210	.0597	.1347	.2527	.4703
	11	.0000	.0000	.0000	.0002	.0012	.0061	.0212	.0576	.1280	.2403
	12	.0000	.0000	.0000	.0000	.0002	.0014	.0062	.0203	.0537	.1189
	13	.0000	.0000	.0000	.0000	.0000	.0003	.0014	.0058	.0183	.0481
	14	.0000	.0000	.0000	.0000	.0000	.0000	.0003	.0013	.0049	.0154
	15	.0000	.0000	.0000	.0000	.0000	.0000	.0000	.0002	.0010	.0038
	16	.0000	.0000	.0000	.0000	.0000	.0000	.0000	.0000	.0001	.0007
	17	.0000	.0000	.0000	.0000	.0000	.0000	.0000	.0000	.0000	.0001
	18	.0000	.0000	.0000	.0000	.0000	.0000	.0000	.0000	.0000	.0000
19	1	.6226	.8649	.9544	.9856	.9958	.9989	.9997	.9999	1.0000	1.0000
	2	.2453	.5797	.8015	.9171	.9690	.9896	.9969	.9992	.9998	1.0000
	3	.0665	.2946	.5587	.7631	.8887	.9538	.9830	.9945	.9985	.9996
	4	.0132	.1150	.3159	.5449	.7369	.8668	.9409	.9770	.9923	.9978
	5	.0020	.0352	.1444	.3267	.5346	.7178	.8500	.9304	.9720	.9904
	6	.0002	.0086	.0537	.1631	.3322	.5261	.7032	.8371	.9223	.9682
	7	.0000	.0017	.0163	.0676	.1749	.3345	.5188	.6919	.8273	.9165
	8	.0000	.0003	.0041	.0233	.0775	.1820	.3344	.5122	.6831	.8204
	9	.0000	.0000	.0008	.0067	.0287	.0839	.1855	.3325	.5060	.6762
	10	.0000	.0000	.0001	.0016	.0089	.0326	.0875	.1861	.3290	.5000
	11	.0000	.0000	.0000	.0003	.0023	.0105	.0347	.0885	.1841	.3238
	12	.0000	.0000	.0000	.0000	.0005	.0028	.0114	.0352	.0871	.1796
	13	.0000	.0000	.0000	.0000	.0001	.0006	.0031	.0116	.0342	.0835
	14	.0000	.0000	.0000	.0000	.0000	.0001	.0007	.0031	.0109	.0318
	15	.0000	.0000	.0000	.0000	.0000	.0000	.0001	.0026	.0028	.0096
	16	.0000	.0000	.0000	.0000	.0000	.0000	.0000	.0001	.0005	.0022
	17	.0000	.0000	.0000	.0000	.0000	.0000	.0000	.0000	.0001	.0004
	18	.0000	.0000	.0000	.0000	.0000	.0000	.0000	.0000	.0000	.0000
	19	.0000	.0000	.0000	.0000	.0000	.0000	.0000	.0000	.0000	.0000
20	1	.6415	.8784	.9612	.9885	.9968	.9992	.9998	1.0000	1.0000	1.0000
	2	.2642	.6083	.8244	.9308	.9757	.9924	.9979	.9995	.9999	1.0000
	3	.0755	.3231	.5951	.7939	.9087	.9645	.9879	.9964	.9991	.9998
	4	.0159	.1330	.3223	.5886	.7748	.8929	.9556	.9840	.9951	.9987
	5	.0026	.4320	.1702	.3704	.5852	.7625	.8818	.9490	.9811	.9941
	6	.0003	.0113	.0673	.1958	.3828	.5836	.7546	.8744	.9447	.9793
	7	.0000	.0024	.0219	.0867	.2142	.3920	.5843	.7500	.8701	.9423
	8	.0000	.0004	.0059	.0321	.1018	.2277	.3990	.5841	.7480	.8684
	9	.0000	.0001	.0013	.0100	.0409	.1133	.2376	.4044	.5857	.7483
	10	.0000	.0000	.0002	.0026	.0139	.0480	.1218	.2447	.4086	.5881
	11	.0000	.0000	.0000	.0006	.0039	.0171	.0532	.1275	.2493	.4119
	12	.0000	.0000	.0000	.0001	.0009	.0051	.0196	.0565	.1308	.2517
	13	.0000	.0000	.0000	.0000	.0002	.0013	.0060	.0210	.0580	.1316
	14	.0000	.0000	.0000	.0000	.0000	.0003	.0015	.0065	.0214	.0577
	15	.0000	.0000	.0000	.0000	.0000	.0000	.0003	.0016	.0064	.0207
	16	.0000	.0000	.0000	.0000	.0000	.0000	.0000	.0003	.0015	.0059
	17	.0000	.0000	.0000	.0000	.0000	.0000	.0000	.0000	.0003	.0013
	18	.0000	.0000	.0000	.0000	.0000	.0000	.0000	.0000	.0000	.0002
	19	.0000	.0000	.0000	.0000	.0000	.0000	.0000	.0000	.0000	.0000
	20	.0000	.0000	.0000	.0000	.0000	.0000	.0000	.0000	.0000	.0000

Table 3

397

Table 3 ·Poisson Distribution Function, $1 - F(x-1) = \sum_{r=z}^{r=\infty} (e^{-a}a^r/r!)$

x	$a = 0.2$	$a = 0.3$	$a = 0.4$	$a = 0.5$	$a = 0.6$
0	1.0000000	1.0000000	1.0000000	1.0000000	1.0000000
1	.1812692	.2591818	.3296800	.393469	.451188
2	.0175231	.0369363	.0615519	.090204	.121901
3	.0011485	.0035995	.0079263	.014388	.023115
4	.0000568	.0002658	.0007763	.001752	.003358
5	.0000023	.0000158	.0000612	.000172	.000394
6	.0000001	.0000008	.0000040	.000014	.000039
7			.0000002	.000001	.000003

x	$a = 0.7$	$a = 0.8$	$a = 0.9$	$a = 1.0$	$a = 1.2$
0	1.0000000	1.0000000	1.0000000	1.0000000	1.0000000
1	.503415	.550671	.593430	.632121	.698806
2	.155805	.191208	.227518	.264241	.337373
3	.034142	.047423	.062857	.080301	.120513
4	.005753	.009080	.013459	.018988	.033769
					.007746
5	.000786	.001411	.002344	.003660	
6	.000090	.000184	.000343	.000594	.001500
7	.000009	.000021	.000043	.000083	.000251
8	.000001	.000002	.000005	.000010	.000037
9				.000001	.000005
10					.000001

x	$a = 1.4$	$a = 1.6$	$a = 1.8$
0	1.000000	1.000000	1.000000
1	.753403	.798103	.834701
2	.408167	.475069	.537163
3	.166502	.216642	.269379
4	.053725	.078813	.108708
5	.014253	.023682	.036407
6	.003201	.006040	.010378
7	.000622	.001336	.002569
8	.000107	.00260	.005602
9	.000016	.000045	.000110
10	.000002	.000007	.000019
11		.000001	.000003

Table 3

x	a = 2.5	a = 3.0	a = 3.5	a = 4.0	a = 4.5	a = 5.0
0	1.000000	1.000000	1.000000	1.000000	1.000000	1.000000
1	.917915	.950213	.969803	.981684	.988891	.993262
2	.712703	.800852	.864112	.908422	.938901	.959572
3	.456187	.576810	.679153	.761897	.826422	.875348
4	.242424	.352768	.463367	.566530	.657704	.734974
5	.108822	.184737	.274555	.371163	.467896	.559507
6	.042021	.083918	.142386	.214870	.297070	.384039
7	.014187	.033509	.065288	.110674	.168949	.237817
8	.004247	.011905	.026739	.051134	.086586	.1333 2
9	.001140	.003803	.009874	.021363	.040257	.068094
10	.000277	.001102	.003315	.008132	.017093	.031828
11	.000062	.000292	.001019	.002840	.006669	.013695
12	.000013	.000071	.000289	.000915	.002404	.005453
13	.000002	.000016	.000076	.000274	.000805	.002019
14		.000003	.000019	.000076	.000252	.000698
15		.000001	.000004	.000020	.000074	.000226
16			.000001	.000005	.000020	.000069
17				.000001	.000005	.000020
18					.000001	.000005
19						.000001

Table 4 ·Ordinates of the Normal Curve,

$$\phi(z) = \frac{1}{\sqrt{2\pi}}\, e^{-1/2 z^2}$$

(TO FOUR DECIMAL PLACES)

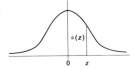

Z	.00	.01	.02	.03	.04	.05	.06	.07	.08	.09
.0	.3989	.3989	.3989	.3988	.3986	.3984	.3982	.3980	.3977	.3973
.1	.3970	.3965	.3961	.3956	.3951	.3945	.3939	.3932	.3925	.3918
.2	.3910	.3902	.3894	.3885	.3876	.3867	.3857	.3847	.3836	.3825
.3	.3814	.3802	.3790	.3778	.3765	.3752	.3739	.3725	.3712	.3697
.4	.3683	.3668	.3653	.3637	.3621	.3605	.3589	.3572	.3555	.3538
.5	.3521	.3503	.3485	.3467	.3448	.3429	.3410	.3391	.3372	.3352
.6	.3332	.3312	.3292	.3271	.3251	.3230	.3209	.3187	.3166	.3144
.7	.3123	.3101	.3079	.3056	.3034	.3011	.2989	.2966	.2943	.2920
.8	.2897	.2874	.2850	.2827	.2803	.2780	.2756	.2732	.2709	.2685
.9	.2661	.2637	.2613	.2589	.2565	.2541	.2516	.2492	.2468	.2244
1.0	.2420	.2396	.2371	.2347	.2323	.2299	.2275	.2251	.2227	.2203
1.1	.2179	.2155	.2131	.2107	.2083	.2059	.2036	.2012	.1989	.1965
1.2	.1942	.1919	.1895	.1872	.1849	.1826	.1804	.1781	.1758	.1736
1.3	.1714	.1691	.1669	.1647	.1626	.1604	.1582	.1561	.1539	.1518
1.4	.1497	.1476	.1456	.1435	.1415	.1394	.1374	.1354	.1334	.1315
1.5	.1295	.1276	.1257	.1238	.1219	.1200	.1182	.1163	.1145	.1127
1.6	.1109	.1092	.1074	.1057	.1040	.1023	.1006	.0989	.0973	.0957
1.7	.0940	.0925	.0909	.0893	.0878	.0863	.0848	.0833	.0818	.0804
1.8	.0790	.0775	.0761	.0748	.0734	.0721	.0707	.0694	.0681	.0769
1.9	.0656	.0644	.0632	.0620	.0608	.0596	.0584	.0573	.0562	.0551
2.0	.0540	.0529	.0519	.0508	.0498	.0488	.0478	.0468	.0459	.0449
2.1	.0440	.0431	.0422	.0413	.0404	.0396	.0387	.0397	.0371	.0363
2.2	.0355	.0347	.0339	.0332	.0325	.0317	.0310	.0303	.0297	.0290
2.3	.0283	.0277	.0270	.0264	.0258	.0252	.0246	.0241	.0235	.0229
2.4	.0024	.0219	.0213	.0208	.0203	.0198	.0194	.0189	.0184	.0180
2.5	.0175	.0171	.0167	.0163	.0158	.0154	.0151	.0147	.0143	.0139
2.6	.0136	.0132	.0129	.0126	.0122	.0119	.0116	.0113	.0110	.0107
2.7	.0104	.0101	.0099	.0096	.0093	.0091	.0088	.0086	.0084	.0081
2.8	.0079	.0077	.0075	.0073	.0071	.0069	.0067	.0065	.0063	.0061
2.9	.0060	.0058	.0056	.0055	.0053	.0051	.0050	.0048	.0047	.0046
3.0	.0044	.0043	.0042	.0040	.0039	.0038	.0037	.0036	.0035	.0034
3.1	.0033	.0032	.0031	.0030	.0029	.0028	.0027	.0026	.0025	.0025
3.2	.0024	.0023	.0022	.0022	.0021	.0020	.0020	.0019	.0018	.0018
3.3	.0017	.0017	.0016	.0016	.0015	.0015	.0014	.0014	.0013	.0013
3.4	.0012	.0012	.0012	.0011	.0011	.0010	.0010	.0010	.0009	.0009
3.5	.0009	.0008	.0008	.0008	.0008	.0007	.0007	.0007	.0007	.0006
3.6	.0006	.0006	.0006	.0005	.0005	.0005	.0005	.0005	.0005	.0004
3.7	.0004	.0004	.0004	.0004	.0004	.0004	.0003	.0003	.0003	.0003
3.8	.0003	.0003	.0003	.0003	.0003	.0002	.0002	.0002	.0002	.0002
3.9	.0002	.0002	.0002	.0002	.0002	.0002	.0002	.0002	.0001	.0001
4.0	.0001	.0001	.0001	.0001	.0001	.0001	.0001	.0001	.0001	.0001
z	.00	.01	.02	.03	.04	.05	.06	.07	.08	.09

Reprinted by permission of Prentice-Hall, Inc., Englewood Cliffs, N.J., © 1961, from Elmer B. Mode, *Elements of Statistics*, 3rd ed.

Table 5 · **Standard Normal Distribution Function** $\Phi(z) = \int_{-\infty}^{z} \frac{1}{\sqrt{2\pi}} e^{-u^2/2} \, du = P(Z \le z)$

Z	0	1	2	3	4	5	6	7	8	9
0.0	0.5000	0.5040	0.5080	0.5120	0.5160	0.5199	0.5239	0.5279	0.5319	0.5359
0.1	0.5398	0.5438	0.5478	0.5517	0.5557	0.5596	0.5636	0.5675	0.5714	0.5753
0.2	0.5793	0.5832	0.5871	0.5910	0.5948	0.5987	0.6026	0.6064	0.6103	0.6141
0.3	0.6719	0.6217	0.6255	0.6293	0.6331	0.6368	0.6406	0.6443	0.6480	0.6517
0.4	0.6554	0.6591	0.6628	0.6664	0.6700	0.6736	0.6772	0.6808	0.6844	0.6879
0.5	0.6915	0.6950	0.6985	0.7019	0.7054	0.7088	0.7123	0.7157	0.7190	0.7224
0.6	0.7257	0.7291	0.7324	0.7357	0.7389	0.7422	0.7454	0.7486	0.7517	0.7549
0.7	0.7580	0.7611	0.7642	0.7673	0.7703	0.7734	0.7764	0.7794	0.7823	0.7852
0.8	0.7881	0.7910	0.7939	0.7967	0.7995	0.8023	0.8051	0.8078	0.8106	0.8133
0.9	0.8159	0.8186	0.8212	0.8238	0.8264	0.8289	0.8315	0.8340	0.8365	0.8389
1.0	0.8413	0.8438	0.8461	0.8485	0.8508	0.8531	0.8554	0.8577	0.8599	0.8621
1.1	0.8643	0.8665	0.8686	0.8708	0.8729	0.8749	0.8770	0.8790	0.8810	0.8830
1.2	0.8849	0.8869	0.8888	0.8907	0.8925	0.8944	0.8962	0.8980	0.8997	0.9015
1.3	0.9032	0.9049	0.9066	0.9082	0.9099	0.9115	0.9131	0.9147	0.9162	0.9177
1.4	0.9192	0.9207	0.9222	0.9236	0.9251	0.9265	0.9278	0.9292	0.9306	0.9319
1.5	0.9332	0.9345	0.9357	0.9370	0.9382	0.9394	0.9406	0.9418	0.9430	0.9441
1.6	0.9452	0.9463	0.9474	0.9484	0.9495	0.9505	0.9515	0.9525	0.9535	0.9545
1.7	0.9554	0.9564	0.9573	0.9582	0.9591	0.9599	0.9608	0.9616	0.9625	0.9633
1.8	0.9641	0.9648	0.9656	0.9664	0.9671	0.9678	0.9686	0.9693	0.9700	0.9706
1.9	0.9713	0.9719	0.9726	0.9732	0.9738	0.9744	0.9750	0.9756	0.9762	0.9767
2.0	0.9772	0.9778	0.9783	0.9788	0.9793	0.9798	0.9803	0.9808	0.9812	0.9817
2.1	0.9821	0.9826	0.9830	0.9834	0.9838	0.9842	0.9846	0.9850	0.9854	0.9857
2.2	0.9861	0.9864	0.9868	0.9871	0.9874	0.9878	0.9881	0.9884	0.9887	0.9890
2.3	0.9893	0.9896	0.9898	0.9901	0.9904	0.0906	0.9909	0.9911	0.9913	0.9916
2.4	0.9918	0.9920	0.9922	0.9925	0.9927	0.9929	0.9931	0.9932	0.9934	0.9936
2.5	0.9938	0.9940	0.9941	0.9943	0.9945	0.9946	0.9948	0.9949	0.9951	0.9952
2.6	0.9953	0.9955	0.9956	0.9957	0.9959	0.9960	0.9961	0.9962	0.9963	0.9964
2.7	0.9965	0.9966	0.9967	0.9968	0.9969	0.9970	0.9971	0.9972	0.9973	0.9974
2.8	0.9974	0.9975	0.9976	0.9977	0.9977	0.9978	0.9979	0.9979	0.9980	0.9981
2.9	0.9981	0.9982	0.9982	0.9983	0.9984	0.9984	0.9985	0.9985	0.9986	0.9986
3.0	0.9987	0.9990	0.9993	0.9995	0.9997	0.9998	0.9998	0.9999	0.9999	1.0000

Table 6

401

Table 6 · Exponential Function

x	e^x	e^{-x}
.00	1.000	1.000
.01	1.010	.990
.02	1.020	.980
.03	1.031	.970
.04	1.041	.960
.05	1.051	.951
.06	1.062	.942
.07	1.073	.932
.08	1.083	.923
.09	1.094	.914
.10	1.105	.905
.11	1.116	.896
.12	1.128	.887
.13	1.139	.878
.14	1.150	.869
.15	1.162	.861
.16	1.174	.852
.17	1.185	.844
.18	1.197	.835
.19	1.209	.827
.20	1.221	.819
.21	1.234	.811
.22	1.246	.802
.23	1.259	.795
.24	1.271	.787
.25	1.284	.779
.26	1.297	.771
.27	1.310	.763
.28	1.323	.756
.29	1.336	.748
.30	1.350	.741
.31	1.363	.733
.32	1.377	.726
.33	1.391	.719
.34	1.405	.712
.35	1.419	.705
.36	1.433	.698
.37	1.478	.691
.38	1.462	.684
.39	1.477	.677

Table 6

x	e^x	e^{-x}
.40	1.492	.670
.41	1.507	.664
.42	1.522	.657
.43	1.537	.651
.44	1.553	.644
.45	1.568	638
.46	1.584	.631
.47	1.600	.625
.48	1.616	.619
.49	1.632	.613
.50	1.649	.607
.51	1.665	.601
.52	1.682	.595
.53	1.699	.589
.54	1.716	.583
.55	1.733	.577
.56	1.751	.571
.47	1.768	.566
.58	1.786	.560
.59	1.804	.554
.60	1.822	.549
.61	1.840	.543
.62	1.859	.538
.63	1.878	.533
.64	1.897	.527
.65	1.916	.522
.66	1.935	.517
.67	1.954	.512
.68	1.974	.507
.69	1.994	.502
.70	2.014	.497
.71	2.034	.492
.72	2.054	.487
.73	2.075	.482
.74	2.096	.477
.75	2.117	.472
.76	2.138	.468
.77	2.160	.463
.78	2.181	.458
.79	2.203	.453

Table 6

403

x	e^x	e^{-x}
.80	2.226	.449
.81	2.248	.445
.82	2.271	.440
.83	2.293	.436
.84	2.316	.432
.85	2.340	.427
.86	2.363	.423
.87	2.387	.419
.88	2.411	.415
.89	2.435	.411
.90	2.460	.407
.91	2.484	.403
.92	2.509	.399
.93	2.535	.395
.94	2.560	.391
.95	2.586	.387
.96	2.612	.383
.97	2.638	.379
.98	2.665	.375
.99	2.691	.372
1.0	2.718	.368
1.1	3.004	.333
1.2	3.320	.301
1.3	3.669	.273
1.4	4.055	.247
1.5	4.482	.223
1.6	4.953	.202
1.7	5.474	.183
1.8	6.050	.165
1.9	6.686	.150
2.0	7.389	.135
2.1	8.166	.122
2.2	9.025	.111
2.3	9.974	.100
2.4	11.023	.091
2.5	12.182	.082
2.6	13.464	.074
2.7	14.880	.067
2.8	16.445	.061
2.9	18.174	.055

Table 6

x	e^x	e^{-x}
3.0	20.086	.050
3.1	22.198	.045
3.2	24.533	.041
3.3	27.113	.037
3.4	29.964	.033
3.5	33.115	.030
3.6	36.598	.027
3.7	40.447	.025
3.8	44.701	.002
3.9	49.402	.020

Table 7 405

Table 7 ·Natural Logarithms of Numbers—0.00 to 4.49 (Base $e = 2.718 \ldots$)

N		0	1	2	3	4	5	6	7	8	9
0.0			5.395	6.088	6.493	6.781	7.004	7.187	7.341	7.474	7.592
0.1		7.697	7.793	7.880	7.960	8.034	8.103	8.167	8.228	8.285	8.339
0.2		8.391	8.439	8.486	8.530	8.573	8.614	8.653	8.691	8.727	8.762
0.3		8.796	8.829	8.861	8.891	8.921	8.950	8.978	9.006	9.032	9.058
0.4		9.084	9.108	9.132	9.156	9.179	9.201	9.223	9.245	9.266	9.287
0.5		9.307	9.327	9.346	9.365	9.384	9.402	9.420	9.438	9.455	9.472
0.6		9.489	9.506	9.522	9.538	9.554	9.569	9.584	9.600	9.614	9.629
0.7		9.643	9.658	9.671	9.685	9.699	9.712	9.726	9.739	9.752	9.764
0.8		9.777	9.789	9.802	9.814	9.826	9.837	9.849	9.861	9.872	9.883
0.9		9.895	0.906	9.917	9.927	9.938	9.949	9.959	9.970	9.980	9.990
1.0	0.0	0000	0995	1980	2956	3922	4879	5827	6766	7696	8618
1.1		9531	*0436	*1333	*2222	*3103	*3976	*4842	*5700	*6551	*7395
1.2	0.1	8232	9062	9885	*0701	*1551	*2314	*3111	*3902	*4686	*5464
1.3	0.2	6236	7003	7763	8518	9267	*0010	*0748	*1481	*2208	*2930
1.4	0.3	3647	4359	5066	5767	6464	7156	7844	8526	9204	9878
1.5	0.4	0547	1211	1871	2527	3178	3825	4469	5108	5742	6373
1.6		7000	7623	8243	8858	9470	* 0078	*0682	*1282	*1879	*2473
1.7	0.5	3063	3649	4232	4812	5389	5962	6531	7098	7661	8222
1.8		8779	9333	9884	*0432	*0977	*1519	*2058	*2594	*3127	*3658
1.9	0.6	4185	4710	5233	5752	6269	6783	7294	7803	8310	8813
2.0		9315	9813	*0310	*0804	*1295	*1784	*2271	*2755	*3237	*3716
2.1	0.7	4194	4669	5142	5612	6081	6547	7011	7473	7932	8390
2.2		8846	9299	9751	*0200	*0648	*1093	*1536	*1978	*2418	*2855
2.3	0.8	3291	3725	4157	4587	5015	5442	5866	6289	6710	7129
2.4		7547	7963	8377	8789	9200	9609	*0016	*0422	*0826	*1228
2.5	0.9	1629	2028	2426	2822	3216	3609	4001	4391	4779	5166
2.6		5551	5935	6317	6698	7078	7456	7833	8208	8582	8954
2.7		9325	9695	*0063	*0430	*0796	*1160	*1523	*1885	*2245	*2604
2.8	1.0	2962	3318	3674	4028	4380	4732	5082	5431	5779	6126
2.9		6471	6815	7158	7500	7841	8181	8519	8856	9192	9527
3.0		9861	*0194	*0526	*0856	*1186	*1514	*1841	*2168	*2493	*2817
3.1	1.1	3140	3462	3783	4103	4422	4740	5057	5373	5688	6002
3.2		6315	6627	6938	7248	7557	7865	8173	847	8784	9089
3.3		9392	9695	9996	*0297	*0597	*0896	*1194	*1491	*1788	*2083
3.4	1.2	2378	2671	2964	3256	3547	3837	4127	4415	4703	4990
3.5		5276	5562	5846	6130	6413	6695	6976	7257	7536	7815
3.6		8093	8371	8647	8923	9198	9473	9746	*0019	*0291	*0563
3.7	1.3	0833	1103	1372	1641	1909	2176	2442	2708	2972	3237
3.8		3500	3763	4025	4286	4547	4807	5067	5325	5584	5841
3.9		6098	6354	6609	6864	7118	7372	7624	7877	8128	8379
4.0		8629	8879	9128	9377	9624	9872	*0118	*0364	*0610	*0854
4.1	1.4	1099	1342	1585	1828	2070	2311	2552	2792	3031	3270
4.2		3508	3746	3984	4220	4456	4692	4927	5161	5395	5629
4.3		5862	6094	6326	6557	6787	7018	7247	7476	7705	7933
4.4		8160	8387	8614	8840	9065	9290	9515	9739	9962	*0185
N		0	1	2	3	4	5	6	7	8	9

$$\log_e 0.10 = 7.69741 \ 49070 - 10$$

N		0	1	2	3	4	5	6	7	8	9
4.5	1.5	0408	0630	0851	1072	1293	1513	1732	1951	2170	2388
4.6		2606	2823	3039	3256	3471	3687	3902	4116	4330	4543
4.7		4756	4969	5181	5393	6041	5184	6025	6235	6444	6653
4.8		6862	7070	7277	7485	7691	7898	8104	8309	8515	8719
4.9		8924	9127	9331	9534	9737	9939	*0141	*0342	*0543	*0744
5.0	1.6	0944	1144	1343	1542	1741	1939	2137	2334	2531	2728
5.1		2924	3120	3315	3511	3705	3900	4094	4287	4481	4673
5.2		4866	5058	5250	5441	5632	5823	6013	6203	6393	6582
5.3		6771	6959	7147	7335	7523	7710	7896	8083	8269	8455
5.4		8640	8825	9010	9194	9378	9562	9745	9928	*0111	*0293
5.5	1.7	0475	0656	0838	1019	1199	1380	1560	1740	1919	2093
5.6		2277	2455	2633	2811	2988	3166	3342	3519	3695	3871
5.7		4047	4222	4397	4572	4746	4920	5094	5267	5440	5361
5.8		5786	5958	6130	6302	6473	6644	6815	6985	7156	7336
5.9		7495	7665	7834	8002	8171	8339	8507	8675	8842	9009
6.0	1.7	9176	9342	9509	9675	9840	*0006	*0171	*0336	*0500	*0665
6.1	1.8	0829	0993	1156	1319	1482	1645	1808	1970	2132	2294
6.2		2455	2616	2777	2938	3098	3258	3418	3578	3737	3896
6.3		4055	4214	4372	4530	4688	4845	5003	5160	5317	5473
6.4		5630	5786	5942	6097	6253	6408	6563	6718	6872	7026
6.5		7180	7334	7487	7641	7794	7947	8099	8251	8403	8555
6.6		8707	8858	9010	9160	9311	9462	9612	9762	9912	*0061
6.7	1.9	0211	0360	0509	0658	0806	0954	1102	1250	1398	1545
6.8		1692	1839	1986	2132	2279	2425	2571	2716	2862	3007
6.9		3152	3297	3442	3586	3730	3874	4018	4162	4305	4448
7.0		4591	4734	4876	5019	5161	5303	5445	5586	5727	5869
7.1		6009	6150	6291	6431	6571	6711	6851	6991	7130	7269
7.2		7408	7547	7685	7824	7962	8100	8238	8376	8513	8650
7.3		8787	8924	9061	9198	9334	9570	9606	9742	9877	*0013
7.4	2.0	0.148	0283	0418	0553	0687	0821	0956	1089	1223	1357
7.5		1490	1624	1757	1890	2022	2155	2287	2419	2551	2683
7.6		2815	2946	3078	3209	3340	3471	3601	3732	3862	3992
7.7		4122	4252	4381	4511	4640	4769	4898	5027	5156	5284
7.8		5412	5540	5668	5796	5924	6051	6179	6303	6433	6560
7.9		6686	6813	6939	7065	7191	7317	7443	7568	7694	7819
8.0		7944	8069	8194	8318	8443	8567	8691	8815	8938	9063
8.1		9186	9310	9433	9556	9679	9802	*9924	*0047	*0169	*0291
8.2	2.1	0413	0535	0657	0779	0900	1021	1142	1263	1384	1505
8.3		1626	1746	1866	1986	2106	2226	2346	2465	2585	2704
8.4		2823	2942	3061	3180	3298	3417	3535	3653	3771	3889
8.5		4007	4124	4242	4359	4476	4593	4710	4827	4943	5060
8.6		5176	5292	5409	5524	5640	5756	5871	5987	6102	6217
8.7		6332	6447	6562	6677	6791	6905	7020	7134	7248	7361
8.8		7475	7589	7702	7816	7929	8042	8155	8267	8380	8493
8.9		8605	8717	8830	8942	9054	9165	9277	9389	9500	8611
9.0		9722	9834	9944	*0055	*0166	*0276	*0387	*0497	*0607	*0717
9.1	2.2	0827	0937	1047	1157	1266	1375	1485	1594	1703	1812
9.2		1920	2029	2138	2246	2354	2462	2570	2678	2786	2894
9.3		3001	3109	3216	3324	3431	3538	3645	3751	3858	3965
9.4		4071	4177	4284	4390	4496	4601	4707	4813	4918	5024
9.5		5129	5234	5339	5444	5549	5654	5759	5863	5968	6072
9.6		6176	6280	6385	6488	6592	6696	6799	6903	7006	7109
9.7		7213	7316	7419	7521	7624	7727	7829	7932	8034	8136
9.8		8238	8340	8442	8544	8646	8747	8849	8950	9051	9152
9.9		9253	9354	9455	9556	9657	9757	9858	9958	*0058	*0158
10.0	2.3	0259	0358	0458	0558	0658	0757	0857	0956	1055	1154
N		0	1	2	3	4	5	6	7	8	9

Table 7

Natural Logarithms of Numbers—10 to 99

N	0	1	2	3	4	5	6	7	8	9
1	2.30259	39790	48491	56495	63906	70805	77259	83321	89037	94444
2	99573	*04452	*09104	*13549	*17805	*21888	*25810	*29584	*33220	*36730
3	3.40120	43399	46574	49651	52636	55535	58352	61092	63759	66356
4	68888	71357	73767	76120	78419	80666	82864	85015	87120	89182
5	91202	93183	95124	97029	98898	*00733	*02535	*04305	*06044	*07754
6	4.09434	11087	12713	14313	15888	17439	18965	20469	21951	23411
7	24850	26268	27667	29046	30407	31749	33073	34381	35671	36945
8	38203	39445	40672	41884	43082	44265	45435	46591	47734	48864
9	49981	51086	52179	53260	54329	55388	56435	47471	58497	59512

$$\log_e 10 = 2.30258\ 50930$$

Table 8 · Binomial Coefficients

n	$\binom{n}{0}$	$\binom{n}{1}$	$\binom{n}{2}$	$\binom{n}{3}$	$\binom{n}{4}$	$\binom{n}{5}$	$\binom{n}{6}$	$\binom{n}{7}$	$\binom{n}{8}$	$\binom{n}{9}$	$\binom{n}{10}$
0	1										
1	1	1									
2	1	2	1								
3	1	3	3	1							
4	1	4	6	4	1						
5	1	5	10	10	5	1					
6	1	6	15	20	15	6	1				
7	1	7	21	35	35	21	7	1			
8	1	8	28	56	70	56	28	8	1		
9	1	9	36	84	126	126	84	36	9	1	
10	1	10	45	120	210	252	210	120	45	10	1
11	1	11	55	165	330	462	462	330	165	55	11
12	1	12	66	220	495	792	924	792	495	220	66
13	1	13	78	286	715	1287	1716	1716	1287	715	286
14	1	14	91	364	1001	2002	3003	3432	3003	2002	1001
15	1	15	105	455	1365	3003	5005	6435	6435	5005	3003
16	1	16	120	560	1820	4368	8008	11440	12870	11440	8008
17	1	17	136	680	2380	6188	12376	19448	24310	24310	19448
18	1	18	153	816	3060	8568	18564	31824	43758	48620	43758
19	1	19	171	969	3876	11628	27132	50388	75582	92378	92378
20	1	20	190	1140	4845	15504	38760	77520	125970	167960	184756

Factorials and Their Reciprocals

n	$n!$	n	$n!$	n	$1/n!$	n	$1/n!$
1	1	11	39916800	1	1.	11	$.25052 \times 10^{-7}$
2	2	12	479001600	2	0.5	12	$.20877 \times 10^{-8}$
3	6	13	6227020800	3	.16667	13	$.16059 \times 10^{-9}$
4	24	14	87178291200	4	$.41667 \times 10^{-1}$	14	$.11471 \times 10^{-10}$
5	120	15	1307674368000	5	$.83333 \times 10^{-2}$	15	$.76472 \times 10^{-12}$
6	720	16	20922789888000	6	$.13889 \times 10^{-2}$	16	$.47795 \times 10^{-13}$
7	5040	17	355687428096000	7	$.19841 \times 10^{-3}$	17	$.28115 \times 10^{-14}$
8	40320	18	6402373705728000	8	$.24802 \times 10^{-4}$	18	$.15619 \times 10^{-15}$
9	362880	19	121645100408832000	9	$.27557 \times 10^{-5}$	19	$.82206 \times 10^{-17}$
10	3628800	20	2432902008176640000	10	$.27557 \times 10^{-6}$	20	$.41103 \times 10^{-18}$

Table 9   **409**

Table 9 ·Basic Integration Formulas

1. $\int x^r \, dx = \dfrac{x^{r+1}}{r+1}, \; r \neq -1$

2. $\int \sin x \, dx = -\cos x$

3. $\int \cos x \, dx = \sin x$

4. $\int \sec^2 x \, dx = \tan x$

5. $\int \csc^2 x \, dx = -\cot x$

6. $\int \sec x \tan x \, dx = \sec x$

7. $\int \csc x \cot x \, dx = -\cos x$

8. $\int e^x \, dx = e^x$

9. $\int \dfrac{1}{x} \, dx = \ln |x|$

10. $\int a^x \, dx = \dfrac{a^x}{\ln a}, \; a > 0, \, a \neq 1$

11. $\int \dfrac{1}{\sqrt{a^2 - x^2}} \, dx = \arcsin(x/a), \; a > 0$

12. $\int \dfrac{1}{a^2 + x^2} \, dx = \arctan (x/a), \; a > 0$

Additional Integration Formulas

13. $\int \dfrac{1}{\sqrt{a^2 - x^2}} \, dx = \ln|x + \sqrt{x^2 + a^2}|$

14. $\int \dfrac{1}{\sqrt{x^2 - a^2}} \, dx = \ln|x + \sqrt{x^2 - a^2}|$

15. $\int \dfrac{1}{a^2 - x^2} \, dx = \dfrac{1}{2a} \ln \left| \dfrac{a+x}{a-x} \right|, \; x^2 < a^2$

16. $\int \dfrac{1}{x^2 - a^2} \, dx = -\dfrac{1}{2a} \ln \left| \dfrac{x+a}{x-a} \right|, \; a^2 < x^2$

17. $\int \dfrac{1}{x\sqrt{a^2 - x^2}} \, dx = -\dfrac{1}{a} \ln \left| \dfrac{a + \sqrt{a^2 - x^2}}{x} \right|, \; 0 < x < a$

Additional Integration Formulas

18. $\int \dfrac{1}{x\sqrt{a^2+x^2}}\,dx = -\dfrac{1}{a}\ln\left|\dfrac{a+\sqrt{a^2+x^2}}{x}\right|$

19. $\int \ln|x|\,dx = x(\ln|x|-1)$

20. $\int \dfrac{x}{ax+b}\,dx = \dfrac{x}{a}-\dfrac{b}{a^2}\ln|ax+b|$

21. $\int \dfrac{x}{(ax+b)^2}\,dx = \dfrac{b}{a^2(ax+b)}+\dfrac{1}{a^2}\ln|ax+b|$

22. $\int \dfrac{1}{x(ax+b)}\,dx = \dfrac{1}{b}\ln\left|\dfrac{x}{ax+b}\right|$

23. $\int \dfrac{1}{x(ax+b)^2}\,dx = \dfrac{1}{b(ax+b)}+\dfrac{1}{b^2}\ln\left|\dfrac{x}{ax+b}\right|$

24. $\int \sqrt{a^2-x^2}\,dx = \dfrac{x}{2}\sqrt{a^2-x^2}+\dfrac{a^2}{2}\arcsin\dfrac{x}{a}$

25. $\int \sqrt{x^2\pm a^2}\,dx = \tfrac{1}{2}x\sqrt{x^2\pm a^2}+\dfrac{a^2}{2}\ln|x+\sqrt{x^2\pm a^2}|$

26. $\int x^2\sqrt{a^2-x^2}\,dx = -\tfrac{1}{4}x(a^2-x^2)^{3/2}+\tfrac{1}{8}a^2x\sqrt{a^2-x^2}+\tfrac{1}{8}a^4\arcsin\dfrac{x}{a}$

27. $\int \tan x\,dx = \ln|\sec x|$

28. $\int \cot x\,dx = \ln|\sin x|$

29. $\int \sec x\,dx = \ln|\sec c+\tan x|$

30. $\int \csc x\,dx = \ln|\csc x-\cot x|$

31. $\int x^n\ln x\,dx = x^{n+1}\left(\dfrac{\ln x}{n+1}-\dfrac{1}{(n+1)^2}\right),\ n\neq -1$

32. $\int x^n e^{ax}\,dx = \dfrac{1}{a}x^n e^{ax}-\dfrac{n}{a}\int x^{n-1}e^{ax}\,dx$

33. $\int e^{ax}\sin bx\,dx = \dfrac{e^{ax}}{a^2+b^2}(a\sin bx-b\cos bx)$

34. $\int e^{ax}\cos bx\,dx = \dfrac{e^{ax}}{a^2+b^2}(b\sin bx+a\cos bx)$

Answers to Selected Problems

Chapter One

Section 1.1

1. (a) Number of pieces of mail that will be handled each day
   (b) Number of cars that will pass a light 8–9 A.M.
   (c) Number of defectives
   (d) Daily stock market sales
   (e) Elapsed time for a rat to run a maze
   (f) Number of television sets sold each week
   (g) Number of days it rains during growing season
5. (a) Yes
   (b) Yes (women unpredictable)
   (c) Yes or no depending on selection process
   (d) Yes
   (e) No

Section 1.2

1. (a) Number of
   (b) Alike or different $\{A, D\}$
3. $\{(3,0), (2,1), (1, 2), (0, 3)\}$
5. Large tribes are A, B; small tribes are a, b, c; some possible alignments are aA abA acA; eight possibilities in all bcB cB bB
7. $\{(0, 0), (0, 1), (0, 2), (1, 0), (1, 1), (1, 2), (2, 0), (2, 1), (2, 2)\}$
9. Set of all pairs (a, b); $a = 1, 2, ..., 6$; $b = 1, 2, ..., 6$; 36 possibilities in all
11. All the integers from 1, 49
13. (a) $\{0, 1, 2, 3, 4, 5\}$ and $\{0, 1, 2, 3\}$ (b) $\{0, 1, ..., 8\}$
15. One space is 27 triples of the form (R, W, W) (i.e., red, white, white); another is 7 sets: $\{R\}$, $\{W\}$, $\{Y\}$, $\{R, W\}$, $\{R, Y\}$, $\{Y, W\}$, $\{R, W, Y\}$; there are others
17. Number of heads for penny, tails for nickel; $S = \{(0, 0), (0, 1), (0, 2), (0, 3), (1, 0), (1, 1), (1, 2), (1, 3), (2, 0), (2, 1), (2, 2), (2, 3)\}$

19. Number of balls in cell 2; $S = \{0, 1, 2\}$
21. $S = \{(x, y, z)\} \mid x, y, z$ are distinct people chosen from given 500$\}$
23. $S = \{0, 1, 2, ...\}$

Section 1.3

1. (a) $E = \{(x, y) \mid x + y = 2, 4, 6, 8, 10, 12\}$
   (b) $E = \{(x, y) \mid x + y = 4\}$
   (c) $E = \{(x, y) \mid x, y = 2, 4,$ or $6\}$
   (d) $E = \{(x, y) \mid x, y = 5$ or $6\}$
   (e) $E = \{(x, y) \mid x = 1, 3, 5; y = 1, 2, ..., 6\}$
3. $S = \{27$ triples of $R, B, Y\}$; $S = \{(x, y, z) \mid x$ color draw 1, y color draw 2, z color draw 3$\}$
   (a) $E = \{(R, B, Y), (R, Y, B), (B, R, Y), (B, Y, R), (Y, R, B), (Y, B, R)\}$
   (b) $E = \{(x, y, z) \mid$ one of x, y, z is Y and the other two are not $Y\}$
   (c) $E = \{(x, y, z) \mid x = z\}$
   (d) $E = \{(x, y, z) \mid x = y, y = z,$ or $x = z\}$
   (e) $E = \{(x, y, z) \mid x, y, z$ are not all different$\}$
5. $S = \{(x, y) \mid x, y =$ weight of item on 1st, 2nd choice$\}$
   (a) $E = \{(5, 10), (10, 20)\}$
   (b) $E = \{(5, 15), (10, 20), (15, 25)\}$
   (c) $E = \{(5, 5), (5, 10), (5, 15), (5, 20), (5, 25), (10, 5), (10, 10),$ $(10, 15), (10, 20), (15, 5), (15, 10), (15, 15), (20, 5), (20, 10),$ $(25, 5)\}$
   (d) $E = \{(25, 5), (20, 10), (15, 15), (10, 20), (5, 25)\}$
   (e) $E = \{(10, 5), (15, 5), (15, 10), (20, 5), (20, 10), (20, 15),$ $(25, 5), (25, 10), (25, 15), (25, 20)\}$

Section 1.4

1. a, b, and d
3. (a) $\{1, 2, 3, 4\}$
   (b) $\{-1, -2, -3, -4\}$
   (c) $\{2, 4, 6, ...\}$
5. $\{a\}, \{b\}, \{c\}, \{a, b\}, \{a, c\}, \{b, c\}, \{a, b, c\}$ $\varnothing$; all but the last two are proper subsets
7. Different
9. (a) $A \cap B' = \{1, 2, 3\}$
   (b) $(A \cap B') \cup (A' \cap B) = \{1, 2, 3, 7, 8, 9\}$
   (c) $(A \cup B)' = \{10, 11, 12\}$
   (d) $A' \cap B' = \{10, 11, 12\}$
   (e) $(A \cap B)' = \{1, 2, 3, 7, 8, 9, 10, 11, 12\}$
   (f) $A' \cup B' = \{1, 2, 3, 7, 8, 9, 10, 11, 12\}$
11. (a) $A_1 \cup A_2$ is the set of all freshmen and sophomores
    (b) $C \cap (A_1 \cup A_2)$ is the set of all freshmen and sophomores who own cars

(c) $B \cap A_3$ is the set of female juniors

(d) $(C \cup B)'$ is the set of male students who do not own cars

(e) $B \cap (A_3 \cup A_4)'$ is the set of female freshmen and sophomores

13. AB   Rh+   AB   Rh−
    A   Rh+   A   Rh−
    B   Rh+   B   Rh−
    O   Rh+   O   Rh−

15. a, b, c, and e

19. 360

21. $12 \cdot 11 \cdot 10 \cdot 9 \cdot 8 \cdot 7$

25. (a) 20   (b) 30   (c) 5

27. His survey revealed that −2 households owned only dish-washers

Section 1.5

1. (a) $E = \{5, 10, 15, 20\}$
   (b) $E = \{3, 5, 7, 11, 13, 17, 19\}$
   (c) $E = \{10, 12, 14, 16, 18, 20\}$
   (d) $E = \{5, 7, 13, 19\}$
   (e) $E = \{2, 3, 4, 6, 8, 9, 10, 12, 14, 15, 16, 18, 20\}$
   (f) $E = \{6, 7, 8, 9, 10, 11, 12, 13, 14, 15, 16, 17, 18, 19, 20\}$

3. $S = \{(H, H, H, H), (H, H, H, T), (H, H, T, H), (H, H, T, T), (H, T, H, H), (H, T, H, T), (H, T, T, H), (H, T, T, T), (T, H, H, H), (T, H, H, T), (T, H, T, H), (T, H, T, T), (T, T, H, H), (T, T, H, T), (T, T, T, H), (T, T, T, T)\}$
   (a) $E = \{(H, H, T, T), (H, T, H, T), (H, T, T, H), (T, H, H, T), (T, H, T, H), (T, T, H, H)\}$
   (b) $E = \{(H, H, T, T), (H, T, H, T), (H, T, T, H)\}$
   (c) $E = \{(H, H, H, H), (H, H, H, T), (H, H, T, H), (H, H, T, T), (H, T, T, T), (T, H, T, T), (T, T, T, T)\}$
   (d) $E = \{(H, H, H, H), (H, H, H, T), (H, H, T, H), (H, T, H, H), (T, H, H, H)\}$
   (e) $E = \{(H, H, H, H), (H, H, T, T), (H, T, H, T), (H, T, T, H), (T, H, H, T), (T, H, T, H), (T, T, H, H), (T, T, T, T)\}$

5. Let $a = A1, b = A2, c = A3, d = A4,$ and $e = A5; S = \{abc, abd, abe, acd, ace, ade, bcd, bce, bde, cde\}$
   (a) $E = \{abc, abd, abe, acd, ace, ade\}$
   (b) $E = \{abc, abd, abe\}$
   (c) $E = \{abc, abd, abe, acd, ace, ade, bcd, bce, bde\}$
   (d) $E = \{acd, ace, ade, bcd, bce, bde, cde\}$

Section 1.6

1. No

3. $(1, a), (2, b), (3, c), (4, a), (5, b)$ and $(1, c), (2, c), (3, c), (4, c), (5, c)$

5. b and c

Problem 7, Section 1.5

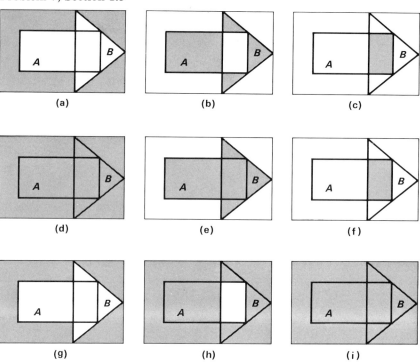

Problem 9, Section 1.5

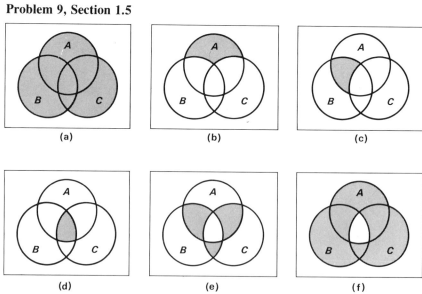

7. (a) 0.1   (b) 1.2   (c) 1.0   (d) 1.0
9. 48
11. (a) All x
   (b) $1, \frac{1}{5}, \frac{1}{101}$
13. (a) $\{0, \frac{1}{4}, \frac{1}{2}, \frac{2}{3}, 1\}$
   (b) $0, \frac{1}{4}, \frac{1}{2}$
   (c) $1 \leqslant x < 3$
15. (a) $x > 0$
   (b) $\frac{1}{5}, 10, \frac{1}{10}$
17. (a) $\{\frac{1}{2}, -2\}$
   (b) $-2 < x < \frac{1}{2}$
19. (a) $0.35, 0 \leqslant x \leqslant \frac{1}{4}$; $0.60, \frac{1}{4} < x \leqslant \frac{1}{2}$; $0.85, \frac{1}{2} < x \leqslant \frac{3}{4}$; $1.10,$
   $\frac{3}{4} < x \leqslant 1$
21. $5; z^{+}; 1$

Chapter Two

Section 2.1

1. B1, B2, B4, PA1, and PA2 are the same; B3: If a finite number of events $A_1, A_2, ..., A_n$ belong to $\mathscr{E}$, then their union belongs to $\mathscr{E}$; B5: If a finite number of events $A_1, A_2, ..., A_n$ belong to $\mathscr{E}$, then their union belongs to $\mathscr{E}$; PA3: The probability of the union of a finite number of mutually exclusive events is the sum of the probabilities of the events

Section 2.2

1. $e_1 = 0$, $e_2 = 1$, $e_3 = 2$, $e_4 = 3$, $e_5 = 4$; $P(E_i) = \frac{1}{5}$, $i = 1, ..., 5$
3. The set of triples of the form (P, O, E), (P, P, E), (E, P, P), etc.; a probability function is $P(E_i) = \frac{1}{27}$, $i = 1, ..., 27$
5. 6 or 2^5 depending on your assumptions
7. In addition to the given probability for E:
   $P(\{e_1, e_2\}) = \frac{3}{8}$ $P(\{e_1, e_2, e_3\}) = \frac{3}{4}$
   $P(\{e_1, e_3\}) = \frac{1}{2}$ $P(\{e_1, e_2, e_4\}) = \frac{5}{8}$
   $P(\{e_1, e_4\}) = \frac{3}{8}$ $P(\{e_1, e_3, e_4\}) = \frac{3}{4}$
   $P(\{e_2, e_3\}) = \frac{5}{8}$ $P(\{e_2, e_3, e_4\}) = \frac{7}{8}$
   $P(\{e_2, e_4\}) = \frac{1}{2}$ $P(\varnothing) = 0$
   $P(\{e_3, e_4\}) = \frac{5}{8}$ $P(S) = 1$
9. (a) $\frac{1}{5}$ (b) $\frac{4}{5}$ (c) $\frac{7}{20}$ (d) $\frac{13}{20}$
11. $\frac{1}{6}$; no
13. $S = \{(3,0,0), (0,3,0), (0,0,3), (2,1,0), (2,0,1), (1,2,0),$
   $(0,2,1), (0,1,2), (1,0,2), (1,1,1)\}$; $P(E_i) = \frac{1}{10}$, $i = 1, ..., 10$
15. $k = \frac{1}{21}, \frac{4}{7}$

17. $S = \{H, TH, TTH, TTT\}$; $P(E_1) = \frac{1}{2}$, $P(E_2) = \frac{1}{4}$, $P(E_3) = \frac{1}{8}$, $P(E_4) = \frac{1}{8}$

Section 2.3

1. $\frac{5}{13}$
3. $\frac{3}{8}$
5. $\frac{4}{11}$
7. (a) $\frac{1}{2}$ (b) $\frac{11}{12}$ (c) $\frac{5}{12}$ (d) $\frac{11}{12}$
9. (a) $\frac{2}{7}$ (b) $\frac{5}{7}$ (c) $\frac{3}{14}$ (d) $\frac{5}{14}$
11. $\frac{17}{20}$
13. $\frac{2}{3}$
15. (a) The set of eight triples $\{(a, b, c) \mid a = H, T; b = H, T; c = H, T)\}$ (b) $\frac{1}{4}$
17. $\frac{1}{3}$
19. $\frac{1}{676}$
21. $\frac{5}{16}$

Section 2.4

1. (a) $\frac{5}{10}$ (b) $\frac{6}{10}$ (c) $\frac{1}{10}$ (d) $\frac{5}{10}$
3. $\frac{29}{30}$
5. $103/52^2$
7. (a) False   (b) True
9. (a) $S = \{1, 2, ..., 10\}$, $P(x) = \frac{1}{10}$, $x = 1, ..., 10$
     (b) $\frac{1}{10}$ (c) $\frac{3}{10}$ (d) $\frac{1}{2}$
11. $\frac{9}{25}$
21. (a) $1 - P(A) - P(B) + P(A \cap B)$
   $P(A) + P(B) - 2P(A \cap B)$
   $P(A \cap B)$
   (b) 1
   $P(A) + P(B) - P(A \cap B)$
   $P(A \cap B)$
   (c) 0
   $1 - P(A \cap B)$
   1
23. (a) $P(A \cap B) + P(A \cap C) + P(B \cap C) - 3P(A \cap B \cap C)$
   (b) $P(A \cap B) + P(A \cap C) + P(B \cap C) - 2P(A \cap B \cap C)$
   (c) $1 - P(A \cap B \cap C)$

Chapter Three

Section 3.1

1. 2^{10}
3. 10
5. 96
7. 144
9. 375
11. $17 \cdot 11 \cdot 32$

Section 3.2

1. (a) 60   (b) 125   (c) 120
3. 30
5. 60
7. 17, 576
9. 720
11. 120
13. 72
15. (a) 12   (b) 32

17. n^k
19. (a) 169   (b) 338   (c) 169
21. (a) $n(n!)$ (b) n (c) $(n + 1)n$
23. 1680
25. (a) $(n)_k$ (b) n^k

Section 3.3

1. (a) 252   (b) 56
3. 21
7. 2502
9. 15,504; 7496
11. 715
13. $\frac{3}{20}$
15. $\dfrac{n(n-1)}{2}$
17. 462
19. 3,719,250

21. 10
23. (a) 1,814,400   (b) 28,800
25. 22
27. 540

35. $n = 18; r = 10$

Section 3.4

1. $\frac{9}{50}$
3. $\frac{1}{105}$
5. $\frac{1849}{6660}$
7. (a) $\frac{1}{9}$ (b) $\frac{37}{108}$
9. $\frac{5}{14}$
11. $(\frac{1}{15})^{25}$
13. (a) $\frac{1}{330}$ (b) $\frac{21}{55}$
15. $\frac{1}{13}$
17. $\frac{91}{228}$
19. $\binom{100}{20} \Big/ \binom{95}{20}$
21. (a) $\frac{1}{6}$ (b) $\frac{1}{30}$ (c) $\frac{3}{10}$
23. $\frac{20}{35}$
25. (a) $\frac{56}{165}$ (b) $\frac{1}{11}$
27. With replacement: (a) $(12/13)^{12}$ (b) $1 - (12/13)^{13}$; without replacement: (a) $(48)_2/(51)_{12}$ (b) $1 - (48)_{13}/(52)_{13}$
29. (a) 0.018   (b) $\frac{123}{288}$
31. 0.1
33. $\frac{505}{1001}$
35. $(10!)^4/((5!)^4(20!))$

Chapter Four

Section 4.1

1. $\frac{2}{3}$

3. (a) $\frac{3}{8}$ (b) $\frac{3}{8}$

5. $\frac{1}{2}$

7. $\frac{86}{350}$

9. (a) $\frac{3}{5}$ (b) $\frac{3}{4}$

11. $\frac{1}{10}$

13. (a) $\frac{7}{25}$ (b) $\frac{4}{15}$ (c) $\frac{12}{23}$

15. (a) $\frac{1}{4}$ (b) $\frac{3}{8}$

17. $\frac{4613}{5807}$; $\frac{562}{651}$; $\frac{299}{1194}$

19. (a) $\frac{5}{12}$ (b) $\frac{1}{36}$

21. $\frac{5}{8}$, $\frac{3}{8}$

23. (a) $\frac{1}{73}$ (b) $\frac{1}{24}$

25. $\frac{3}{4}$

31. (a) $\frac{1}{2}$ (b) $\frac{1}{2}$

Section 4.2

1. $\frac{3}{200}$

3. $\frac{71}{216}$

5. 0.65

7. $\frac{4}{15}$

9. $\frac{15}{23}$

11. $\frac{4}{15}$

13. $\frac{11}{108}$

15. $\frac{1}{40}$

17. $\frac{8}{9}$

19. $\frac{8}{11}$

21. $\frac{147}{341}$

25. $\frac{8}{9}$

27. $\frac{1}{2}$, $\frac{1}{3}$, $\frac{1}{6}$

29. 0.45; 0.2, freshman; 0.16, sophomore; 0.3, junior; 0.34, senior

31. (a) $\frac{2}{105}$ (b) $\frac{2}{5}$

Section 4.3

3. Yes; no

5. (a) $\frac{3}{8}$ (b) $\frac{5}{25}$ (c) $\frac{23}{24}$

7. $\frac{1}{4}$

9. (a) 0.02 (b) 0.26

11. $S = \{e_1 e_2 e_3 e_4\}$, where $e_1 = $ HH, $e_2 = $ HT, $e_3 = $ TH, and $e_4 = $ TT; the events are $\varnothing$, $\{e_1\}$, $\{e_2\}$, $\{e_3\}$, $\{e_4\}$, $\{e_1, e_2, e_3, e_4\}$, $\{e_1, e_2\}$, $\{e_1, e_3\}$, $\{e_1, e_4\}$, $\{e_2, e_3\}$, $\{e_2, e_4\}$, $\{e_3, e_4\}$, $\{e_1, e_2, e_3\}$, $\{e_1, e_2, e_3\}$, $\{e_1, e_3, e_4\}$, $\{e_2, e_3, e_4\}$; the sets $\{e_i, e_j\}$, $\{e_j, e_k\}$ are independent of $j \neq i, j \neq k$; no other pairs are independent

13. (a) $\frac{3}{100}$ (b) $\frac{1}{24}$

15. Yes; no

17. $\frac{81}{100}$, $\frac{899}{1110}$

19. (a) $\frac{1}{4}$ (b) $\frac{2}{3}$

21. $\frac{25}{28}$

23. (a) 0.1375 (b) 0.32 (c) 0.19

25. 0.72

Section 4.4

1. (a) $\frac{9}{64}$ (b) $\frac{15}{64}$

3. $\frac{49}{5000}$

5. (a) 0.33 (b) 0.41

7. (a) 0.216   (b) 0.108
9. $\frac{19}{64}$
11. (1) 1:6:9   (2) 1:2:1
13. (a) $\frac{1}{8}$ (b) $\frac{1}{16}$
15. (a) $(\frac{2}{3})^5$ (b) $(\frac{5}{18})^5$ (c) $\frac{40}{243}$
17. $31/7^5$
19. (1) 9:6:1   (2) 1:2:1
21. Two heads
23. 0.7; 0.66

Section 4.5

1. $\frac{15}{65}$; $\frac{57}{64}$
3. $\frac{280}{2187}$; $\frac{2088}{2187}$
5. (a) 0.00065   (b) $9^4/10^{10}$
7. $3^6 \cdot 71/4^8$
9. (a) 0.24   (b) 0.26   (c) 0.988
11. (a) $30/4^{10}$ (b) 0.0781
13. 0.1201
15. 0.2304
17. $3125/6^5$
19. 160/421
21. $3^4/2^8$
23. $\frac{1}{64}$, $\frac{9}{64}$, $\frac{27}{64}$, $\frac{27}{64}$
25. $1 - (0.8)^7$
27. $200/3^7$

Chapter Five

Section 5.1

1. $S_x = \{4, 2, 0, -2, -4\}$; $P_x(4) = \frac{1}{16}, P_x(2) = \frac{1}{4}, P_x(0) = \frac{3}{8}, P_x(-2) = \frac{1}{4},$ $P_x(-4) = \frac{1}{16}$
3. $S_x = \{0, 1, 2\}$; $P_x(0) = \frac{3}{14}, P_x(1) = \frac{4}{7}, P_x(2) = \frac{3}{14}$
5. $S_x = \{3, 4, 5, 6\}$; $P_x(3) = \frac{1}{10}, P_x(4) = \frac{2}{5}, P_x(5) = \frac{2}{5}, P_x(6) = \frac{1}{10}$
7. $S_x = \{1, 2, 3, ..., n, ...\}$; $P_x(n) = (\frac{1}{4})^{n-1}(\frac{3}{4})$, $n = 1, 2, ...$
9. $\frac{23}{60}$
11. $S_x = \{-2, -1, 1, 2\}$; $P_x(-2) = \frac{1}{8}, P_x(-1) = \frac{1}{8}, P_x(1) = \frac{1}{4}, P_x(2) = \frac{1}{2}$
13. $S_x = \{0, 1, 2, ..., n\}$; $P_x(k) = \binom{n}{k}p^k(1 - p)^{n-k}, k = 0, 1, ..., n$
15. $\frac{14}{15}$
17. 0.080, 0.4, 0.45

Section 5.2

1. (b) $F(x) = 0, x < 0$; $F(x) = \frac{1}{16}, 0 \leqslant x < 1$; $F(x) = \frac{5}{16}, 1 \leqslant x < 2$; $F(x) = \frac{11}{16}, 2 \leqslant x < 3$; $F(x) = \frac{15}{16}, 3 \leqslant x < 4$; $F(x) = 1, 4 \leqslant x$

3. (b) $F(x) = 0$, $x < 2$; $F(x) = \frac{1}{36}$, $2 \leqslant x < 3$; $F(x) = \frac{3}{36}$, $3 \leqslant x < 4$;
$F(x) = \frac{6}{36}$, $4 \leqslant x < 5$; $F(x) = \frac{10}{36}$, $5 \leqslant x < 6$; $F(x) = \frac{15}{36}$, $6 \leqslant x < 7$;
$F(x) = \frac{21}{36}$, $7 \leqslant x < 8$; $F(x) = \frac{26}{36}$, $8 \leqslant x < 9$; $F(x) = \frac{30}{36}$, $9 \leqslant x < 10$;
$F(x) = \frac{33}{36}$, $10 \leqslant x < 11$; $F(x) = \frac{35}{36}$, $11 \leqslant x < 12$; $F(x) = 1$, $12 \leqslant x$

5. (b) $F(x) = 0$, $x < 0$; $F(x) = \frac{10}{56}$, $0 \leqslant x < 1$; $F(x) = \frac{40}{56}$, $1 \leqslant x < 2$;
$F(x) = \frac{55}{56}$, $2 \leqslant x < 3$; $F(x) = 1$, $3 \leqslant x$

7. $P_X(k) = \binom{6}{k} (0.2)^k (0.8)^{6-k}$

9. (a) $k = \frac{3}{20}$

(b) $F(x) = 0$, $x < 0$; $F(x) = \frac{2}{20}$, $0 \leqslant x < 1$; $F(x) = \frac{11}{20}$, $2 \leqslant x < 3$;
$F(x) = \frac{17}{20}$, $3 \leqslant x < 4$; $F(x) = 1$, $4 \leqslant x$

(c) $\frac{15}{20}$

(d) $\frac{11}{20}$

11. $F(x) = 0$, $x < 1$; $F(x) = \frac{1}{20}$, $1 \leqslant x < 2$; $F(x) = \frac{9}{20}$, $s \leqslant x < 3$; $F(x) = \frac{11}{20}$, $3 \leqslant x < 4$; $F(x) = \frac{16}{20}$, $4 \leqslant x < 5$; $F(x) = \frac{18}{20}$, $5 \leqslant x < 6$; $F(x) = 1$, $6 \leqslant x$

13. (a) 0.40   (b) 0.6   (c) 0.3   (d) 0.45

15. $P_X(\cdot) = \binom{4}{k}\binom{48}{4-k} \Big/ \binom{52}{4}$; $k = 0, 1, 2, 3, 4$; $\binom{13}{k}\binom{39}{4-k} \Big/ \binom{52}{4} = P_X(k)$

19. (a) $\frac{1}{2900}$ (b) $\frac{49}{100}$ (c) $\frac{43}{125}$ (d) $\frac{73}{245}$

Chapter Six

Section 6.1

1. (a) 1.8   (b) 0.56   (c) 0   (d) 0.52
3. (a) 4.95   (b) 4.14   (c) 0   (d) 1.65
5. Mean, 82.5; median, 85.5; mode, 86
7. 192, 250

Section 6.2

1. 2
3. 3.8
5. 1
7. 7
9. $\sum_{k=0}^{13} k \binom{13}{k}\binom{39}{13-k} \Big/ \binom{52}{13} = \frac{13}{4}$
11. $\frac{6}{5}$; $\frac{6}{5}$
13. 5 cents
15. 3
17. $3.00
19. $1.75; $2.00; $2.83
21. Mine 2

Section 6.3

1. 0.1
3. 1.7
5. 29.8
7. $2250
9. $\frac{107}{6}$; no
11. 1
13. $E(x) = \sum_{k=0}^{100} 20kp(k) + \sum 2000 + 25(k - 100)p(k)$, where $p(k) = \binom{125}{k}(0.02)^k(0.98)^{125-k}$
15. Use old process; expected profit, old method $= \$0.40$; expected profit, new method $= \$0.21$
17. $(k_2 - k_1 + k_3)4.15 - nk_3$

Section 6.4

1. 2.24
3. 2.64
5. 0.7, 0.21
7. 5, $\frac{15}{4}$
9. 2.32
11. 5, $\frac{5}{2}$
13. $\frac{40}{9}$
15. 11,875
17. 0, 0.8, 0.894, 0.8
19. -2.4, 15.84, 3.97, 3.52
21. $\frac{47}{64}$

Chapter Seven

Section 7.1

1.

x_i \ y_i	0	1	2	3
0	$\frac{1}{8}$	$\frac{2}{8}$	$\frac{1}{8}$	0
1	0	$\frac{1}{8}$	$\frac{2}{8}$	$\frac{1}{8}$

3.

X \ Y	0	1
0	$\frac{2}{20}$	$\frac{6}{20}$
1	$\frac{6}{20}$	$\frac{6}{20}$

5.

X \ Y	2	3	4	5	6	7	8	9	10	11	12
1	$\frac{1}{36}$										
2		$\frac{2}{36}$	$\frac{4}{36}$								
3			$\frac{2}{36}$	$\frac{2}{36}$	$\frac{1}{36}$						
4				$\frac{2}{36}$	$\frac{2}{36}$	$\frac{2}{36}$	$\frac{1}{36}$				
5				$\frac{2}{36}$	$\frac{2}{36}$	$\frac{2}{36}$	$\frac{2}{36}$	$\frac{2}{36}$	$\frac{1}{36}$		
6						$\frac{2}{36}$	$\frac{2}{36}$	$\frac{2}{36}$	$\frac{2}{36}$	$\frac{2}{36}$	$\frac{1}{36}$

0's elsewhere

7.

Y \ X	1	2	3	$P_Y(\cdot)$
2	$\frac{2}{66}$	$\frac{4}{66}$	$\frac{6}{66}$	$\frac{12}{66}$
4	$\frac{4}{66}$	$\frac{8}{66}$	$\frac{12}{66}$	$\frac{24}{66}$
5	$\frac{5}{66}$	$\frac{10}{100}$	$\frac{15}{66}$	$\frac{30}{66}$
$P_X(\cdot)$	$\frac{11}{66}$	$\frac{22}{66}$	$\frac{33}{66}$	

9.

X_i	-1	0	1
$P_X(\cdot)$	$\frac{1}{3}$	0	$\frac{2}{3}$

Y_i	-1	0	1
$P_Y(\cdot)$	$\frac{1}{4}$	$\frac{1}{4}$	$\frac{1}{2}$

; $\frac{1}{3}$

11.

X \ Y	0	1	2	3	4	5	6
1	0	$\frac{1}{36}$	$\frac{1}{36}$	$\frac{1}{36}$	$\frac{1}{36}$	$\frac{1}{36}$	$\frac{1}{36}$
2	0	$\frac{1}{36}$	$\frac{1}{36}$	$\frac{1}{36}$	$\frac{1}{36}$	$\frac{1}{36}$	$\frac{1}{36}$
3	0	$\frac{1}{36}$	$\frac{1}{36}$	$\frac{1}{36}$	$\frac{1}{36}$	$\frac{1}{36}$	$\frac{1}{36}$
4	0	$\frac{1}{36}$	$\frac{1}{36}$	$\frac{1}{36}$	$\frac{1}{36}$	$\frac{1}{36}$	$\frac{1}{36}$
5	$\frac{1}{48}$	$\frac{3}{48}$	$\frac{3}{48}$	$\frac{1}{48}$	0	0	0
6	$\frac{1}{48}$	$\frac{3}{48}$	$\frac{3}{48}$	$\frac{1}{48}$	0	0	0

$P(X = 2, Y < 2) = \frac{1}{18}$

13.

X \ Y	0	1	2	3
1				$\frac{5}{70}$
2			$\frac{30}{70}$	
3		$\frac{30}{70}$		
4	$\frac{5}{70}$			

0's elsewhere

 15. (a) $\frac{14}{27}$ (b) $\frac{11}{27}$ (c) $\frac{1}{27}$ (d) $\frac{21}{27}$
17. $P(X > Y) = \frac{6}{25}$

Section 7.2

1.

X \ Y	1	2
1	$\frac{2}{12}$	$\frac{1}{12}$
2	$\frac{2}{12}$	$\frac{1}{12}$
3	$\frac{4}{12}$	$\frac{2}{12}$

5. No
9. No

11.

X \ Y	A	B	C	D	$P_X(\cdot)$
21–31	0.045	0.078	0.08	0.02	0.25
31–41	0.054	0.093	0.120	0.033	0.30
41–51	0.054	0.093	0.120	0.033	0.30
Over 51	0.027	0.0465	0.06	0.0165	0.15
$P_X(\cdot)$	0.18	0.31	0.40	0.11	

Probability, 0.0495

13. (a)

y_i	−1	0	1
$P(Y = y \mid X = 1)$	$\frac{2}{3}$	$\frac{1}{3}$	0

(b)

x_i	−1	1
$P(X = x \mid Y = -1)$	$\frac{1}{5}$	$\frac{4}{5}$

Section 7.3

1. $z_i = 2, 3, 4, 5, 6$; $p_Z(z_i) = 0.1, 0.2, 0.3, 0.4, 0$; $E(x + y) = 4$
5. 15
7. $E(xy) = \frac{5}{12}$; $E(x) = \frac{3}{4}$; $E(Y) = \frac{2}{3}$
9. (a) $\frac{5}{2}$, $\frac{31}{16}$ (b) $\frac{71}{16}$ (c) $\frac{47}{8}$ (d) No
11. $\frac{40}{3}$
13. $E(x + y) = 1.75$; $E(xy) = 0.245$ (assume independence)
15. Put $\binom{52}{3} = \frac{1}{q}$

X Y	0	1	2	3
0	$\binom{44}{3}q$	$\binom{44}{2}\binom{4}{1}q$	$\binom{44}{1}\binom{4}{2}q$	$\binom{4}{3}q$
1	$\binom{42}{2}\binom{4}{1}q$	$\binom{44}{1}\binom{4}{1}\binom{4}{1}q$	$\binom{4}{2}\binom{4}{1}q$	0
2	$\binom{42}{1}\binom{4}{2}q$	$\binom{4}{2}\binom{4}{1}q$	0	0
3	$\binom{4}{3}q$	0	0	0

$E(x + y) = \frac{102}{221}$

17. $80, -20$; $a \cdot 30 + b \cdot 50$

19.

z_i	1	2	3
$p_Z(z_i)$	0.1	0.4	0.5

$E(Z) = 0.1 + 0.8 + 1.5 = 2.4$

w_i	1	2	3
$p_W(w_i)$	0.4	0.6	0

$E(W) = 0.4 + 1.2 = 1.6$

Section 7.4

1.

z_i	1	2	3	4	5	6
$p_Z(z_i) =$	$\frac{1}{24}$	$\frac{4}{24}$	$\frac{7}{24}$	$\frac{7}{24}$	$\frac{4}{24}$	$\frac{1}{24}$

2. $\frac{17}{12}$
3. $\frac{60}{27}$
5. $\frac{95}{4}$

	$E(\ )$	$(V(\ ))^{1/2}$
7. $x + y$	75	136
$x - y$	5	136

9. 250, 145
11. 4595, 63,348.75
13. 4.05, 4.05
15. 45, $\frac{205}{6}$
23. $\sqrt{0.102}$

Section 7.5

1. (a) $(0,0)$ $(0,1)$ $(0,2)$ $(1,0)$ $(1,1)$ $(1,2)$ $(2,0)$ $(2,1)$ $(2,2)$
    (b) $1, \frac{1}{4}$
3. (a) $\frac{7}{10}$, 0.605
    (b) $\frac{7}{10}$, 0.121
5. $\frac{14}{9}, \frac{38}{81}$
7. Since $V(\bar{X})$ is small, then $\bar{x} = 0.0979$ is a reasonable approximation of $E(\bar{X})$
9. (a) $\frac{10}{25}$ (b) $\frac{1}{10}$
11. $\frac{200}{3}, \frac{200}{9}; \frac{200}{3}, \frac{4}{9}$
13. 225
15. About 63% of the chips are red
17. $\bar{x} = 74.5$ – raise one eyebrow
19.

L	-1	0	1
$p_L(\cdot)$	$\frac{1}{125}$	$\frac{63}{125}$	$\frac{61}{125}$

Section 7.6

1. $\frac{2}{9}$
3. 4000
5. $\frac{25}{9}$

	Exact probability	Estimate by Chebyshev
7. (a)	$\frac{54}{64}$	>0
(b)	$\frac{63}{64}$	$>\frac{3}{4}$
(c)	1	$>\frac{8}{9}$

9. 6
11. $\frac{1}{10}$
13. $P(|r/n - P| < 11/100) \geqslant 1 - 100^2/4n^3$
15. (a) $\frac{24}{25}$ (b) 40
17. $h > 0.22$
21. $\frac{1}{10}$
23. $k \geqslant \sqrt{10}$

Chapter Eight

Section 8.1

1. Let S denote a 2 or 3, F a 1, 4, 5, 6: {S; FS; FFS; ...; FF ... FS; ...}.
3. {FM: MF; FFM; MMF; FFFM; MMMF; ...}.
5. $(\frac{5}{6})^6$
7. $\frac{30}{51}$
9. 0.0837
11. $\frac{1}{3}$
13. $A: \frac{4}{7}, B: \frac{2}{7}, C: \frac{1}{7}$

Section 8.2

1. (a) $(\frac{1}{3})^7$ (b) $\frac{1}{13}$
5. 1550
7. 0, no
9. $\frac{1}{2}$
11. $x = \frac{3}{4}$, $(\frac{1}{4})^m$
13. $p(k) = (k - 1)\ (\frac{3}{4})^{k-2}\ (\frac{1}{4})^2$

Section 8.3

1. (a) 0.3840 (b) 0.4232 (c) 0.1404 (d) 0.0067
3. 0.0343
5. (a) 0.0839 (b) e^{-9}
7. (a) 0.1338 (b) 0.8488 (c) 0.1512 (d) 0.1339
9. 0.0638
11. 0.5768
13. $1 - (1 - p)^4$, where $p = 0.5768$
15. (a) 0.6767 (b) 0.2391
17. $E(Y) = 2\lambda$; $V(Y) = 4\lambda$
21. $p = 0.2642$; $\Sigma_{k=0}^{10} \binom{500}{k} p^k (1 - p)^{500-k}$
23. $P(X \text{ even}) = \frac{1}{2}(e^\lambda + e^{-\lambda})$

Chapter Nine

Section 9.1

1. $S_X = \{x \mid 0 \le x \le 4\}$; $P_X(x \le a) = a/4$
3. $S_X = \{x \mid 0 \le x \le 2\}$; $P_X(x \le a) = (\pi a^2/4)$
5. $S_Z = \{z \mid 0 \le z \le 6\}$; $P_Z(z < 1) = (1 + \ln \frac{2}{3})/6$
7. $S_X = \{x \mid 1 \le x \le \sqrt{5}\}$; $P_X(x \le 2) = \sqrt{3}/3$
9. a^2; $\frac{3}{4}$

Section 9.2

1. $F(x) = \begin{cases} 0, & x < 2 \\ x^2 - 4x + 4, & 2 \le x < 3 \\ 1, & x \ge 3 \end{cases}$

3. $F(x) = \begin{cases} 0, & x < -2 \\ -x^2/8 + \frac{1}{2}, & -2 \le x < 0 \\ x^2/8 + \frac{1}{2}, & 0 \le x < 2 \\ 1, & x \ge 2 \end{cases}$

5. $F(x) = \begin{cases} 0, & x < 0 \\ -\frac{1}{18}(x-3)^2 + \frac{1}{2}, & 0 \leq x < 3 \\ (x-3)^2/8 + \frac{1}{2}, & 3 \leq x < 5 \\ 1, & x \geq 5 \end{cases}$

7. (a) $A = 6$

(b) $F(x) = \begin{cases} 0, & x < 0 \\ x^2(3 - 2x), & 0 \leq x < 1 \\ 1, & x \geq 1 \end{cases}$

(c) $\frac{11}{32}$

9. (a) $A = 20$

(b) $F(x) = \begin{cases} 0, & x < 0 \\ x^4(5 - 4x), & 0 \leq x < 1 \\ 1, & x \geq 1 \end{cases}$

(c) $\frac{11}{81}$

13. $f(x) = \begin{cases} \frac{3}{2}x^2, & -1 \leq x \leq 1 \\ 0, & \text{otherwise} \end{cases}$

15. $f(x) = \begin{cases} 3e^{-3x}, & 0 < x \\ 0, & \text{otherwise} \end{cases}$

17. $\frac{1}{2}$

19. $a = \frac{1}{2} + \sqrt{3}/6$

21. (b) $F(x) = \begin{cases} 0, & x < \frac{1}{2} \\ x - \frac{1}{2}, & \frac{1}{2} \leq x < 1 \\ \frac{1}{2}, & 1 \leq x < 2 \\ x/2 - \frac{1}{2}, & 2 \leq x < 3 \\ 1, & 3 \leq x \end{cases}$

(d) $\frac{1}{2}$

Section 9.3

1. $f_Y(\cdot) = \begin{cases} \frac{2}{9}(2 - y - y^2), & -2 < y < 1 \\ 0, & \text{otherwise} \end{cases}$

$F_Y(\cdot) = \begin{cases} 0, & y < -2 \\ \frac{2}{9}(\frac{10}{3} + 2y - y^2/2 - y^3/3), & -2 \leq y < 1 \\ 1, & y \geq 1 \end{cases}$

3. $f_Y(\cdot) = 1, \quad \frac{7}{2} < y < \frac{9}{2}$

$F_Y(\cdot) = \begin{cases} 0, & y < \frac{7}{2} \\ y - \frac{7}{2}, & \frac{7}{2} \leq y < \frac{9}{2} \\ 1, & y \geq \frac{9}{2} \end{cases}$

5. $f_Y(y) = \frac{1}{3} e^{-y^{1/3}} y^{-2/3}, \quad y > 0$

7. $f_Y(\cdot) = \begin{cases} (9/2500)(9y - 168), & 60/9 < y \leq 90/9 \\ -(9/2500)(9y - 268), & 90/9 < y \leq 140/9 \\ 0, & \text{otherwise} \end{cases}$

9. $f_Y(\cdot) = 1/(2\sqrt{y - 1}), \quad 1 < y < 2$

11. $f_Y(y) = (\frac{1}{2})1/\sqrt{y}, \quad 0 < y < 1$

13. $f_V(\cdot) = (\frac{3}{4}\pi)^{2/3}v^{-1/3}, \quad 0 < v < 1$

15. $P(Y < \sqrt{3}) = \frac{1}{2}$

17. $f_Y(\cdot) = e^{-e^y} \cdot e^y, \quad -\infty < y < \infty$

Section 9.4

1. $E(X) = \frac{8}{3}, V(X) = \frac{1}{18}$

3. $E(X) = \frac{4}{3}, V(X) = \frac{731}{36}$

5. $E(X) = 1, V(X) = 1$

7. $E(Y) = -\frac{1}{2}, V(Y) = \frac{99}{40}$

9. $E(Y) = \frac{4}{3}, V(Y) = \frac{4}{45}$

11. $E(X) = \frac{13}{8}, V(X) = \frac{59}{84}$

13. $E(Y) = (-2/\pi^2); V(Y) = \frac{1}{2} - 4/\pi^4$

15. $E(Y) = 0.17$

17. $E(Y) = \$200$

19. $E(5 \text{ throws}) = 13.50$

21. $a = \sqrt{5}/5, b = -2\sqrt{5}$

Chapter Ten

Section 10.1

1. $\frac{2}{3}$

3. $\frac{1}{3}$

5. $\frac{1}{3}$

7. $1 - (1/\pi)^{1/2}$

9. $f_Y(y) = e^y, \ln 2 < y < \ln 3$

11. (a) $\frac{4}{3}$　(b) 1　(c) 3　(d) $\frac{3}{4}$

13. $\frac{192}{243}$

15. $a = \alpha + \sqrt{3}\beta, b = \alpha - \sqrt{3}\beta$

17. $f_Y(y) = y/(y^2 - 1)^{1/2}, \quad 1 < y < \sqrt{2}$

$F_Y(y) = \begin{cases} 0, & y < 1 \\ (y^2 - 1)^{1/2}, & 1 \leq y < \sqrt{2} \\ 1, & \sqrt{2} \leq y \end{cases}$

Section 10.2

1. 0.3413

3. 0.0556

5. 0.8413

7. 0.5899

9. $a = 51.12$

11. (a) 0.6010 (b) 0.6298
13. (67.325, 70.675)
15. $f_Y(y) = (1/0.4\pi\sqrt{2}\sqrt{y})\, e^{-(1/2)((\sqrt{y}-\sqrt{\pi})/0.2\sqrt{\pi})^2}$
17. 65

19. $F_Y(y) = \dfrac{\displaystyle\int_{-\infty}^{y} e^{-1/2((t-3.2)/0.1)^2}\, dt}{\displaystyle\int_{-\infty}^{3} e^{-1/2((t-3.2)/0.1)^2}\, dt}$, $-\infty < y < 3$

$\qquad\qquad 1,\qquad\qquad\qquad\qquad y \geq 3$

21. $E(Y) = 0.68c_1 - 0.16(c_2 + c_3)$
23. 0.39
25. (a) 0.32 (b) 0.58

Section 10.3

 1. $E(X) = 2,\ V(X) = 4$
 3. $(1 - e^{-240})$
 5. $(e^{-6} - e^{-12})$
 7. $\Sigma_{r=3}^{6} \binom{6}{r}(e^{-2})^r(1 - e^{-2})^{6-r}$
13. $R(X) = e^{-\alpha x}$
15. Process 2

Section 10.4

 1. 0.424, 1.0
 3. 3, $\frac{1}{300}$
 5. 0.95
 7. $1 - (V(X)/nc^2)$
 9. 0.9997
11. 0.1314
13. 11
15. 0.8944
17. 0.7062
19. 0.5
21. 0.383
23. 0.9999

Appendix A

1. (a) $\frac{5}{6} < x$
   (b) $x < 0$ or $x > 3$
2. (a) $-1 \leq x \leq 1$
   (b) $6 < x$
3. (a) $x \leq 0$ or $x > 1$
   (b) $x < -2$ or $x > 3$

4. (a) $-1 < x < 1$
   (b) $x < 0$ or $x > 4$
5. (a) All real numbers
   (b) $-2 < x < 4$
15. (a) $-5 < x < 9$
   (b) $-3 < x < 3$
16. (a) $x < 1$ or $x > 5$
   (b) $-5 \leqslant x \leqslant 7$
17. (a) $-3 < x < -1$
   (b) $x < -2$ or $x > 5$

Appendix B

1. (a) $x_2 + x_3 + x_4 + x_5 + x_6 + x_7$
   (b) $1 + 2 + 3 + 4 + 5$
2. (a) $x^2 + x^3 + x^4$
   (b) $2 + 5 + 8 + 11 + 14 + 17$
3. (a) $1 + 4 + 27$
   (b) $1 + \frac{1}{4} + \frac{1}{9} + \frac{1}{16}$
4. $1 + 2 + 3 + 4 + 6 + 12$
5. $15, 4, k$
6. (a) $\frac{25}{12}$ (b) $\frac{7}{12}$
7. (a) $\Sigma_{n=1}^{4} 2n$ (b) $\Sigma_{n=6}^{10} n^2$ (c) $\Sigma_{k=1}^{n} (2k - 1)$ (d) $\Sigma_{k=1}^{n} k^3$
8. (a) 0 (b) 0, n even; -1, n odd
24. $(n + 1)/2n$

Appendix C

Section C.1

1. $1, 2, 7\frac{1}{3}, -1, 4$
2. $\frac{47}{70}, \frac{107}{140}$
6. 4
7. $-3.2, -0.1$
8. B lies between A and C
9. $-13 + 10\sqrt{2}$
11. $(-\frac{1}{2}, \frac{3}{2})$
12. $2\sqrt{5}$
13. $\sqrt{74}/2$
15. Yes
16. No
17. No
18. No
19. (a) $(2, 2)$ (b) $3\sqrt{2}$

20. $(5 - \sqrt{7}, 1)$
22. 15
23. $(0, 1)$
25. $\sqrt{149}/2$
27. $(\frac{37}{9}, 0)$
29. $(\frac{58}{21}, \frac{29}{21})$
30. $|x + 3|$

Section C.2

1. $x - 2y + 9 = 0$
2. $y + 2 = 0$
3. $x - 4y + 5 = 0$
4. $3x + y - 9 = 0$
5. $2x - y - 1 = 0$
6. $3x + 4y - 12 = 0$
7. $3x - 2y - 1 = 0$
8. $3x - 5y - 21 = 0$
9. $4x + y = 0$
10. $y + 5 = 0$
11. $m = -\frac{2}{3}, b = -\frac{5}{3}$
12. $m = \frac{3}{8}, (-\frac{4}{3}, 0)$
13. $(-\frac{1}{2}, 0), (0, -\frac{1}{7})$
14. $x + 5y - 11 = 0$
15. $7x + 3y + 23 = 0$
16. $(-\frac{17}{7}, -\frac{3}{7})$
17. (a) Parallel (b) neither
18. $5x - 2y - 24 = 0$
19. $3x + 8y + 27 = 0$
20. $12x - 2y - 25 = 0$
22. No
24. $2\sqrt{17}$

Section C.3

1. Intercepts $(\pm 2, 0), (0, \pm 2)$; symmetric to x-axis and y-axis
2. Intercepts $(0, 0)$; no symmetry
3. Intercepts $(1, 0)$; symmetric to origin
4. Intercepts $(\pm 1, 0)$ $(0, 0)$; symmetric to origin
5. Intercepts $(-\frac{1}{2} \pm \sqrt{5}/2, 0)$; symmetric to x-axis
6. Intercepts $(-1, 0)$; no symmetry; $x = 0$ vertical asymptote
7. Intercepts $(-1, 0)$ $(0, -1)$; $x = 1$ vertical asymptote; $y = 1$ horizontal asymptote
8. Intercepts $(0, 0)$; $x = \pm \sqrt{3}/3$ vertical asymptotes; $y = 0$ horizontal asymptote; symmetric to origin
9. Intercept $(0, 0)$; symmetric to y-axis
10. Intercepts $(0, 0)$ $(\pm 1, 0)$; symmetric to y-axis

11. $x = 1$ vertical asymptote; $y = 0$ horizontal asymptote
12. Intercept $(0, 0)$; symmetric to origin; $y = 0$ horizontal asymptote
13. Intercept $(0, 0)$; symmetric to x- and y-axes
14. $x = 0$ vertical asymptote; $x = \pm 1$ vertical asymptotes; $y = 0$ horizontal asymptote; symmetric to origin
15. Intercept $(0, 3)$; symmetric to y-axis
16. Intercept $(0, 0)$; $x = \pm 2$ vertical asymptotes; $y = 1$ horizontal asymptote; symmetric to y-axis
17. Intercept $(0, 1)$; $x = -2$ vertical asymptote
18. Intercept $(0, 1)$; symmetric to y-axis; $x = 0$ vertical asymptote
19. Intercepts $(0, 0)$ $(\pm 2, 0)$; symmetric to origin
20. Intercepts $(4^{1/3}, 0)$ $(0, 4)$
21. Intercept $(0, 0)$; $x = 1$ vertical asymptote; $x = 2$ vertical asymptote; $y = 0$ horizontal asymptote
22. Intercepts $(-2, 0)$ $(0, -2)$; $y = 0$ horizontal asymptote; $x = \pm 1$ vertical asymptote; $x + -1$ vertical asymptote
23. Intercepts $(\pm 1, 0)$; $x = 0$ vertical asymptote; $x = -2$ vertical asymptote; $y = 1$ horizontal asymptote
24. Intercepts $(0, 1)$, $(-1, 0)$

Section C.4
1. $x^2 = -12y$
2. $y^2 = (\frac{16}{3})x$
3. $(x - 1)^2 = 8y$
4. $(y - 1)^2 = 6(x - \frac{9}{2})$
5. Focus $(\frac{3}{4}, 0)$; directrix $x = -\frac{3}{4}$
6. Focus $(0, \frac{1}{16})$; directrix $y = -\frac{1}{16}$
7. Focus $(3, 3)$; directrix $x = 1$
8. Focus $(-1, 3)$; directrix $y = -1$
9. $x^2/25 + y^2/9 = 1$
10. $x^2/25 + y^2/1 = 1$
11. $x^2/81 + y^2/45 = 1$
12. $(x - 6)^2/9 + (y - 2)^2/5 = 1$
13. Foci $(\pm 2\sqrt{3}, 0)$; vertices $(\pm 5, 0)$
14. Foci $(0, \pm\sqrt{7}/2)$; vertices $(0, \pm 2)$
15. Foci $(-4, -1 \pm \sqrt{5}/6)$; vertices $(-4, -1 \pm \frac{1}{2})$
16. Foci $(1 \pm 1, -1)$; vertices $(1 \pm 2, -1)$
17. $x^2/9 - y^2/16 = 1$
18. $y^2/9 - x^2/9 = 1$
19. $x^2/9 - y^2/27 = 1$
20. $3(x + 2)^2/16 - 3(y - 2)^2/64 = 1$
21. Vertices $(\pm 3, 0)$; foci $(\pm 5, 0)$; asymptotes $y = +\frac{4}{3}x$
22. Vertices $(\pm 1, 0)$; foci $(\pm\sqrt{10}, 0)$; asymptotes $y = \pm 3x$
23. Vertices $(-4, 3 \pm 3)$; foci $(-4, 3 \pm \sqrt{13})$; asymptotes $y = \pm\frac{2}{3}x + \frac{17}{3}$
24. Vertices $(2, 2 \pm 1)$; foci $(2, 2 \pm \sqrt{5})$; asymptotes $y = \pm 2x - 2$

Index